Guide to Organic Stereochemistry

Sheila R. Buxton

and

Stanley M. Roberts

 LONGMAN

Addison Wesley Longman

Addison Wesley Longman Limited,
Edinburgh Gate, Harlow,
Essex CM20 2JE, England
and associated companies throughout the world

First published 1996

British Library Cataloguing in Publication Data
A catalogue entry for this title is available from the British Library

ISBN 0-582-23932-X

Library of Congress Cataloging-in-Publication Data
A catalog entry for this title is available from the Library of Congress

Set by 40 in 10/12 Times Roman
Produced by Longman Singapore Publishers (Pte) Ltd.
Printed in Singapore

Contents

Preface

Stereochemistry is one of the most important scientific concepts and also one of the hardest areas of chemistry to grasp. Its importance is manifest not only in chemistry but also in many other of the life sciences. To acquire a good understanding of stereochemistry and its effects, one must be able to think and visualize in three dimensions and become familiar with the extensive and often confusing terminology.

Guide to Organic Stereochemistry is not intended to be a comprehensive text. However, we envisaged a book that would be a step-by-step guide for the beginner and also one that would provide some insight into how a knowledge of stereochemistry is used in planning synthetic strategy or rationalizing the outcome of some reactions.

The early chapters were constructed with the first-year undergraduate chemistry student in mind, and also possibly those engaged in the biochemical or medical disciplines, where an awareness of stereochemistry and its effects is necessary, but not the complexities. These chapters cover the identification and assignment of various chiral features of organic molecules, and, while assuming a level of understanding of general chemistry, do not require any prior knowledge of stereochemistry.

The middle chapters are orientated more to the advanced undergraduate chemistry student, where an understanding of the predictive capabilities of stereochemical models, Cram's rule for example, and of the more advanced stereochemical terminology, such as topicity or stereoselectivity, becomes more important.

After much thought, we decided also to discuss towards the end some of the more difficult and thought-provoking stereochemical topics of asymmetric synthesis, with the final chapter illustrating how these techniques have been used to make two complex, chiral, natural products. As this was thought likely to be of interest largely to early postgraduate chemists, the text is written more in the usual research style.

As mentioned above, *Guide to Organic Stereochemistry* is not intended to be comprehensive; for those interested enough to wish to read further around this subject, we strongly recommend Eliel and Wilen's excellent text *Stereochemistry of Organic Compounds*.

Whatever the level of the reader of this book, there is one thing that we do advocate unreservedly, and that is to read the text in conjunction with a molecular modelling kit. Building representations of chiral molecules and comparing their various stereoisomeric forms is invaluable for a proper understanding of stereochemical concepts. It is often astonishing how something that is almost impossible to visualize in two dimensions on a sheet of paper becomes crystal clear when constructed into a three-dimensional model. During the writing of this book, we tried a number of different modelling kits, and the best ones in our opinion, in terms of cost and effectiveness, were the Darling models. However, even a basic inexpensive kit, such as the Orbit set, can be an enormous aid to comprehension.

Finally, there are a number of questions set in each chapter, which sometimes present points not otherwise exemplified in the chapter. They are intended to make the reader extrapolate from what has been learned from the text, and thus it is important that the problems be attempted.

It is appropriate here to acknowledge the many people who have offered us assistance and encouragement during the preparation of the text. For practical work (preparation of Mosher's esters, determining $[\alpha]_D$ values and running NMR spectra) we thank Dr Vladimir Sik and Dr Richard Hufton and for critical reading of all or parts of the manuscript we are indebted to Dr Gordon Read, Dr Brian Ridge, Dr Elena Lasterra, Dr Alex Drake, Professor Mark Baird, Professor Don Bethell, Professor Wes Borden, Miss Kathryn Wright, Miss Alison Murphy, Mr Stephen Hermitage, Dr Nazir Bashir, Mr Thierry Guyot, Mr Pierre Kary, Mr David Varley and Mr Daniel Watson. Special thanks go to Professor Milos Hudlicky for many ideas and suggestions.

The photographs of crystal structures of selected proteins, featured on the front cover and on pages 173 and 174, were kindly provided by Dr Jenny Littlechild (Exeter University). The Institut Pasteur (Paris) provided the photographs of Louis Pasteur's notebook and models (front cover and page 100).

Professor Ray Abraham and Drs Andrew Carnell, Rick Cosstick, Ian O'Neil and Dick Storr assisted at the proof-reading stage and Miss May-Britt Nielsen, Mr Pierre Kary, Mr Stephen Hermitage and Mr Daniel Watson helped in the compilation of the index.

Sheila R. Buxton
Stanley M. Roberts
1996

Foreword

We live in a three-dimensional world at every level from the molecular to the macroscopic. Nowhere is the third dimension more important than in the field of organic chemistry. The study of chemistry in three dimensions is called 'stereochemistry'. In 1860, Pasteur described his famous resolution of tartaric acid, thereby giving birth to the field. A major step forward was taken by van't Hoff and Le Bel who recognised that a carbon with four bonds was tetrahedral, thereby providing the foundation for the modern concepts of stereoisomerism. But it was not until later that the study of the conformations of organic molecules brought the field of organic chemistry off the flat, two dimensional pages of chemistry journals forever and added the fourth dimension of time to the three geometric dimensions.

In the 1990s, stereochemistry pervades every aspect of the field of organic chemistry, and it is simply not possible to study any aspect of this field without a solid knowledge of stereochemistry. But the principles and players in stereochemistry — configuration, conformation, enantiomerism, diastereoisomerism, asymmetric synthesis, chiral auxiliaries, and so on — are not at all self-evident to aspiring organic chemists. Indeed, stereochemistry and its associated chemical con-sequences must surely be one of the reasons why many otherwise very bright students find introductory courses in organic chemistry so difficult. As the field continues to grow, the gap between the basic principles taught in introductory courses and the sophisticated applications of these principles in advanced books, review articles and original research papers grows ever wider.

The book by Roberts and Buxton steps boldly into this gap by providing a bridge for students. It helps them cross from the introductory level over to the real world with illustrations showing the principles and applications of stereochemistry. Sixteen easy-to-read chapters take students all the way from the shape of methane through the sophisticated and beautiful world of enzymes, to the elegant field of asymmetric synthesis. The book conveys the importance and the excitement of the field of stereochemistry along with its principles. Students taking the time and effort to cross the bridge with the aid of Roberts and Buxton's book will marvel at what they begin to see on the other side as the field of stereochemistry begins to unfold for them.

Dennis P. Curran
Distinguished Service Professor of Chemistry
University of Pittsburgh
July 1996

1 Shapes of simple molecules

1.1 Methane

The hydrocarbon methane is the simplest organic molecule. It contains four equivalent carbon–hydrogen bonds and its shape has been shown by various spectroscopic methods to be that of a regular tetrahedron (Fig. 1.1). The carbon sits in the centre of the tetrahedron and hydrogen atoms take up positions at the vertices.

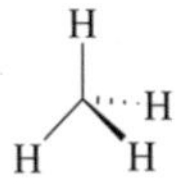

Fig. 1.1

The shape of a molecule is determined by the electrons in the constituent atoms, both those which participate in chemical bonds and those which are non-bonding. The electron configurations of a number of atoms commonly encountered in organic chemistry are given in Table 1.1. If we consider the lowest energy (ground state) electron configuration of carbon, $1s^2\,2s^2\,2p^2$, the very formation of methane would appear to be anomalous. The three 2p orbitals $(2p_x\,2p_y\,2p_z)$ are able to accommodate a maximum of six electrons as three pairs of opposite spin. However, according to Hund's rule, when there are three or fewer electrons to accommodate spin pairing does not occur. In the case of carbon with two 2p electrons the two electrons enjoy sole occupation of two of the three p orbitals. From this it would look as if carbon should be divalent

Table 1.1 Electron configuration of some elements.

Element	Electron configuration
H	$1s^1$
C	$1s^2\,2s^2\,2p^2$
N	$1s^2\,2s^2\,2p^3$
O	$1s^2\,2s^2\,2p^4$
P	$1s^2\,2s^2\,2p^6\,3s^2\,3p^3$
S	$1s^2\,2s^2\,2p^6\,3s^2\,3p^4$

(since it apparently has only two electrons available for covalent bonding), but we know from the chemistry of carbon that it is in fact tetravalent, as in methane.

1.2 Hybridization

The explanation proposed to account for the tetravalency of carbon is based on the concept of hybridization. In this model the four valence electrons of carbon occupy four energetically equivalent (degenerate) sp^3 orbitals which, as the name suggests, are 'hybrids' of the 2s and 2p originals. The hybrids are directed towards the vertices of a regular tetrahedron, and the angle between each pair of hybrids is 109.5°, which coincidentally is the angle of maximum separation.

Unlike the p orbitals, which are symmetrical in shape and which can accommodate the electron in either lobe with equal probability, the hybrid orbitals are highly unsymmetrical with a much greater probability of accommodating the electron in the large lobe. For convenience, conventional representations usually show only the large lobes (Fig. 1.2).

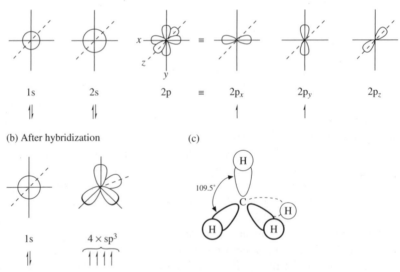

Fig. 1.2 Representation of orbitals before and after hybridization and the bonding orbitals of methane.

1.3 Ethene and ethyne

sp^3 Hybridization accounts for the tetrahedral shape of the carbon atoms in all saturated organic compounds. But what about compounds with double and triple bonds such as ethene (ethylene) and ethyne (acetylene)? Physical techniques have demonstrated that these molecules are planar and linear, respectively (Fig. 1.3).

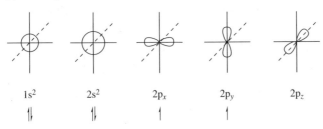

ethene   ethyne

Fig. 1.3

Again, recourse is taken to the concept of hybridization but this time not all of the p orbitals are involved. In the ethene molecule, each of the carbon atoms has undergone sp^2 hybridization; that is, the 2s and only two of the 2p orbitals of the original configuration have combined to give three degenerate sp^2 orbitals and one remaining 2p orbital, Fig. 1.4. The calculated arrangement of the new orbitals is trigonal (angle = $120°$), which also provides the maximum separation.

(a) Before hybridization

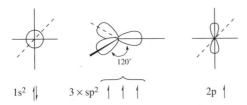

(b) sp^2 hybridization

Fig. 1.4

The sp^2 orbitals again have characteristics of both 'parent' s and p orbitals but there are differences between the shapes of the sp^2 and sp^3 lobes: sp^2 orbitals have more 's' character than their sp^3 counterparts (33.3% compared with 25%) and so have rounder, shorter lobes.

The term 'double bond' is a poor descriptor of the alkene unit, particularly when used with conventional structural formulae which tend to show alkenic bonds as two equally contributing bonds. This is not the case: a double bond comprises a strong σ-bond and a weaker π-bond. Figure 1.5 shows how a double bond is constructed.

The σ-bond is formed by direct end-on overlap between one of the sp^2 orbitals on each of the carbon atoms. The π-bond, on the other hand, is formed by weaker, sideways overlap between the two p orbitals, one on each carbon. Such

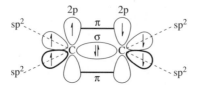

Fig. 1.5

lateral overlap is all that the p orbitals can manage, because the p orbital is placed orthogonally (perpendicularly, zero overlap) to the sp^2 hybrids in the plane and the result is a molecular orbital consisting of a 'π-cloud' above and below the plane of the σ-bond. The remaining sp^2 orbitals σ-bond with hydrogen or other substituents.

Finally, similar treatment gives us the structure of ethyne, which has one strong σ-bond and two weak π-bonds. Hybridization in this case involves only one of the 2p orbitals and the 2s orbital, Fig. 1.6 (sp hybridization). The sp

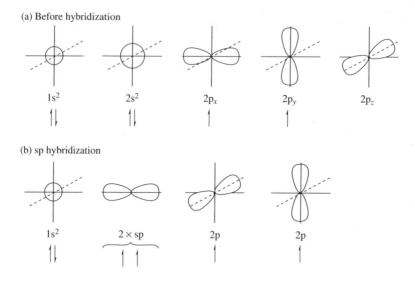

Fig. 1.6

orbitals have 50% 's' character and are rounder and shorter still, and, as before, engage in end-on overlap to make a strong σ-bond, Fig. 1.7. As there are only two sp orbitals, each containing an electron, they become linearly disposed and the 180° angle affords maximum separation. The two remaining

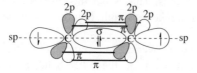

Fig. 1.7

orthogonal p orbitals on each carbon atom participate in weaker sideways over-lap above and below, and in front and behind the axis of the σ-bond. The decrease in bond length as we proceed from ethane (1.54 Å) through ethene (1.33 Å) to ethyne (1.20 Å) reflects the increased amount of bonding between the carbons and the increase in 's' character of the central σ-bond.

1.4 Molecules containing elements other than carbon

The hybridization model can also be applied to other elements that are frequently encountered in organic chemistry. Amines, ethers and water, although possessing less than four ligands, nevertheless are approximately sp³ hybridized species.

The electron configuration of nitrogen is $1s^2\,2s^2\,2p^3$, therefore the sp³ orbitals resulting from hybridization of the 2s and 2p orbitals are required to accommo-date five electrons. Nitrogen does this by placing a single electron in three of the four lobes and a pair of electrons in the remaining lobe. Thus the lone pair becomes the fourth 'ligand' and the tetrahedral arrangement is again assumed. The effectively large volume required by the lone pair, compared with the bond-ing electrons, means that the ligands are not disposed in quite a regular tetra-hedral fashion and there is a certain amount of deviation from the normal 109.5° angle (for example, the H—N—H angle in NH_3 is 106.8°, Fig. 1.8).

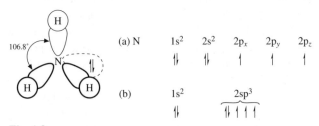

Fig. 1.8

The tetrahedral arrangement in amines and ammonia is by no means static: the lone pair is able to migrate to the other face of the molecule causing the config-uration to flip or invert to give the mirror image (Fig. 1.9). However, the energy barrier associated with this process is very small and the inversion occurs much too rapidly for either of the two forms to be isolated (for example, ammonia is estimated to undergo 2×10^{11} inversions per second, and has an inversion energy E_{inv} of 23 kJ mol^{-1}).

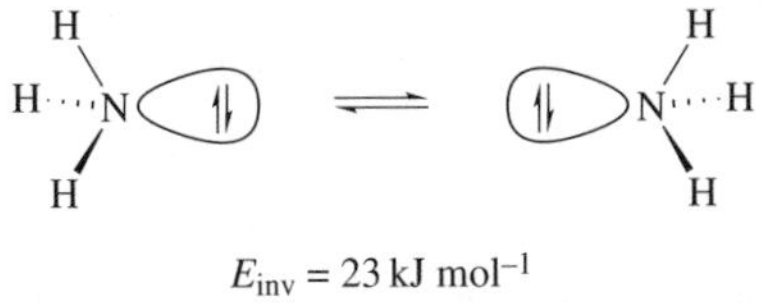

$$E_{inv} = 23\,\text{kJ mol}^{-1}$$

Fig. 1.9

This is not so with the phosphorus analogues, called phosphines. Phosphorus, a third-row element, has the electronic configuration $1s^2\,2s^2\,2p^6\,3s^2\,3p^3$ and here hybridization between the 3s and 3p orbitals is not very effective. Therefore we have three singly occupied p orbitals, which form bonds with other ligands, and one doubly occupied s orbital (the lone pair). The barrier to inversion in phosphines is much higher than in amines, so much so that inversion takes place only at higher temperatures (E_{inv} = 113 kJ mol^{-1}).

Ethers and water are also approximately tetrahedral: in oxygen (electron configuration $1s^2\,2s^2\,2p^4$) the valence shell is hybridized to $2sp^3$ with two of the lobes occupied by two lone pairs of electrons. The situation for H_2S and sulfides is akin to that for phosphines. Thus bonds are made to singly occupied p orbitals, lone pairs of electrons occupying the 3s and orthogonal 3p orbital.

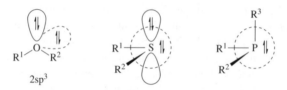

Fig. 1.10 The approximately tetrahedral structures of ethers and water (the H—O—H bond angle in water is 104.5°) and the shape of sulfides and phosphines.

The electronegative oxygen atom exerts a very tight grip on the two pairs of non-bonding electrons. However, nitrogen and phosphorus, being less electronegative, are able to undergo oxidation by donation of one of the lone pairs to give, for example, amine oxides and phosphine oxides, respectively. Each species now has four separate ligands, which lock the tetrahedral structure and prevent inversion. Sulfur can also donate one of its lone pairs to form sulfoxides, in which the remaining lone pair is one of the ligands. If each of the ligands is different, i.e. $R^1 \neq R^2 \neq R^3$ (Fig. 1.11), then the oxides display chirality, the stereochemical consequences of which can be very important. A fuller discussion of chiral compounds is given in Chapter 2.

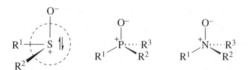

Fig. 1.11 Representations of sulfoxides, phosphine oxides and amine oxides. In phosphine oxides the requirement to form a fourth bond overcomes the reluctance of 3s and 3p orbitals to hybridize (cf. phosphines). The hybridization in the phosphine oxides approximates to sp^3 and the bond angles are close to tetrahedral.

This sp^2 hybridization accounts for the shapes of ketones, thiones and other carbonyl-containing compounds, imines, oximes and azo-containing molecules. Nitriles are sp hybridized (Fig. 1.12).

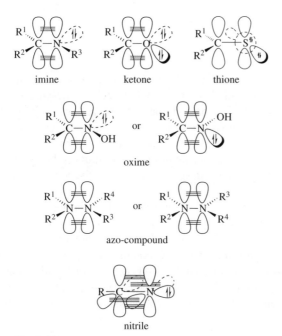

Fig. 1.12

It must be stressed at this point that the discussion of the hybridization model presented here is a very qualitative description of a complex mathematical treatment. The important points to remember are that the number of orbitals must be the same before and after hybridization [one s and three p orbitals give (a) four sp^3, (b) three sp^2 + one p or (c) two sp + two p orbitals] and that the arrangement affording maximum separation of the electrons in the hybridized orbitals determines the shape of the molecule (sp^3 tetrahedral, sp^2 trigonal planar and sp linear) (Table 1.2).

Table 1.2 Some characteristics of hybridization.

No. of ligands	Hybridization	Geometry	Internuclear angle	% s character	Example
4	sp^3	tetrahedral	109.5°	25	CH$_4$
3	sp^2	trigonal	120°	33.3	CH$_2$=CH$_2$
2	sp	linear	180°	50	CH≡CH

Q. 1.1 Determine the electron configuration of the carbon atoms CH$_3^+$ and CH$_3^-$. What type of hybridization will be associated with each species and what shape will they be?

1.5   Conformation

1.5.1   Ethane

Ethane, like methane, is made up of tetrahedral sp³ carbon atoms. Unlike methane, however, its overall shape is not fixed because the sigma bond that arises from end-on overlap of one of the sp³ orbitals on each of the two constituent carbon atoms, although strong, is not rigid and can freely rotate around its axis. This fact leads us on to considerations of conformation, that is, the spatial array of atoms in a molecule of given constitution. Let us focus, first of all, on the spatial array of the hydrogen atoms around the central carbon atoms of ethane.

The rotation of the CH_3 groups around the carbon–carbon single bond in ethane is very fast. Nevertheless, distinct conformations do exist, some of which are favourable low-energy arrangements and others that are higher in energy and unfavourable. These conformations were commonly represented by the so-called 'sawhorse' projection in which the C—C bond is artificially elongated so that the relative orientations of the C—H bonds can be displayed (Fig. 1.13); however, this has been largely superseded by the use of the Newman projection.

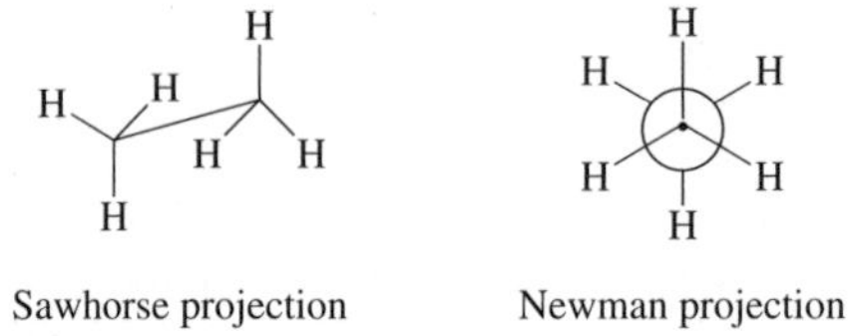

Sawhorse projection   Newman projection

Fig. 1.13

The Newman projection is a stylized view along the bond undergoing rotation, from front to back. The circle is merely a visual aid to differentiate between the substituents (in this case hydrogen atoms) on the nearest carbon, which is placed in the centre of the circle, whereas for the carbon directly behind it, the hydrogens are drawn from the circumference rather than the centre. The particular conformations depicted in Fig 1.13 are the lowest-energy ones and therefore most favourable, because the electrons in the C—H bonds experience maximum separation in this 'staggered' arrangement. If the carbon atom at the back is rotated through 60° in either direction the C—H bonds on the two carbons

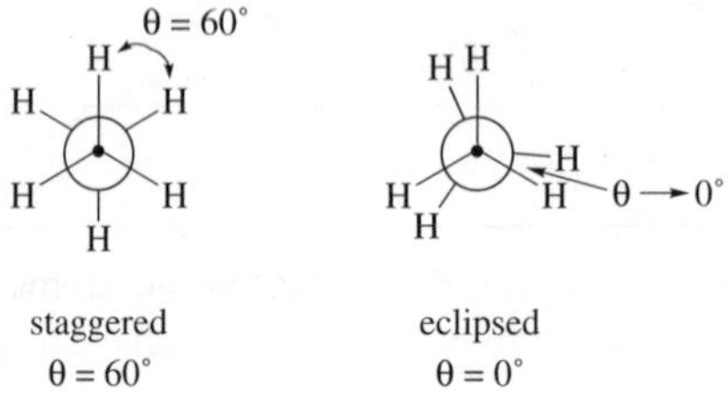

staggered   eclipsed
$\theta = 60°$ $\theta = 0°$

Fig. 1.14

become adjacent, forcing the bonding electrons closer together. This 'eclipsed' conformation is higher in energy and therefore less favourable (Fig. 1.14).

In Fig. 1.14 theta (θ) is the dihedral or torsion angle and is 60° in the staggered conformation and 0° in the eclipsed conformation. The staggered and eclipsed conformations are interconvertible provided that sufficient energy is supplied to overcome the energy barrier to rotation. For ethane, with only C—H bonds to contend with, the energy barrier between the staggered and eclipsed conformations is relatively low (14 kJ mol^{-1}) and normal room temperature (about 25 ° C) provides enough thermal energy for the C—C bond to rotate freely and give both conformers.

1.5.2 Butane

Staggered and eclipsed are, unfortunately, not the only terms in common usage to describe conformation: two other sets also exist, namely, *cis–trans–gauche* and synperiplanar–synclinal–anticlinal–antiperiplanar. These terms are used to describe more complex systems where the substituents are not all the same. Consider, for example, the conformation of butane about the C2—C3 bond (Fig. 1.15). Four limiting conformations are possible.

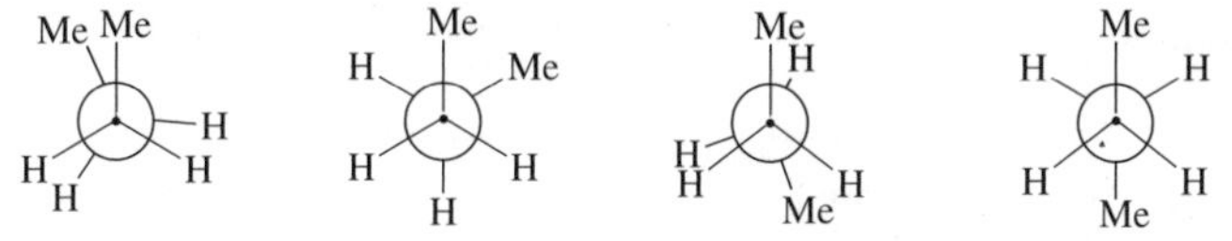

synperiplanar (*sp*)	synclinal (*sc*)	anticlinal (*ac*)	antiperiplanar (*ap*)
cis (*c*)	*gauche*	$\theta = 120°$	*trans* (*t*)
$\theta = 0°$	(g^+ or g^-)		$\theta = 180°$
	$\theta = 60°$		

Fig. 1.15

Here we have two eclipsed (*sp* and *ac*) and two staggered (*sc* and *ap*) conformations. The synperiplanar conformer is the highest energy and therefore least favoured owing to the close proximity of the bulky methyl groups, each with their four pairs of mutually repulsive bonding electrons; at the other end of the scale, the antiperiplanar conformer is lowest energy and most favoured as it affords the maximum distance between all the electron pairs. The *sc* and *ac* conformers are of intermediate energy as illustrated in the potential energy diagram in Fig. 1.16. The important destabilizing conformational interactions for butane are listed in Table 1.3.

At 25°C approximately 75% of a given population of butane molecules assume the antiperiplanar conformation with most of the remaining 25% taking up the synclinal conformation.

Q. 1.2 Using the values given in Table 1.3, calculate the energy gap $\Delta E = b$ in Fig. 1.16.

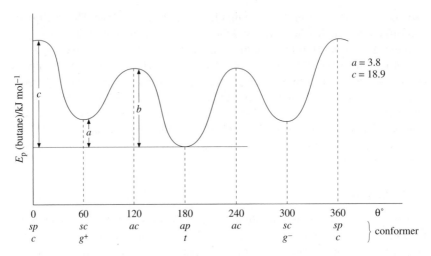

Fig. 1.16

Table 1.3 Destabilizing conformational interactions in butane.

Interaction		Energy of destabilization kJ mol^{-1}
H H	eclipsed	4.2
H CH$_3$	eclipsed	5.8
CH$_3$ CH$_3$	eclipsed	10.5
CH$_3$ CH$_3$	gauche	3.8

1.5.3   Unsymmetrically substituted alkanes

It is relatively simple to recognize the various conformations adopted by ethane and butane since they are symmetrical and it is obvious which atoms or groups define the torsion angle. This becomes less obvious when the substituents on the carbons at each end of the bond being examined are all different. To address this problem a number of criteria have been devised. These are listed and illustrated in Fig. 1.17.

(a) 2-chloropropanamide

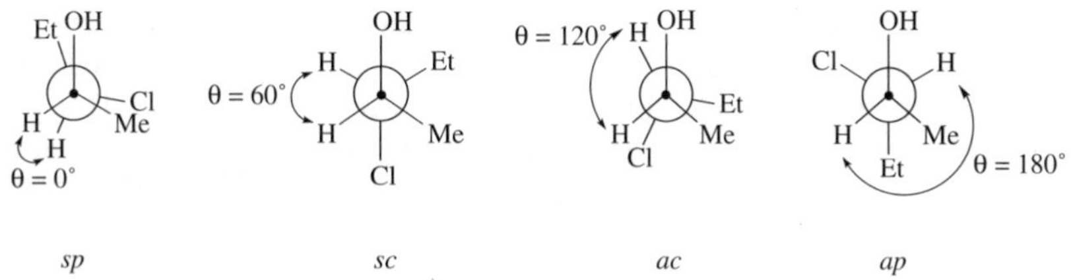

Et OH   OH $\theta = 120°$ H OH   OH
         $\theta = 60°$ Et   Cl   H
    Cl   Et
H   Me   H   Me   H   Et   H   Me
 H   Cl   Cl   Me   Et $\theta = 180°$
$\theta = 0°$

   sp   *sc*   *ac*   *ap*

(b) 3-chloropentan-2-ol

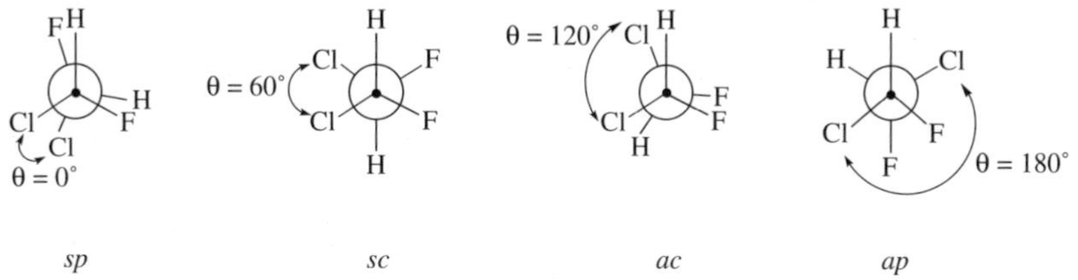

F H   H $\theta = 120°$ Cl H   H
         $\theta = 60°$ Cl   F   H   Cl
   H   Cl   F   F
Cl   F   Cl   F   Cl   F
 Cl   H   H   F $\theta = 180°$
$\theta = 0°$

   sp   *sc*   *ac*   *ap*

(c) 1,2-dichloro-1,2-difluoroethane

Fig. 1.17

(a) When all the atoms or groups are different, the angle theta (θ) is measured
   from the one in each set that is preferred by the sequence rule (i.e. is of
   highest priority; for a detailed discussion of the sequence rule, see Chapter
   2, p. 25). For example, consider 2-chloropropanamide [Fig. 1.17(a)]. The
   sequence-rule-preferred atoms are those of highest atomic number, so in this
   case θ is defined by the oxygen atom of the carbon at the front, and the
   chlorine on the carbon behind. Remember, because of the planar, trigonal
   nature of the carbonyl carbon arising from sp^2 hybridization, the doubly
   bonded oxygen and the NH$_2$ group appear, from this view, to be 180° apart.

(b) If there is an atom or group common to both ends of the bond, θ is measured
   between these two, even if the remaining substituents are different and of
   higher priority, for example in 3-chloropentan-2-ol [Fig. 1.17(b)].

(c) If there are two atoms or groups common to both ends of the bond, that pair
   of highest priority defines θ, for example in 1,2-dichloro-1,2-difluoroethane
   [Fig. 1.17(c)].

1.6 Conformations of cyclic molecules

1.6.1 Cyclohexane

Saturated carbon atoms are sp^3 hybridized and therefore tetrahedral and the
saturated carbon atoms in cyclic molecules are no exception. A planar arrangement of the six methylene groups of cyclohexane does not give the required
tetrahedral shape for every carbon atom and this is only achieved by puckering

of the ring. Cyclohexane does this mainly by adopting two conformations: the 'chair' and the 'boat' (Fig. 1.18). The ring is very flexible and cyclohexane freely converts from one conformer to the other at room temperature. This conversion involves no bond breakage at all, only bond rotation, and if you construct a cyclohexane molecule from a modelling kit you will get an idea of the flexibility of this six-membered ring by manipulating it from one conformation to another.

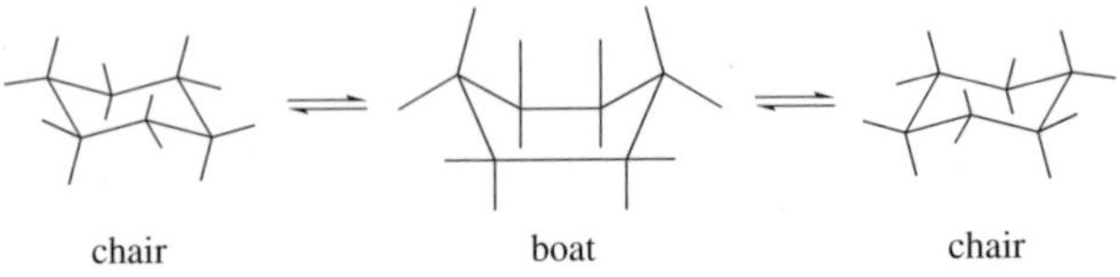

chair   boat   chair

Fig. 1.18

Having said that the two conformations are interconvertible, the chair form is energetically much more favourable than the boat. The reason for this can also be seen by inspecting the model you have just made. If you hold the chair model so that the downward-pointing carbon is at the front and the carbon pointing up is at the back and look along the two parallel bonds, you will see that all the bonds are staggered ($\theta = 60°$); the so-called double Newman projection for this is illustrated in Fig. 1.19. In fact if you look along *any* of the carbon–carbon bonds in the chair conformation of cyclohexane you will find that they are *all* staggered.

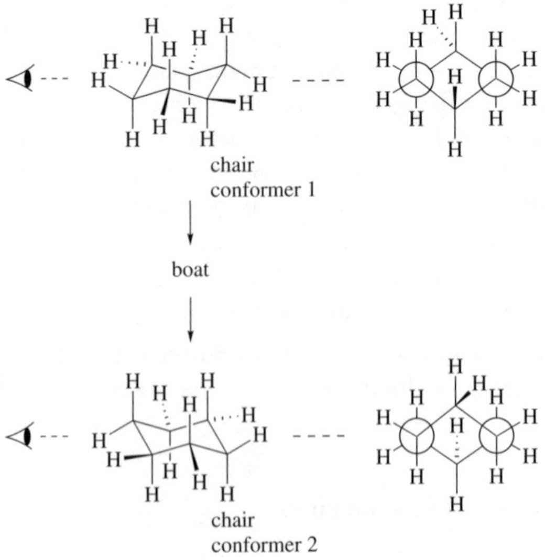

chair
conformer 1

boat

chair
conformer 2

Fig. 1.19

Now flip the model from the chair into the boat and look along the two parallel carbon–carbon bonds; this time you will see that the carbon–hydrogen and carbon–carbon bonds are in an eclipsed arrangement (Fig. 1.20). In addition to this, there are other energetically unfavourable interactions between the hydrogens on the two carbons defining the stern and prow of the 'boat', which, in this conformation, are brought into uncomfortably close proximity (1,4-transannular interaction).

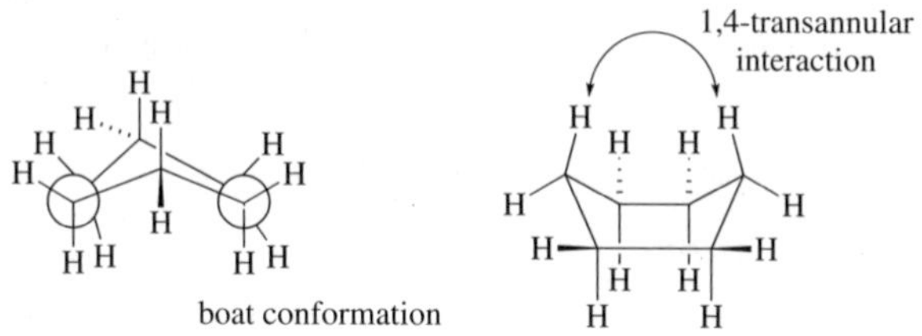

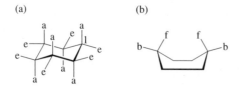

Fig. 1.20

1.6.2 Substituents on cyclohexane

Having established that the ring skeleton in cyclohexane is (a) not planar and (b) not rigid, we can now consider the implications of this for the hydrogens attached to the ring. You should be able, from Fig. 1.21(a) or better still from the model, to discern two different types of hydrogen in the chair conformation. From a side view, starting from the uppermost carbon (C1 in Fig. 1.21) there are six vertical C—H bonds which alternate up or down as you progress round the ring. These six hydrogens are described as occupying the axial positions (labelled a in Fig. 1.21). The remaining six hydrogens take up positions around the equator of the ring and are called, not surprisingly, equatorial hydrogens (labelled e in Fig. 1.21). These too have an alternating pattern.

Fig. 1.21

What you should also notice from your manipulations of the model is that when the chair is converted through the boat into the mirror image chair, the hydrogens which were axial to begin with become equatorial in the mirror image and those that were equatorial become axial. This is an important feature of the chemistry of molecules containing cyclohexanes, as larger substituents frequently show very distinct preferences for axial or equatorial occupancy when they are introduced into the ring during synthesis.

For example, the methyl group in methylcyclohexane prefers to be equatorial because, in the axial position, it experiences unfavourable interactions with the other two axial hydrogens on the same side of the ring (1,3-diaxial interactions). However, the methyl group is small enough for such transannular interactions to be relatively slight and at around 25 °C there is rapid interconversion between the two conformers. In contrast with this, *tert*-butylcyclohexane can have the *tert*-butyl group only in an equatorial position because the transannular inter-actions between the other axial hydrogens and this very bulky substituent are just too unfavourable for it to be axial (Fig. 1.22). This effect can be useful in synthesis when, for example, it is necessary to lock the cyclohexane into a particular chair conformation, perhaps to force the introduction of further substituents into less-preferred orientations.

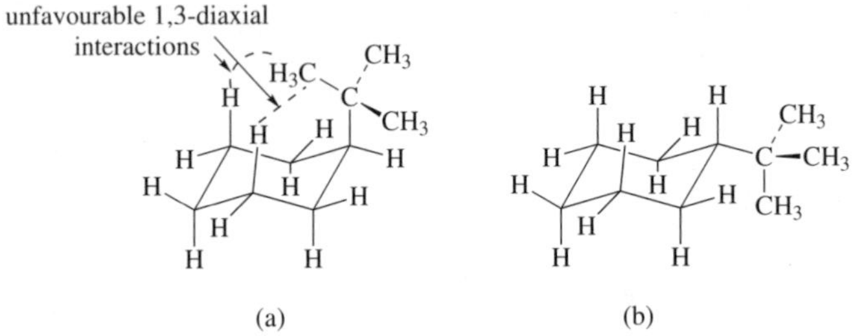

unfavourable 1,3-diaxial interactions

(a)   (b)

Fig. 1.22

The hydrogens on the boat conformation are also of different types. On the two carbons raised above the plane of the middle four carbons (C1 and C4), two of the C—H bonds project into the ring (causing the unfavourable 1,4-trans-annular interactions mentioned earlier) and the other two project away from the ring. In keeping with the nautical theme, these have been labelled 'flagpole' (f) and 'bowsprit' (b) positions, respectively [Fig. 1.21(b)].

Q. 1.3 Draw, using the types of cyclohexane structures featured in Fig. 1.21, the preferred conformations of 1,3-dimethylcyclohexanes.

1.6.3   Cyclohexene and cyclohexanone

Cyclohexene and cyclohexanone molecules contain double bonds and therefore sp^2 hybridized carbon atoms. The skeleton of cyclohexanone has only one trig-onal carbon atom to accommodate and is able to do this without too much diffi-culty by adopting the usual chair conformation. There will be some variations on the tetrahedral angle but the ring is flexible enough to cope with this (Fig. 1.23).

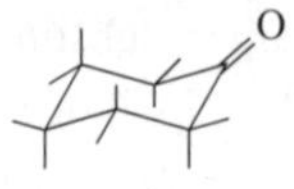

cyclohexanone

Fig. 1.23

Cyclohexene is slightly more of a problem as two trigonal, planar sp^2-hybridized carbons must be accommodated within the ring. The ring is prevented from adopting the chair/boat conformations because the endocyclic double bond requires four contiguous carbons to be coplanar; however, the remaining two sp³ carbon atoms have sufficient flex to be able to rotate into positions that give as near a tetrahedral arrangement as possible. The two resulting, interconvertible conformers are called half-chair and half-boat and are illustrated in Fig. 1.24. The hydrogens attached to the sp³ carbons of the cyclohexene ring are said to be 'pseudoaxial' and 'pseudoequatorial'.

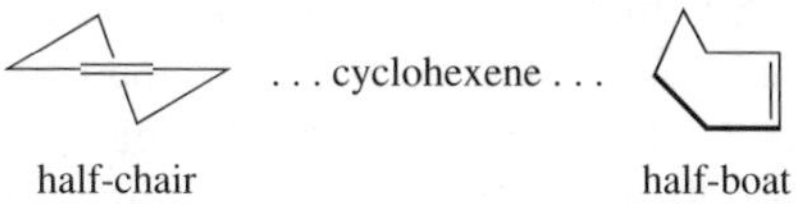

... cyclohexene ...

half-chair   half-boat

Fig. 1.24

1.6.4 Smaller rings

The deviation from the normal hybridization angle value (109° 20′ for sp^3, 120° for sp^2 and 180° for sp hybridization) is known as angle strain or Baeyer strain. Two other types of strain have already been encountered in this chapter: eclipsing strain (or Pitzer strain) and transannular strain (such as 1,3-diaxial interactions, Fig. 1.22).

Rings smaller than five- and six-membered are particularly susceptible to angle strain and attempts to relieve this strain account for much of the reactivity of some of these molecules.* Four- and five-membered rings adopt puckered conformations to relieve strain and achieve as near as possible the tetrahedral angle. Cyclopentane, for example, exists in the envelope conformation in which four of the carbons are coplanar and the fifth carbon is placed above or below the plane. This interconverts with another conformation, the twist, in which there are three coplanar carbons and one carbon above and one below the plane (Fig. 1.25). The differences between these two conformations are particularly difficult to visualize in two dimensions and construction of a model is recommended.

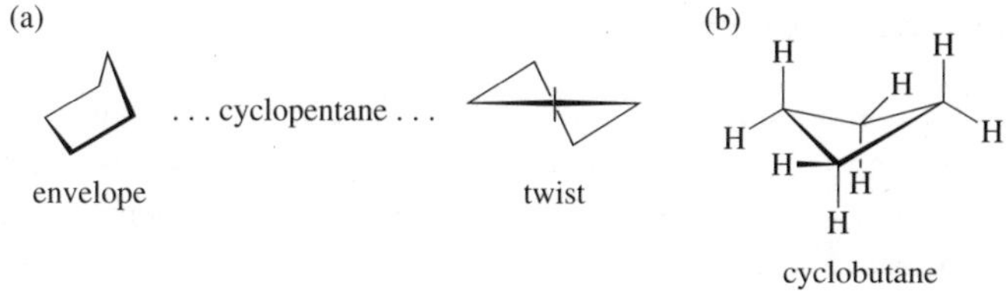

(a)

... cyclopentane ...

envelope   twist

(b)

cyclobutane

Fig. 1.25

The cyclobutane ring is bent along the diagonal to give the optimum conformation. The hydrogens are staggered as far as possible along the C—C bonds, which compensates, to a degree, the transannular 1,3-interactions.

---

* Note that strain energy is defined as the experimentally determined energy (enthalpy) of a strained structure (e.g. cyclobutane) over and above the corresponding unstrained structure.

1.6.5 Cyclopropane

Cyclopropane is a special case since, being a three-membered ring, it is unable to undergo planar deformation in order to overcome its angle strain. As the three carbons form the vertices of an equilateral triangle, the C—C—C bond angle must be 60°, which represents severe deviation for sp^3-hybridized carbons from the normal tetrahedral angle.

In order to account for this, a modification of the hybridization model already discussed has been proposed, which differs in two respects. The first is that hybridization does not occur equally in all the bonds of cyclopropane. From Table 1.2 it is clear that the more p character an orbital has, the smaller the bonding angle will be (compare sp^3 109° 20′ and sp 180°); therefore in cyclopropane the C—C bonds must have a higher p character than the normal sp^3 bond. Consequently the C—H bonds must have less p character and therefore more s character. In quantitative terms, this has led to the concept of 'percentage s character' and, based on ^{13}C NMR experiments, values of 33% for the C—H bonds of cyclopropane and 17% for C—C bonds have been suggested.

The second difference is that the orbitals of the ring bonds do not arise from end-on overlap but are 'bent'; that is, overlap takes place to a certain extent from the side (Fig. 1.26). For most strain-free molecules, the internuclear angle and the interorbital angle are the same (for example, in methane both are 109.5°). However, cyclopropane is a case where the two are different: the internuclear angle is 60° but the interorbital angle is approximately 104°.

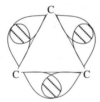

Fig. 1.26

1.6.6 Polycyclic compounds

In polycyclic compounds, the individual rings continue to take up, as far as possible, their preferred conformations. Figure 1.27 illustrates the preferred shape of the tetracyclic steroid nucleus, containing three six-membered rings and one five-membered ring.

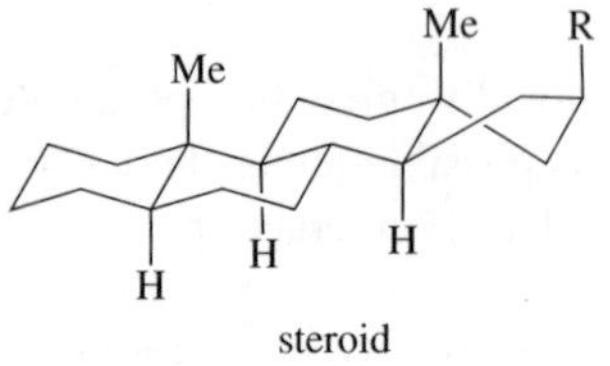

steroid

Fig. 1.27

Q. 1.4 Consider the structure of $CH_2=C=CH_2$. Its name is allene and it is the simplest member of a series of molecules known as cumulenes. What levels of hybridization are likely to occur for each carbon atom and how does the central carbon atom differ from the two end carbon atoms? Once you have answered these questions, what do you notice about the orientations of the C—H bonds and what shape is the molecule as a whole?

The unusual shape of allene has stereochemical implications when substituents other than hydrogen are introduced into the molecule and this will feature later on when we come to discuss chirality. The next homologue in this series, $CH_2=C=C=CH_2$, is called cumulene. Determine the hybridization of each of the carbon atoms and the general shape of the molecule: how does it compare with that of allene? What is the relationship between the shape of a cumulene and the number of carbons in the skeleton?

1.6.7 Heterocyclic ring systems

A six-membered ring containing an oxygen (tetrahydropyran) assumes a chair conformation (Fig. 1.28). Alkyl substituents at all positions prefer an equatorial arrangement; in contrast an α-alkoxy substituent shows enhanced stability in the axial position. This effect, called the anomeric effect, is due, at least in part, to unfavourable dipolar interactions of the oxygen atoms when the alkoxy group is in the equatorial situation.

For *N*-ethylpiperidine the ethyl substituent prefers the equatorial position while the lone pair occupies the axial slot (Fig. 1.28).

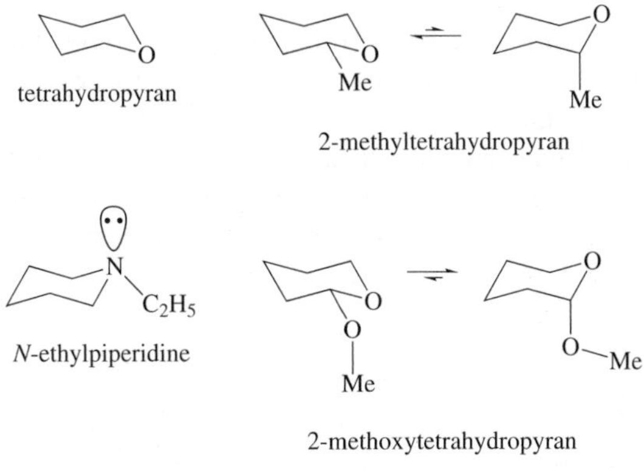

tetrahydropyran

2-methyltetrahydropyran

N-ethylpiperidine

2-methoxytetrahydropyran

Fig. 1.28

Answers to questions

Q. 1.1   For CH_3^+ the ground state electron configuration of the carbon atom is $1s^2 2s^2 2p^1$. This becomes sp^2 hybridized to give (after bonding to hydrogen atoms) a flat, trigonal arrangement with a vacant p orbital.

For CH_3^- the ground state electron configuration of the carbon atom is $1s^2 2s^2 2p^3$: sp^3 hybridization and bonding to hydrogen atoms gives a pyramidal arrangement of carbon and hydrogen atoms (energy barrier to pyramidal inversion 63 kJ mol^{-1}). The 'invisible' lone pair of electrons makes up the fourth arm of the tetrahedron.

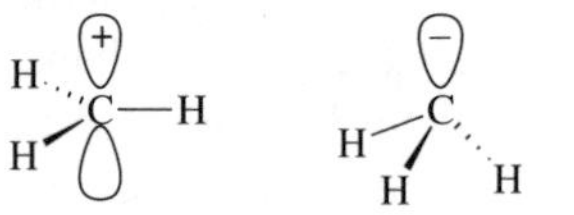

Q. 1.2   15.8 kJ mol^{-1}. The reader may also wish to correlate values $\Delta E = a$ and $\Delta E = c$ with values listed in Table 1.3.

Q. 1.3   The preferred conformations of two 1,3-dimethylcyclohexanes are shown below.

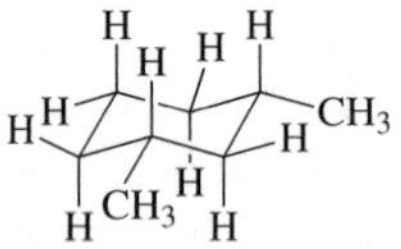

Both substituents equatorial: the arrangement with the two methyl groups axial is less stable.

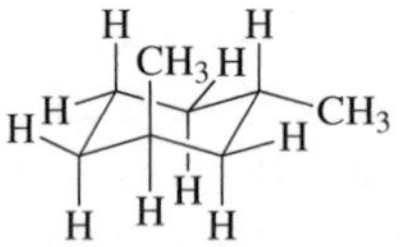

One methyl group is axial, the other equatorial. 'Flipping' to the other chair form does not change the overall situation.

Q. 1.4   The central carbon is sp hybridized, the terminal carbon atoms are sp^2 hybridized. The plane containing the carbon and two hydrogens to the left is orthogonal to the plane containing the carbon and two hydrogens to the right, i.e. the overall structure is an elongated tetrahedron.

For cumulene two carbons are sp hybridized and two carbons are sp^2 hybridized. The hydrogen atoms are contained in the same plane and therefore the overall structure is flat.

allene    cumulene

In general, for $H_2CC_nCH_2$ when n is odd the pairs of terminal hydrogen atoms will be contained in orthogonal planes. When n is even the hydrogen atoms will be in the same plane.

2 Chirality in molecules containing asymmetrically substituted tetrahedral centres

Of the many notable characteristics of carbon, its tetrahedral nature, discussed at length in Chapter 1, is one of the most remarkable. Unsymmetrical substitution of tetrahedral carbon gives rise to the phenomenon of chirality, a property which is fundamental to much of the extensive chemistry of this extraordinary element.

Chirality is a term that describes the 'handedness' of a molecule and is the property to which this chapter is devoted. Figure 2.1 shows a series of simple molecules that are chiral.

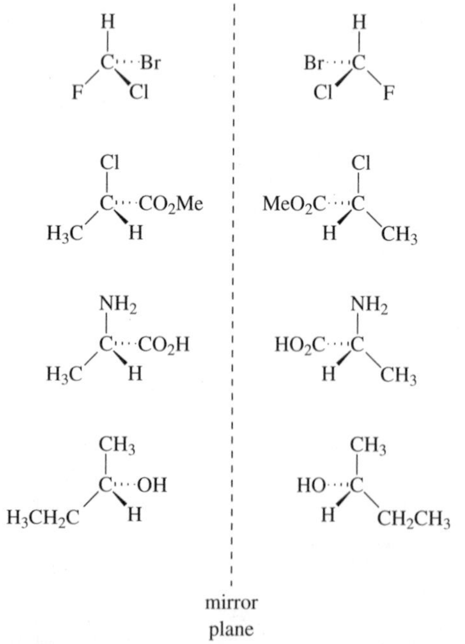

mirror
plane

Fig. 2.1

Two things should be apparent: one is that the pairs of molecules shown are mirror images of each other and the other is that the mirror images cannot be superimposed upon each other. If you find this difficult to visualize from the two-dimensional illustration, construct a model of the two forms of bromo(chloro)fluoromethane and demonstrate for yourself the non-super-imposability, or chirality, of the mirror-image related pair.

Chirality is not restricted to molecular structures: in nature there are many examples of non-superimposable mirror images, your hands and feet being probably your nearest examples. Indeed nature displays a high degree of 'handedness' and on a molecular level many of the most important biological processes take place only with one chiral form of a particular substrate. A truly amazing example of this is the transfer of genetic information from one generation to the next: only the form of DNA with a right-handed spiral (see Chapter 13) performs this most fundamental life process.

Returning to simpler molecules, such as those in Fig. 2.1, each contains a carbon atom with four different substituents. Such carbon atoms are variously referred to as chiral centres, asymmetric centres or stereogenic centres. Although all three alternatives are in current use, in this text the term 'chiral' is applied to the whole molecule to mean 'being capable of existing in non-superimposable, mirror-image forms', and 'stereogenic centre' is used to describe the carbon atom giving rise to the observed chirality.

2.1 Chiral molecules with one stereogenic centre: enantiomers

Figure 2.2 illustrates the two forms of butan-2-ol. The two forms are termed enantiomers or enantiomorphs. The physical and chemical properties of the pair of enantiomers are identical, with the important exceptions of their interactions with (a) plane-polarized light (optical activity) and (b) chiral reagents.

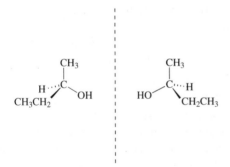

Fig. 2.2

2.1.1 Optical activity

When a beam of transmitted plane-polarized light is passed through a cell containing a sample of a single enantiomer, the plane of polarization will be bent or rotated to either the right or the left. The other enantiomer under exactly the same conditions (temperature, solvent, concentration) will rotate the plane of the light to exactly the same extent but in the opposite direction. This phenomenon is referred to as optical activity.

By convention, the enantiomer that causes rotation to the right (clockwise) when viewed looking towards the light source is described as dextrorotatory (*d*) or the (+)-enantiomer. That which rotates the plane to the left (anticlockwise) is laevorotatory (*l*) or the (–)-enantiomer The optical rotation is measured by a polarimeter and a sketch of the apparatus is given in Appendix 1.

2.1.2 Specific rotation

The extent and sign of the optical activity of an enantiomer are customarily conveyed as the specific rotation (or, more fully, the specific optical rotatory power). This is equal to $\alpha/\gamma l$, where α is the angle of optical rotation in degrees, γ is the mass concentration in g cm^{-3} and *l* is the length of the path of the light through the solution in dm. This information is usually given in the form $[\alpha]_\lambda^\theta$, where λ is the wavelength of the polarized light and is normally the sodium D-line (589 nm) and θ is the Celsius temperature at which the measurement is made. Many specific rotations are measured at or around room temperature and the possibly more familiar expression is $[\alpha]_D^{20}$. As an example, the specific rotation of dextrorotatory butan-2-ol is $[\alpha]_D^{20} = +13$ (neat). The units of specific rotation are deg cm^3 g^{-1} dm^{-1}, but by convention these are not expressed. It is, however, necessary to state the solvent (if any), the temperature and the concentration (c), as variation of any of these parameters will cause a change in the numerical value. Thus for laevorotatory (–)-2-chloropropanoic acid the following values are observed: $[\alpha]_D^{32} = -27$ (c, 1.0 hexane), $[\alpha]_D^{32} = -35$ (c, 1.0 chloroform), $[\alpha]_D^{32} = -17$ (neat), $[\alpha]_D^5 = -24$ (c, 1.0 hexane).

The dependence of the magnitude and (occasionally) the sign of $[\alpha]_\lambda^\theta$ on solvent is due to the fact that molecules in solution are solvated to a greater or lesser extent and that light interacts with the whole molecular assembly, the solute *and* its solvent shell. The dependence of $[\alpha]_\lambda^\theta$ on temperature is due to (i) change in density or concentration with temperature, (ii) change in association equilibrium constants and (iii) change in population of different (asymmetric) conformations with temperature.

Q. 2.1 Dextrorotatory α-pinene has a specific rotation $[\alpha]_D^{20} = +51.3$. What percentage of each enantiomer is present in a sample of α-pinene that shows a specific rotation of $[\alpha]_D^{20} = +30.8$? All specific rotations measured in the same solvent at the same concentration.

2.1.3 Racemates

Because the optical activities of a pair of enantiomers are equal but opposite, it follows that a sample containing equal proportions of both enantiomers will be optically inactive, that is, give rise to no observed rotation. This is because the observed rotation of one enantiomer is cancelled by that of the other.

Such a mixture is called a racemate and, in general, the formation of a compound containing a stereogenic centre from non-chiral starting material and reagents will lead to a racemate. For example, the reduction of butan-2-one with sodium borohydride gives a 50:50 mixture of the two enantiomers of butan-2-ol (Fig. 2.3), designated (±)-butan-2-ol.

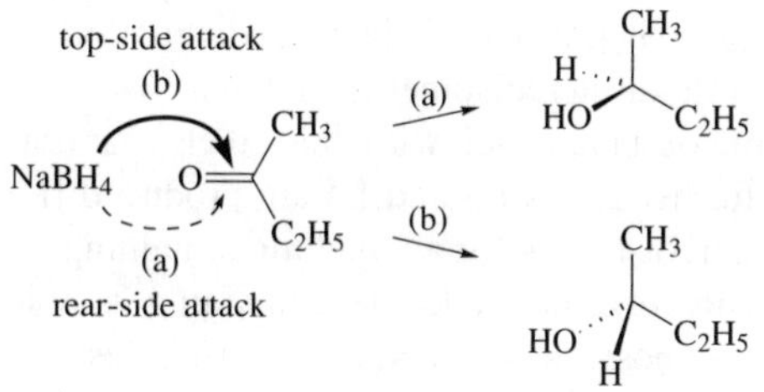

Fig. 2.3

2.2 Chiral molecules with two stereogenic centres: diastereoisomers

2.2.1 Reaction of enantiomers with chiral reagents

Continuing with our example of butan-2-ol, we can now examine its reaction with one enantiomer of a chiral reagent containing one stereogenic centre. Such a reagent is 2-chloropropanoic acid and the two products arising from its reaction with (±)-butan-2-ol are the esters E1 and E2 (Fig. 2.4).

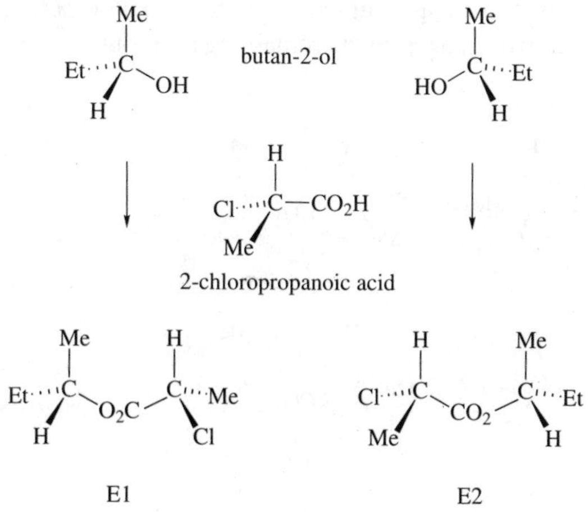

Fig. 2.4

E1 and E2 are still chiral and still isomeric, but are not enantiomeric because they are no longer mirror images of each other. (As with many concepts in stereochemistry, this may be easier to visualize if you construct a model of each product.) The two esters, which now each contain two stereogenic centres, are referred to as diastereoisomers or diastereomers. Diastereoisomers have a major advantage over enantiomers from a practical point of view in that they have different physical properties such as m.p., b.p., solubility, retention times and R_f

values, and standard techniques such as crystallization, distillation or chromatography can be used to separate diastereoisomeric mixtures.

Now consider the reaction of butan-2-ol with the other enantiomer of 2-chloropropanoic acid. Two further esters E3 and E4 are produced [Fig. 2.5(a)] which gives a total of four stereoisomers for a structure containing two stereogenic centres. In fact, for any organic molecule containing n non-identical stereogenic centres, there are 2^n possible stereoisomeric structures.

Q. 2.2   Place asterisks by the stereogenic centres in the following molecules:

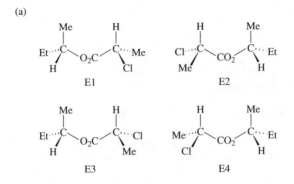

Figure 2.5(b) illustrates the relationship between E1, E2, E3 and E4. The pairs E1, E4 and E2, E3 are mirror images and are therefore enantiomeric pairs. No other pairing produces mirror images so the remaining relationships (E1, E2; E1,

(a)

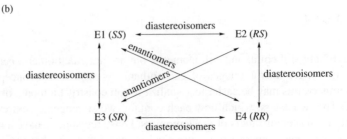

(b)

Fig. 2.5

E3; E2, E4; E3, E4) are diastereoisomeric. It is important to note that, in general, a molecule can have only one enantiomer, but it may have many diastereoisomers.

2.2.2 The Cahn–Ingold–Prelog (CIP) sequence rule

For molecules containing even more stereogenic centres it becomes necessary to distinguish between the large number of possible stereoisomers. To this end, the sequence rule was devised by Cahn, Ingold and Prelog to enable individual stereoisomers to be readily discussed. This system is one of priority of substituents around a stereogenic centre, which in turn is based on atomic number.

To take a simple example, in bromo(chloro)fluoromethane, CHBrClF, with a single stereogenic centre, the two enantiomers are distinguished as follows. First the substituents are placed in order of preference (or priority): the higher the atomic number the higher the preference. Therefore, for our example Br > Cl > F > H, where > denotes 'is preferred to' (or 'has priority over').

The next step is to view the molecule from the side opposite to the least-preferred substituent, in this case hydrogen (Fig. 2.6). What you will see is a trigonal arrangement of the three remaining ligands. The arrangement in which the priority sequence follows a clockwise direction is known as the *R*-enantiomer (from the Latin *rectus*, right). The other arrangement, where the ligands are in an anticlockwise sequence of priority, is known as the *S*-enantiomer (*sinister*, left).

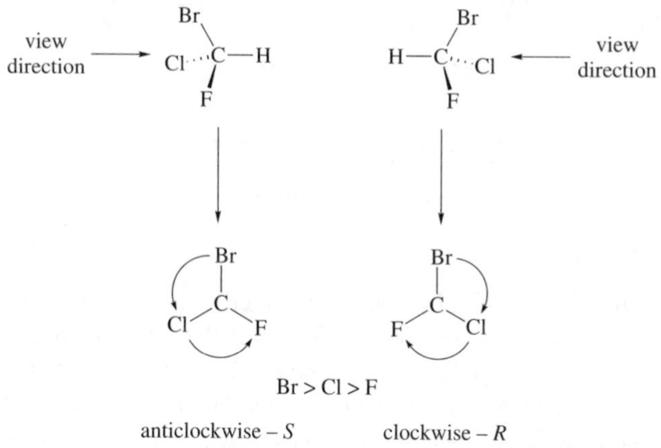

Br > Cl > F

anticlockwise – *S*   clockwise – *R*

Fig. 2.6

The stereodescriptors *R* and *S* are placed in front of the name of the structure, within parentheses. Thus the two enantiomers identified above are (*S*)-bromo-(chloro)fluoromethane and (*R*)-bromo(chloro)fluoromethane.

Butan-2-ol has a stereogenic centre, but unlike the example above, the ligands around it are mainly groups rather than single atoms, and two of them are carbon-containing groups. The most-preferred and least-preferred ligands are easy to identify: oxygen has the highest atomic number of the four atoms attached directly to the stereogenic centre, and so the OH group has the highest priority. The hydrogen has the lowest atomic number and is therefore least preferred.

To decide between the CH_3 and the CH_2CH_3 groups we must apply the priority rule again, but this time to the set of atoms directly attached to the ligand carbon atoms. Each set of atoms is written in order of preference and the two sets are compared. Priority is then assigned at the first point of difference. A simple way of doing this is as shown.

Methyl group – three hydrogens, equal priority   C($\underline{H}$HH)
Ethyl group – one carbon (higher priority), two hydrogens C($\underline{C}$HH)

The first point of difference is underlined, carbon is preferred to hydrogen, therefore the ethyl group has priority over the methyl group. The *R* and *S* stereodescriptors can now be assigned and the two enantiomers described (Fig. 2.7).

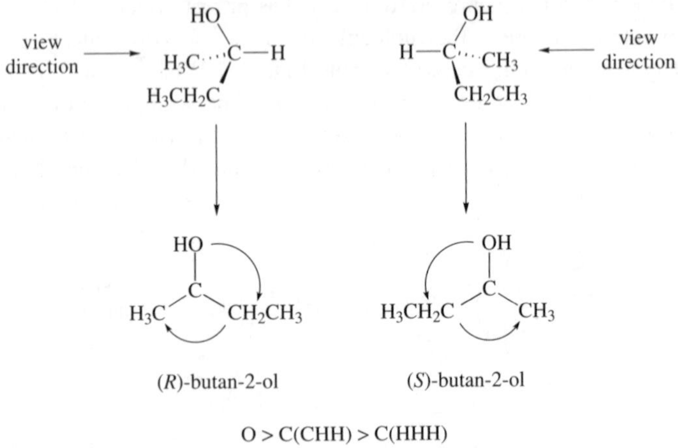

(*R*)-butan-2-ol   (*S*)-butan-2-ol

O > C(CHH) > C(HHH)

Fig. 2.7

For more complex molecules a number of criteria exist to determine the priority of substituents. One important criterion is that multiple bonds are considered as the equivalent number of single bonds. For example, a carbonyl group is treated as two C—O bonds, an alkene double bond as two C—C bonds, an alkyne triple bond as three C—C bonds and a nitrile group as three C—N bonds.

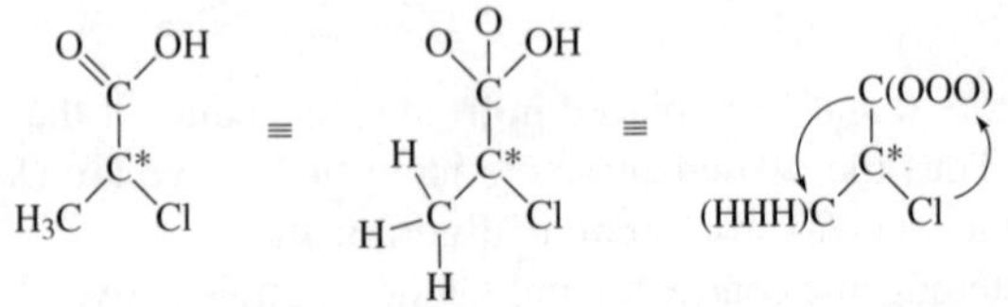

Fig. 2.8 The stereogenic carbon is indicated by an asterisk. Since we are viewing it away from the least-preferred ligand, the hydrogen is at the back.

To illustrate this, the enantiomer of 2-chloropropanoic acid giving rise to the esters E1 and E2 can be represented as in Fig. 2.8, looking down the C—H bond.

The order of priority is Cl > C(OOO) > C(HHH), which in this case is anti-clockwise, therefore the isomer is *S*. The enantiomer giving E3 and E4, being the mirror image, is (*R*)-2-chloropropanoic acid (Fig. 2.9).

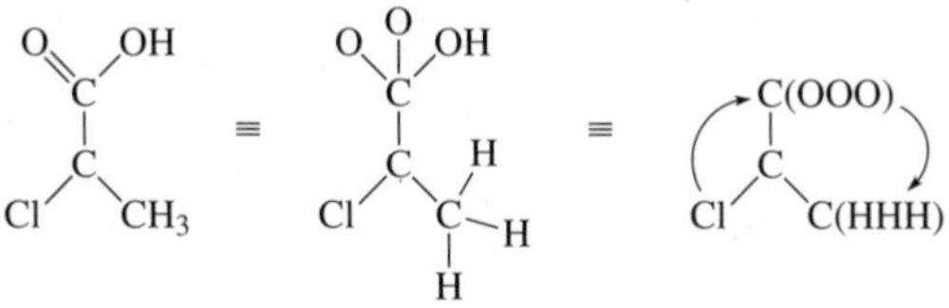

Fig. 2.9

You should now be able to assign *R* and *S* stereodescriptors to the two stereo-genic centres in the esters E1 and E4, as follows:

E1 = (*S*)-butan-2-yl (*S*)-2-chloropropanoate
E2 = (*R*)-butan-2-yl (*S*)-2-chloropropanoate
E3 = (*S*)-butan-2-yl (*R*)-2-chloropropanoate
E4 = (*R*)-butan-2-yl (*R*)-2-chloropropanoate

Now relate this to Fig. 2.5 and you will find that *RR* and *SS* are enantiomers, as are *RS* and *SR*, but that the pairs *RR*, *RS*; *RR*, *SR*; *SS*, *RS* and *SS*, *SR* are diastereoisomers.

The CIP sequence rule may seem cumbersome to begin with, but it is an extremely important concept in organic stereochemistry. Its application becomes easier with practice as you become used to dealing with structures in three dimensions.

Q. 2.3 Draw the enantiomeric forms for each compound listed in Fig. 2.10, then determine the least-preferred ligand for each example. Determine the sequence of the remaining ligands as viewed from the side opposite to the least-preferred. Assign *R* and *S* stereodescriptors.

2-hydroxyphenylethanoic acid

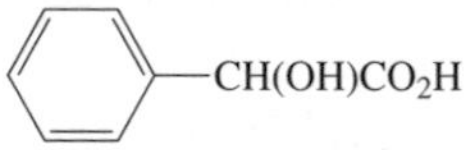

2-bromomalonaldehydic acid

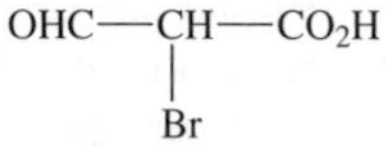

$$OHC\text{—}\underset{\underset{Br}{|}}{CH}\text{—}CO_2H$$

3-bromo-3-methylcyclopentene

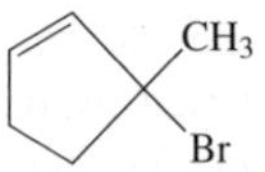

Fig. 2.10

It is important to remember that the descriptors must be determined for individual structures and that seemingly similar molecules may have opposite priority sequences. For example, in the 3-bromo-3-methylcyclopentene (Fig. 2.10) the methyl group is the least-preferred substituent. If the methyl group is replaced by a trifluoroethyl group, the least-preferred ligand becomes the ring CH_2 group and the stereodescriptors are different even though the spatial arrangement is still apparently the same (Fig.2.11).

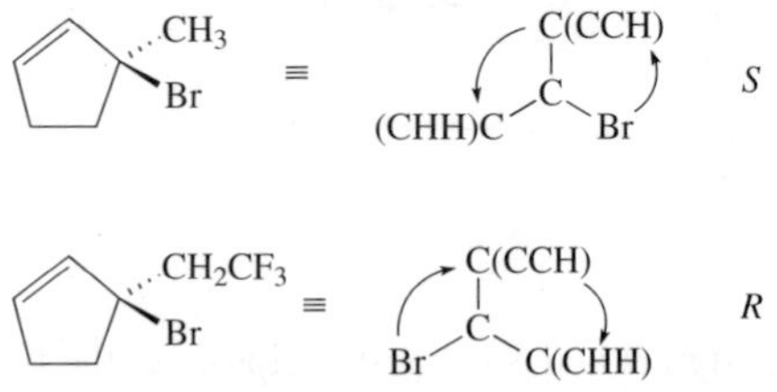

Fig. 2.11

The criteria described above are applicable to stereogenic centres with a ligancy of four, for which hydrogen is almost always the group of lowest priority. However, in certain instances it is possible to have a stereogenic centre with a ligancy of three. An example of this is when a nitrogen atom is incorporated into a ring skeleton and is thus unable to invert (see Chapter 1). The structure shown in Fig. 2.12 is (1*R*,6*R*,7*S*,9*S*,11*R*,16*S*)-sparteine. In order to determine the stereodescriptors of the nitrogen atoms at positions 1 and 16 the lone pair is considered as a ligand and is the least preferred.

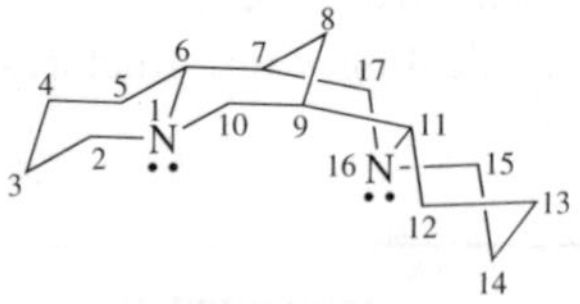

Fig. 2.12 (–)-sparteine.

2.3 Pseudoasymmetric centres and *meso*-compounds

Application of the CIP sequence rule becomes less straightforward when the ligands around a stereogenic centre occur in pairs and are themselves enantiomorphic.

2.3.1 meso-*Compounds*

The three compounds illustrated in Fig. 2.13, butane-2,3-diol, tartaric acid and 1,2-dichloro-1,2-dihydroacenaphthylene have two features in common: (i) a plane of symmetry bisecting the molecule (best seen in the sawhorse-type projection) and (ii) two stereogenic centres with opposite stereodescriptors. The latter condition renders the molecules optically inactive because there is internal compensation: the rotation caused by the *R* stereogenic centre is cancelled by that caused by the *S*. Molecules like these are usually referred to as *meso* forms. Figure 2.14 summarizes the possible forms of butane-2,3-diol.

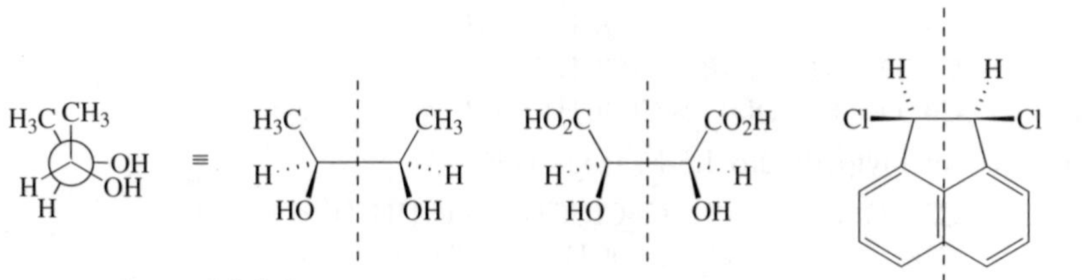

butane-2,3-diol

tartaric acid

1,2-dichloro-1,2-
dihydroacenaphthylene

- - - - - - denotes plane of symmetry

Fig. 2.13

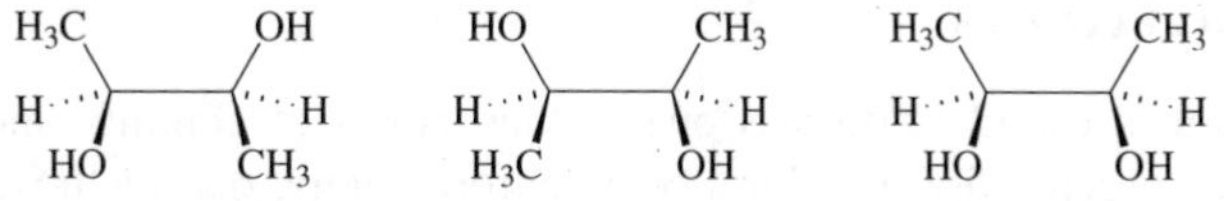

S,S-enantiomer   *R,R*-enantiomer   *meso-S,R*-diastereoisomer
        optically active   optically inactive

Fig. 2.14 Forms of butane-2,3-diol.

2.3.2 *Pseudoasymmetric centres*

A stereogenic centre which has two identical enantiomorphic ligands plus two other non-identical ligands, but which actually lies on a plane of symmetry, is termed a pseudoasymmetric centre. The stereodescriptors in this case are given as lower case italic letters to differentiate such centres from true stereogenic centres. Ribaric acid, xylaric acid and 2-methylperhydro-1,3-benzodioxole each have a pseudoasymmetric centre (Fig.2.15).

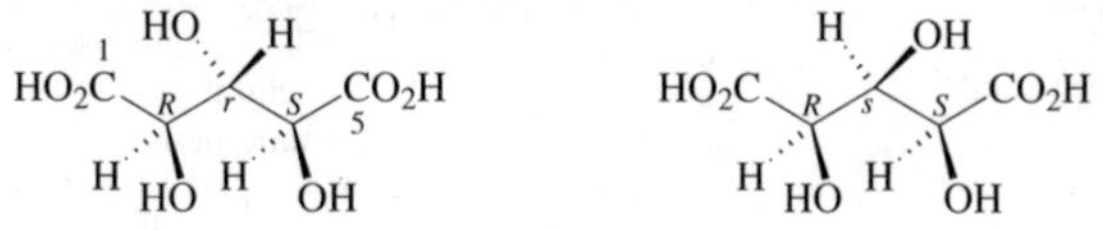

ribaric acid
3*r*

xylaric acid
3*s*

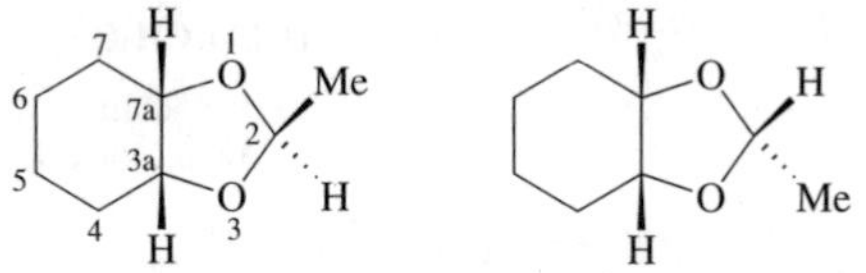

(2*r*,3a*R*,7a*S*)   (2*s*,3a*R*,7a*S*)

2-methylperhydro-1,3-benzodioxole

Fig. 2.15

Because two of the ligands on the pseudoasymmetric centre are the same, in terms of atomic number, a further criterion is invoked and that is that an *R* stereogenic centre has priority over an *S* one (*R* before *S* in the alphabet). Just to recap, the order of preference for ribaric and xylaric acid is as follows:

C2 O > C(OOO) > C(OCH) > H
C3 O > C(OCH)R > C(OCH)S > H
C4 O > C(OOO) > C(OCH) > H

and for 2-methylperhydro-1,3-benzodioxole:

C2   O–C(CCH)R > O–C(CCH)S > C(HHH)
C3a   O > C(OCH) > C(CHH)
C7a   O > C(OCH) > C(CHH)

Atomic number is thus the criterion applied in the first instance and in cases where there is no difference in atomic number the stereodescriptor is used.

2.4 Prochiral centres

At this point it is useful to discuss one further aspect of chirality and that is the concept of prochirality. Prochirality is, unfortunately, one of those stereo-chemical terms that has acquired several meanings. One such definition concern-ing trigonal systems is discussed in Chapter 8 (p. 107). In the current context

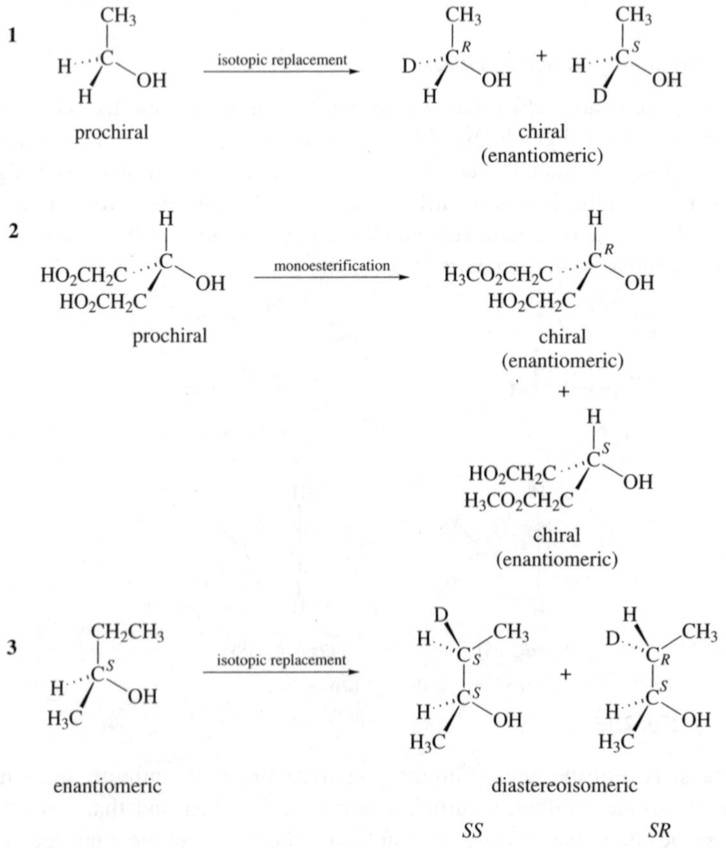

Fig. 2.16

of identifying chiral elements in molecular structures, however, a second application is as follows.

A group $Cabc_2$, where a, b and c are different substituents, can be made chiral by a single desymmetrization step. Simple examples are given in Fig. 2.16. The operations illustrated are the replacement of one of the groups c with a different group d so that the symmetrical entity $Cabc_2$ is converted into Cabcd, thus introducing chirality or, in example 3, further chirality.

The groups c (the hydrogens of the methylene group in **1**, the carboxymethyl groups in **2** and the methylene hydrogens of the ethyl group in **3**) are termed stereoheterotopic groups. In examples **1** and **2** these are enantiotopic because the chiral entities formed are enantiomers, and those in example **3** are diastereotopic because replacement of one of them leads to diastereoisomers.

2.4.1 pro-R, pro-S

The two groups c in $Cabc_2$ are distinguished as *pro-R* and *pro-S*: this is done by notionally assigning CIP priority to one group c over the other. If the resulting prochirality centre (C*) is R, the group c assigned higher priority is *pro-R*. If an S asymmetric centre is formed the group c is *pro-S*.

A simple example is $CHBrCl_2$ (Fig. 2.17). The two chlorines (denoted Cl^1 and Cl^2) are enantiotopic. If Cl^1 is considered to be of higher priority than Cl^2, then a clockwise preference sequence (a) is established. Cl^1 is therefore *pro-R*. If, on the other hand, Cl^2 is considered to be preferred over Cl^1, the priority sequence is anticlockwise and Cl^2 is thus *pro-S*.

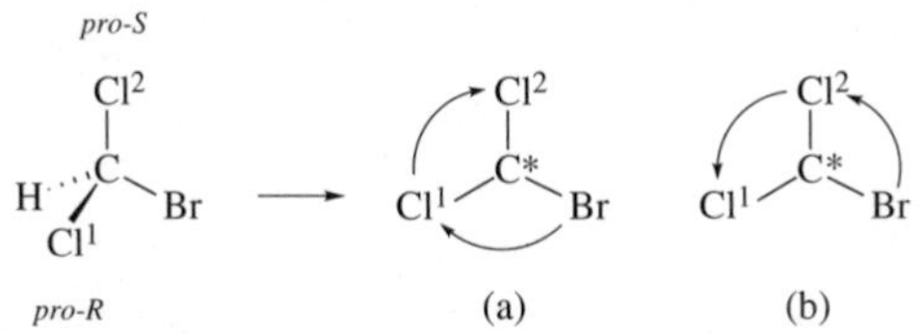

Fig. 2.17

We have already come across an instance of prochirality: the two oxygens in 2-methylperhydro-1,3-benzodioxole are enantiotopic (Fig. 2.18).

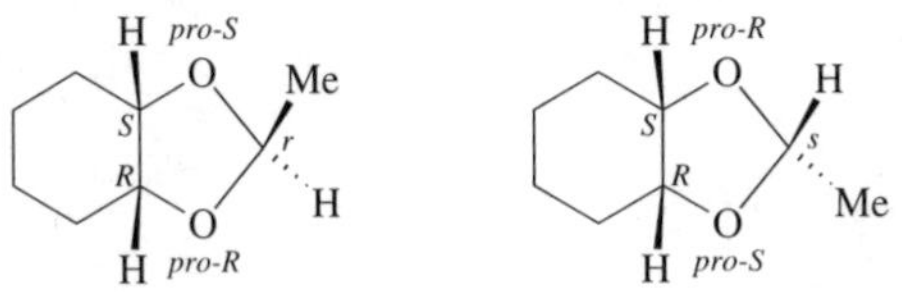

Fig. 2.18

Q. 2.4 State the stereochemical relationship (e.g. identical molecules, enantiomers, diastereoisomers) of the pairs of structures in each of the following sets:

(a)

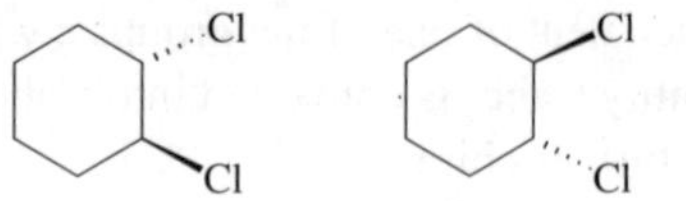

(b)

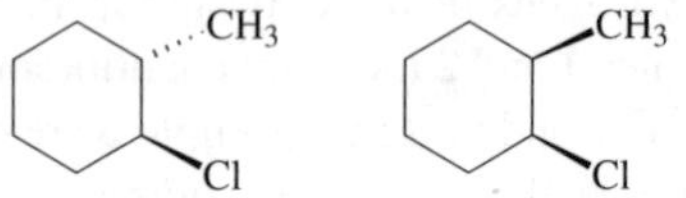

(c)

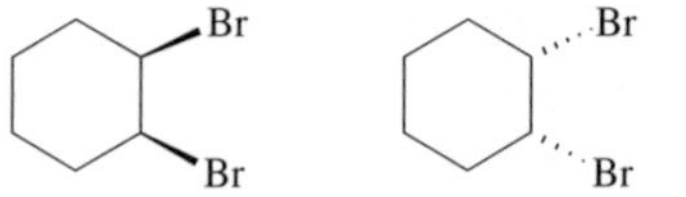

2.5 Axes of symmetry

The *meso*-form of tartaric acid has a plane of symmetry and is optically inactive. (*S,S*)-tartaric acid (Fig. 2.19) *is* asymmetric, *is* optically active but does possess

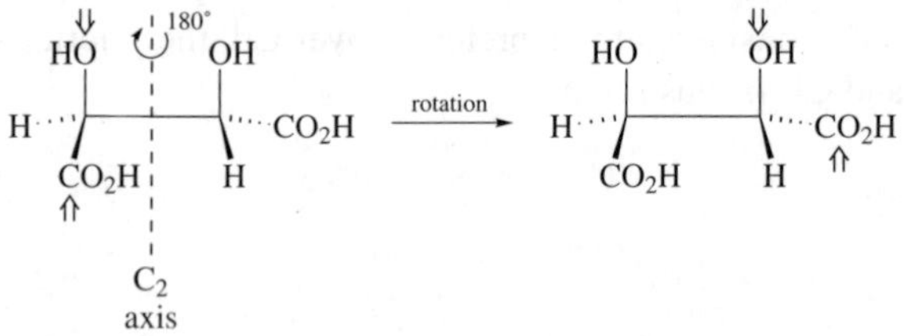

Fig. 2.19

an element of symmetry, namely rotational symmetry, with a C_2 axis. Imagine a vertical axis through the centre of the molecule. Rotation through 180° about this axis gives the same structure: the compound is said to have a 360/180 = two-fold

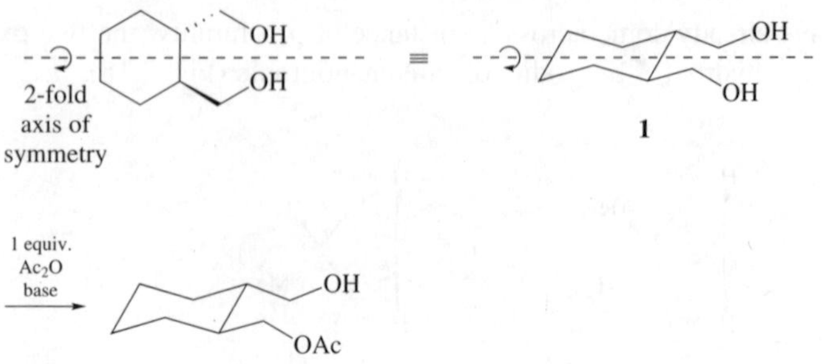

Fig. 2.20

axis of symmetry. The designation C_2 comes from group theory, a concept used more extensively by inorganic chemists than by organic chemists. Similarly, the diol **1** (Fig. 2.20) has a two-fold axis of symmetry. This means that the two hydroxy groups are equivalent: mono-acetylation gives just one compound. More generally, organic molecules can have n-fold axes of symmetry where n is an integer. For example, amongst other elements of symmetry cyclopropane (Fig. 2.21) has a three-fold axis of symmetry.

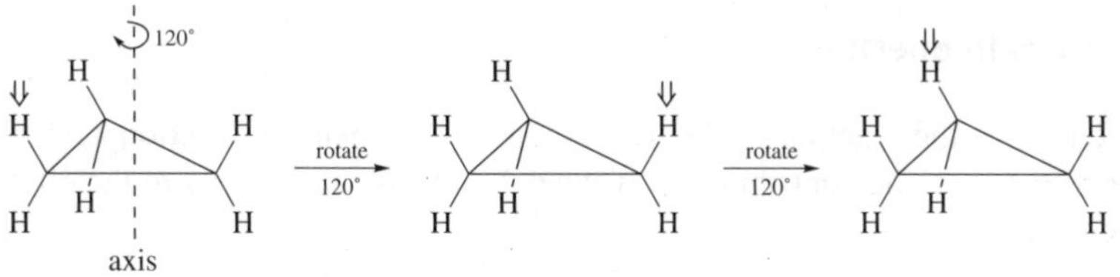

Fig. 2.21

Q. 2.5 The amino acid alanine $H_2NCH(CH_3)CO_2H$ can be dehydrated to give *cis*-(**A**) and *trans*-(**B**) dioxopiperazines. The *cis* isomer can exist in two enantiomorphic forms. The *trans*-isomer is optically inactive because it has a certain symmetry. Can you deduce the particular symmetry present in this case?

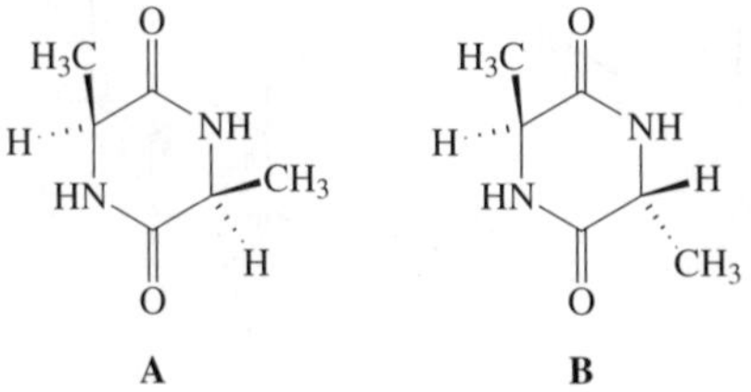

2.6   Representing three-dimensional molecules in two dimensions

In this chapter we have represented chiral molecules in the manner recommended by Maehr. Thus to represent substituents projecting in front of the plane of the paper, a wedge is employed. For substituents projecting below the plane of the paper a tapering broken line is used. In Fig. 2.22(a) (1*R*,2*R*)-dichloro-cyclohexane is depicted. If we wish to represent *trans*-1,2-dichlorocyclohexane

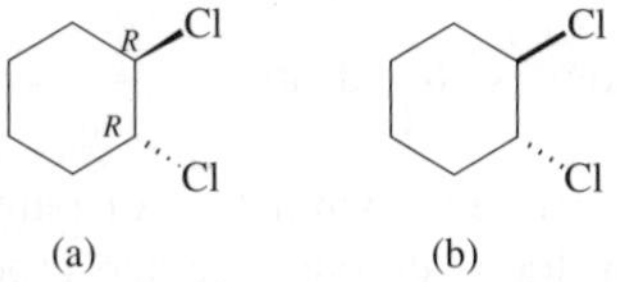

Fig. 2.22

without specifying the absolute configuration then a non-tapering wedge and broken line are used. Thus Fig. 2.22(b) represents a sample of *trans*-dichloro-cyclohexane for which the absolute configuration is unknown or which is made up of a mixture of enantiomers (for example, the racemate).

We will continue using this system for the three-dimensional representation of molecules throughout the rest of this book.

Answers to questions

Q. 2.1   Let the fraction of (+)-(α)-pinene = a. Then the fraction of (–)-(α)-pinene = $1 - a$. The contributions of the (+) and (–) enantiomers to the observed specific rotation are:

$$+51.3a + [-51.3 (1 - a)] = +30.8$$

Simplifying: $51.3 (2a - 1) = 30.8$
$$102.6a - 51.3 = 30.8$$
$$102.6a = 82.1$$
$$a = 0.8$$

Therefore the sample contains 80% (+)-pinene and 20% (–)-pinene.

Q. 2.2

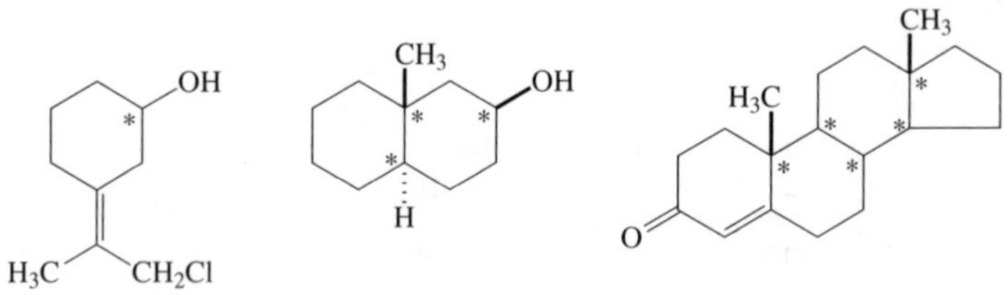

Q. 2.3

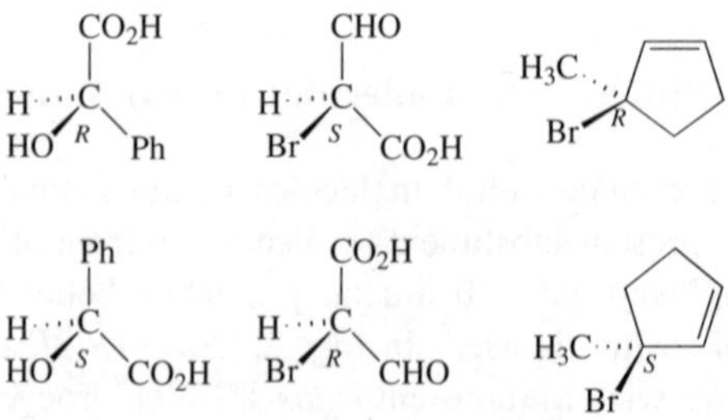

Q. 2.4   (a) Enantiomers; (b) diastereoisomers; (c) identical.

Q. 2.5   *trans*-Dioxopiperazine (**B**) has a centre of symmetry. A centre of symmetry is a point from which lines, when drawn on one side and produced an equal distance on the other side, will meet exactly similar points in the molecule.

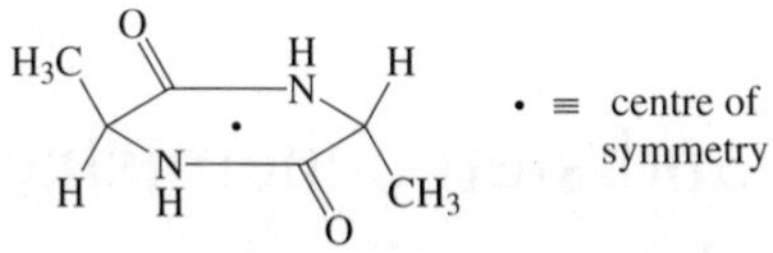

B

• ≡ centre of
      symmetry

Note that isomer **A** is not superimposable on its mirror image whereas mole-cule **B** is. Superimposability of object and mirror image is the ultimate test as to whether a compound will exhibit optical activity or not. Non-superimposability of object and mirror image means that a compound is capable of existing in optically active forms.

3 Nomenclature and stereochemistry of amino acids and some simple carbohydrates

There are three reasons why it is sensible to consider the stereochemical features of amino acids and carbohydrates separately. First, the molecules themselves are important, amino acids being the components of protein and carbohydrates being essential foodstuffs (for example, glucose), feedstocks (see the 'chiral pool', Chapter 14) and components of the genetic code (for example, ribose). Secondly, a type of nomenclature is still used for amino acids and carbohydrates that is no longer used for most other molecules. Thirdly, carbohydrates are often pictorially represented in a way which is peculiar to that type of compound.

3.1 Stereochemical descriptors: D and L notation, and Fischer projections

The simplest of the carbohydrates [general formulae $(CH_2O)_n$] is glyceraldehyde. The R-configuration of glyceraldehyde is shown in Fig. 3.1. The molecule can be viewed conventionally (left-hand formula, Fig. 3.1) or from a slightly

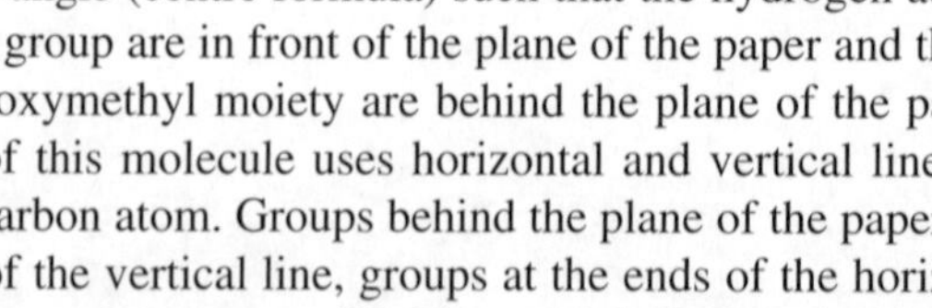

Fig. 3.1

different angle (centre formula) such that the hydrogen atom and the secondary hydroxy group are in front of the plane of the paper and the aldehyde group and the hydroxymethyl moiety are behind the plane of the paper. The Fischer projection of this molecule uses horizontal and vertical lines passing through the central carbon atom. Groups behind the plane of the paper are put at the top and bottom of the vertical line, groups at the ends of the horizontal line are in front of the plane of the paper. (S)-Glyceraldehyde has the Fischer projection shown in Fig. 3.2. By convention the carbon chain is on the vertical axis and, for carbohydrates and related molecules, the more highly oxidized centre is positioned at the top of the diagram.*

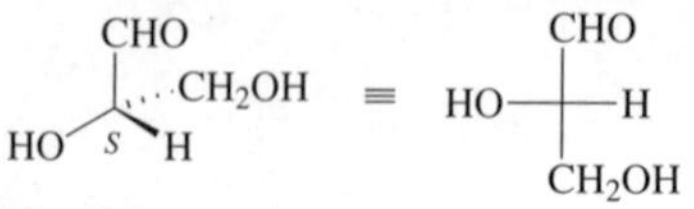

Fig. 3.2

* Uronic acids [OHC—$(CHOH)_n$—CO_2H] are an exception to this rule.

Glyceraldehyde was used as the cornerstone, the key reference compound, in a nomenclature which pre-dates the *RS* system by many years. (*R*)-Glyceraldehyde was called D-glyceraldehyde and (*S*)-glyceraldehyde was termed L-glyceraldehyde. Please note that the descriptors D and L have nothing at all to do with the direction of rotation of plane-polarized light (dextrorotatory may be represented by '*d*', laevorotatory by '*l*', *not* the capital letters).

At the time that the L and D notation was put forward the absolute configuration of glyceraldehyde was unknown. Luckily the proponents of the L and D nomenclature chose correctly, labelling dextrorotatory (+)-glyceraldehyde as D-glyceraldehyde. If they had chosen incorrectly all the old stereochemistry books would have been wrong.

There are 20 commonly occurring amino acids called the essential amino acids. One is achiral (glycine); all the others have a stereogenic centre. A selection of these common amino acids is shown in Fig. 3.3. All have the *S*-configuration except cysteine, which because of the high priority of the sulfur atom gets the stereodescriptor *R* . It is worth noting that this exception is one that results from the CIP rules and not an oddity of nature.

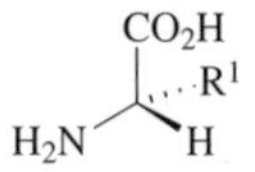

	Side chain (R^1)	Name	Configuration
	H	glycine	–
	CH$_3$	alanine	S
	CH(CH$_3$)$_2$	valine	S
	CH$_2$OH	serine	S
	CH$_2$Ph	phenylalanine	S
	CH$_2$SH	cysteine	R

Fig. 3.3

Figure 3.4 shows (*S*)-alanine in the conventional and the Fischer projections. By keeping the carbon chain on the vertical line and the carbon atom of the higher oxidation state to the top it can be seen in the Fischer projection that

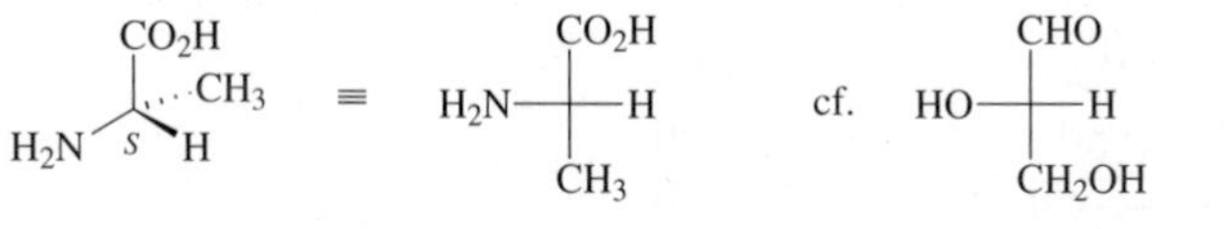

L-alanine

L-glyceraldehyde
[(*S*)-glyceraldehyde]

Fig. 3.4

the heteroatom (in this case nitrogen) lies on the left of the horizontal line and so the molecule can also be designated L-alanine. Thus attaching stereodescriptors to the amino acids is a fairly straightforward business; the situation with regard to carbohydrates is somewhat more complicated.

Q. 3.1 Draw (*R*)-cysteine as the Fischer projection. What is the effect of turning this Fischer projection through (a) 90° (b) 180° (c) 270°?

Let us return to the key compound D-glyceraldehyde. Any compound that can be derived from D-glyceraldehyde without disturbing the stereogenic centre is also labelled as a D-species (Appendix 2). Addition of HCN to the aldehyde group of glyceraldehyde followed by hydrolysis of the nitrile group yields the corresponding acids (Fig. 3.5). Both products have retained the configuration at the stereogenic centre at the bottom of the 'tree' and hence both compounds are stipulated to be, by convention, in the D-series. The two acids shown in Fig. 3.5 are diastereoisomers; partial reduction affords the corresponding aldehydes (Fig. 3.6).

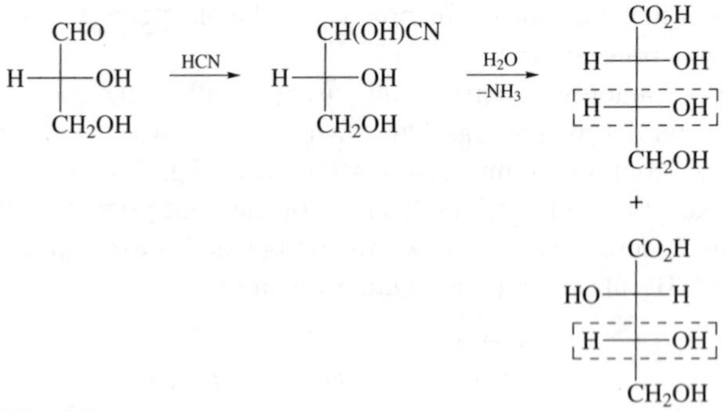

CHO — H—OH — CH₂OH
HCN →
CH(OH)CN — H—OH — CH₂OH
$\xrightarrow[-NH_3]{H_2O}$
CO₂H — H—OH — [H—OH] — CH₂OH
+
CO₂H — HO—H — [H—OH] — CH₂OH

Fig. 3.5

3.2 Nomenclature and stereochemistry of C₄ carbohydrates and tartaric acid

The C₄ aldehydes (Fig. 3.6) are carbohydrates labelled D-threose and D-erythrose. The corresponding L-sugars are shown in Fig. 3.7 and the inter-relationships are summarized in Fig. 3.8.

Tartaric acid is sometimes seen referred to as L-tartaric acid or D-tartaric acid and the relevant formulae are shown in Fig. 3.9 (see also Appendix 2). In this case the two carbon atoms at the end of the vertical chain have the same

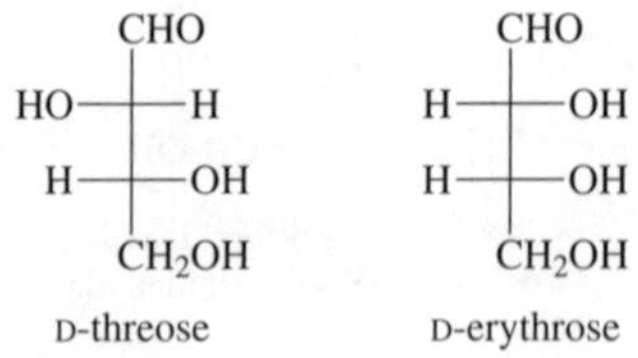

CHO — HO—H — H—OH — CH₂OH
D-threose

CHO — H—OH — H—OH — CH₂OH
D-erythrose

Fig. 3.6

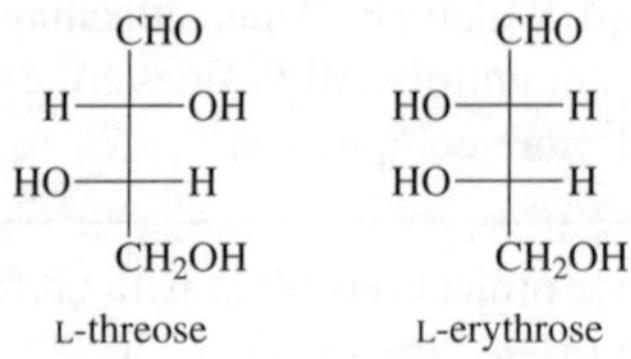

CHO — H—OH — HO—H — CH₂OH
L-threose

CHO — HO—H — HO—H — CH₂OH
L-erythrose

Fig. 3.7

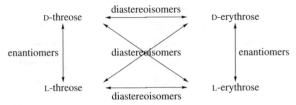

Fig. 3.8

oxidation state. However, it does not matter which carbon atom is situated at the bottom of the chain. Turn D-tartaric acid through 180° about the z-axis shown in Fig. 3.9 and the same arrangement of atoms and groups is found. Note that you can turn a Fischer projection through 180° without creating problems. However, you cannot turn the projection through 90° or 270° to represent the same compounds: in fact turning the Fischer projection through 90° or 270° gives the mirror image (see question 3.1).

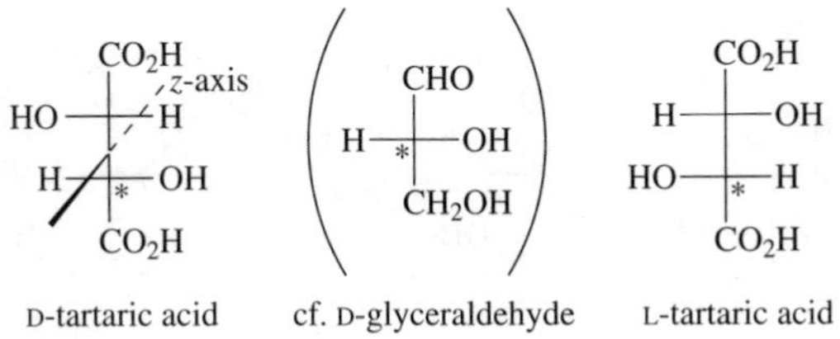

D-tartaric acid   cf. D-glyceraldehyde   L-tartaric acid

Fig. 3.9

3.3 C₅ and C₆ carbohydrates

Adding one extra carbon atom with hydrogen and hydroxy substituents to the threose/erythrose duo gives eight compounds (2^3 for three stereogenic centres). D-Ribose is shown in Fig. 3.10 together with D-arabinose (a diastereoisomer).

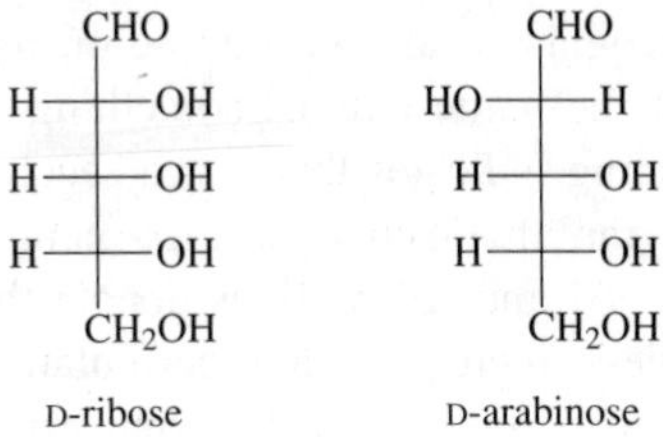

D-ribose   D-arabinose

Fig. 3.10

Ribose is a key constituent of RNA (see Chapter 13). Sugars like ribose with five or more carbon atoms in the backbone can exhibit ring–chain tautomerism, that is they can exist as a chain or a ring as illustrated in Fig. 3.11. The ring form is preferred, as shown by physical methods such as NMR spectroscopy. The wavy line in the ring formula indicates that there is a mixture of two compounds, comprising the two possible diastereoisomers produced by forming the new stereogenic centre (Fig. 3.12). (A wavy line is often used in this way to

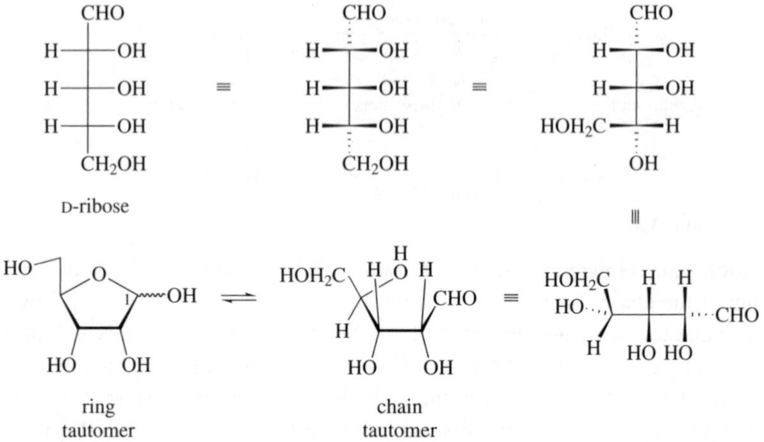

D-ribose

Fig. 3.11

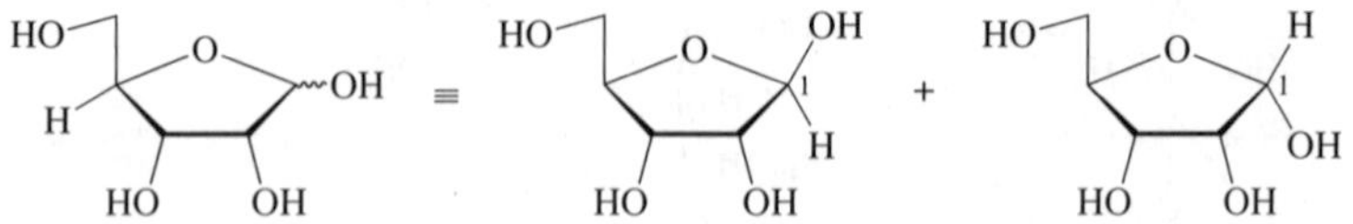

ring tautomer   chain tautomer

Fig. 3.12

represent a mixture of stereoisomers). Carbon atom 1 of the ring form of ribose is given a special name – an anomeric centre (see Chapter 1).

Q. 3.2 Draw L-arabinose in the five-membered ring form.

The structure shown in Fig. 3.13 is that of (1R,2R,3S,4R,5R)-5-hydroxymethyl-tetrahydropyran-1,2,3,4-tetraol, but, drawn in this particular way, in the ring form, most organic chemists would recognise it as α-D-glucose or, more fully, α-D-*gluco*-hexopyranose. Each part of the name tells us something about the structure of the molecule: the 'ose' ending indicates that it is a saccharide, the 'hexo' that it has six carbons and the 'pyran' that it contains a six-membered ring incorporating an oxygen. The '*gluco*', 'α' and 'D' prefixes denote the stereochemistry and, as you can probably guess, there are many permutations of the same molecular formula. So it is vital to know exactly the arrangement of the groups of atoms in space to be able to understand the chemistry of these very important compounds. The three prefixes (*gluco*, α and D) are discussed in detail below. Unless you are used to dealing with saccharide structures, it is difficult from the structure in Fig. 3.13 to grasp the significance of the stereochemical descriptors. To make their application more obvious, we need to examine the alternative straight-chain form with which the cyclic form exists in equilibrium in solution.

Conventionally, the linear form of a carbohydrate is drawn as the Fischer projection, with the backbone of carbon atoms drawn vertically and with the

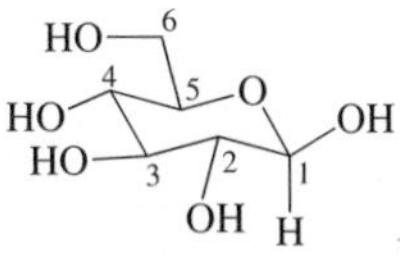

Fig. 3.13

most oxidized carbon at the top and the hydroxymethyl group at the bottom (Fig. 3.14). The cyclic structure is formed by a nucleophilic attack by the C5-hydroxy group on the aldehyde (C1) and is shown, by convention, as a Haworth projection in which the ring is considered to be at right angles to the plane of the paper with the substituents pointing up or down in the plane of the paper. The carbonyl carbon of the original Fischer projection now becomes part of a hemi-acetal group, the 'glycosidic' or 'anomeric' centre. It also retains its original numbering, that is, it is still C1, and it is to the disposition of the newly created hydroxy group that the α and β descriptors refer.

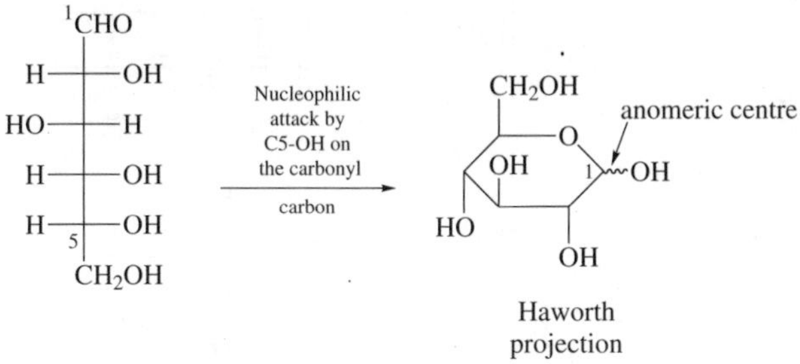

Fig. 3.14

In order to determine the correct anomeric prefix, we need to know the reference point and this is defined as the highest numbered asymmetric carbon in the Fischer projection; in this case C5 is at the bottom of the 'tree'. The reference carbon also enables us to assign the correct D or L descriptor. We must use the reference carbon in order to decide which anomer (alpha or beta) a particular structure is. We do this by constructing a Fischer representation of the cyclic form, Fig. 3.15. The term 'representation' is apt, as the picture is a completely unrealistic, stylized version, which, however, enables us to compare the configuration of the newly formed hemiacetal carbon with the reference carbon. Again, by convention, if the anomeric carbon has the same orientation as the reference carbon, that is the new OH group is on the same side as the C_{ref}—O bond, the structure is an alpha anomer, and if the new OH group is on the opposite side to the C_{ref}—O bond the structure is a beta anomer.

Armed with these facts we should now be able to make sense of the conversion process in Fig. 3.15. Beginning with arguably the most familiar saccharide, the linear Fischer projection depicted on the left is D-glucose [D because the OH group attached to the reference carbon (C5) is on the right-hand side]. The first step is a cyclization between C5—OH and the aldehyde group and both possible results are illustrated: the top structure is α-D-glucopyranose, alpha because

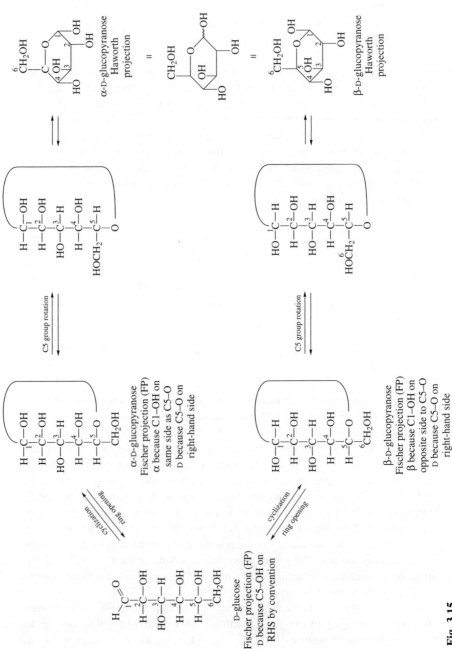

Fig. 3.15

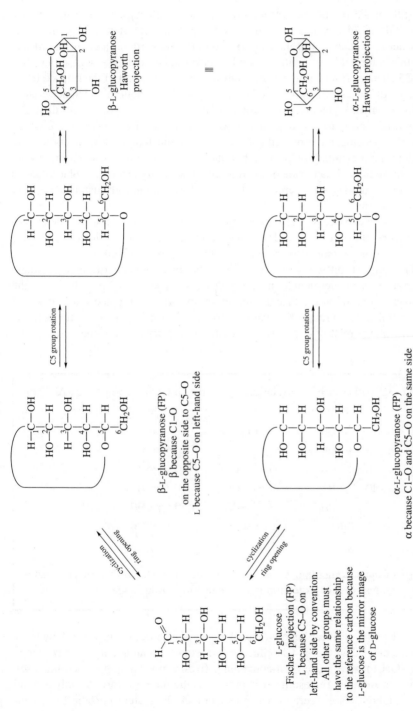

Fig. 3.16

C1—OH is on the same side as C5—O, D because C5—O (formerly C5—OH) is still on the right. The bottom structure is β-D-glucopyranose, beta because C1—OH is on the opposite side to C5—O. The next step is to convert the Fischer representation of the ring form into the Haworth projection. In order to do this, a formal bond rotation must be made because the oxygen atom attached to C5 changes from being part of a substituent to being part of the ring skeleton. The modified structure now has the ring oxygen at the bottom which necessarily shifts the hydroxymethyl group round to the left and the hydrogen to the right. The final step is to construct the Haworth projection by drawing the conventional six-membered ring with the oxygen at the top right-hand corner and adding the remaining hydroxymethyl and hydroxy groups: those groups which appear on the left-hand side of the modified Fischer projection are placed above the ring in the Haworth representation and those on the right-hand side appear below the ring.

The reverse process is also given and is likely to be more useful. The stereochemistry of cyclic saccharides is not always immediately obvious and conversion to the linear isomer is often required for it to be assigned.

The biggest difficulty in visualizing the Fischer–Haworth–Fischer transformation is the reorientation of C5 described above. Figure 3.16 gives the conversion between the Fischer and Haworth forms of L-glucose and it would be useful to go through the exercise for this isomeric pair as well. Note the mirror image relationship between the D and L pairs in both linear and cyclic forms.

Q. 3.3 (i) Describe the relationship of the following compounds **A–C** to D-glucose.

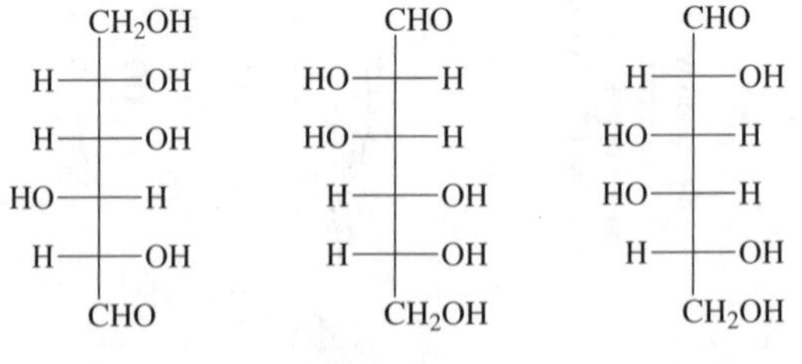

CH$_2$OH	CHO	CHO
H——OH	HO——H	H——OH
H——OH	HO——H	HO——H
HO——H	H——OH	HO——H
H——OH	H——OH	H——OH
CHO	CH$_2$OH	CH$_2$OH
A	**B**	**C**

(ii) Depict compound **C** as the α-anomer in a Haworth projection and translate this projection into a chair form of this sugar.

Cyclization of linear hexopyranosides can, of course, take place with hydroxy groups other than that at C5. Glucose (for example) can undergo cyclization at C4—OH to give an oxygen-containing, five-membered ring (glucofuranose). While in solution glucose itself is present as the furanose form to only a small extent (about 0.4%), certain reactions can lock the system into the five-membered ring form. Thus, reaction of glucose with acetone under acid catalysis gives a di-acetonide with the structure shown in Fig. 3.17. In contrast with glucose, the structural isomer called D-fructose forms significant amounts of the furanose form in aqueous solution (Fig. 3.18).

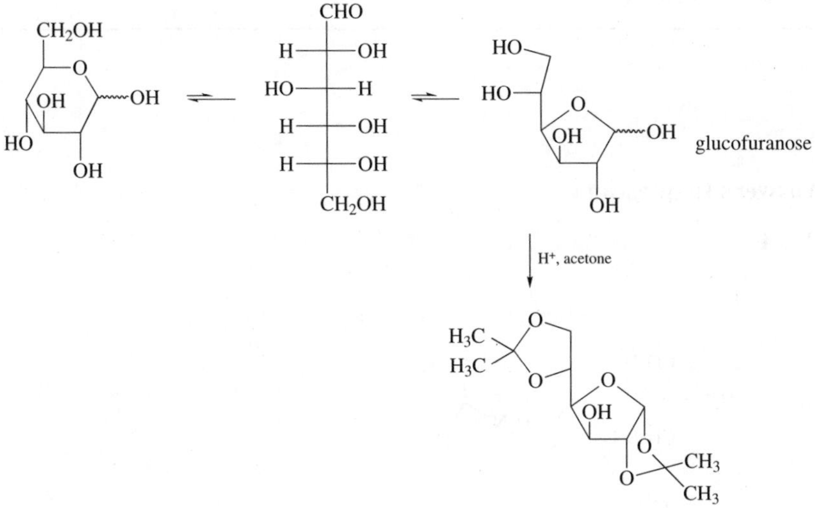

glucofuranose

H⁺, acetone

Fig. 3.17

56%

3%

32%

9%

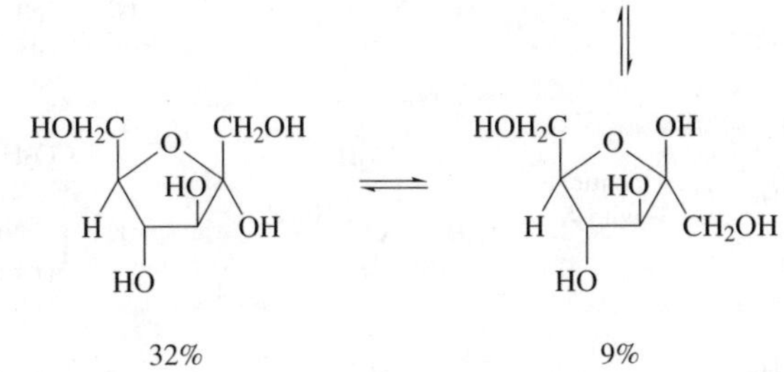

Fig. 3.18

Q. 3.4 When pure α-D-glucopyranose is dissoved in water it has a specific rotation +112. With time the rotation of the solution decreases to +52.7. When pure β-D-glucopyranose is dissolved in water it has a specific rotation of +18.7 which increases over time to +52.7. Explain this phenomenon.

Answers to questions

Q. 3.1

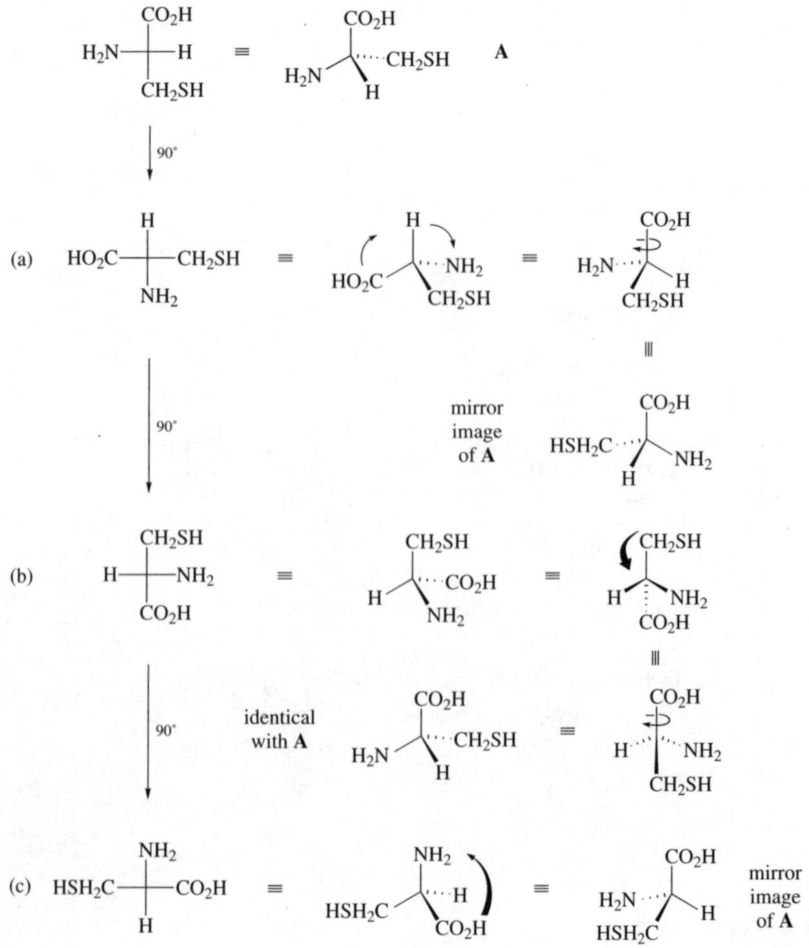

(a)

(b)

(c)

identical
with A

mirror
image
of A

mirror
image
of A

Q. 3.2

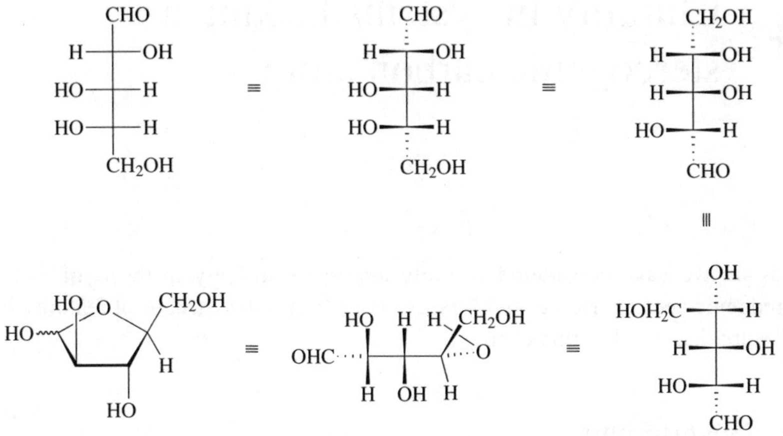

CHO
H——OH
HO——H
HO——H
CH₂OH

≡

CHO
H——OH
HO——H
HO——H
CH₂OH

≡

CH₂OH
H——OH
H——OH
HO——H
CHO

‖‖

OH
HOH₂C——H
H——OH
HO——H
CHO

HO O CH₂OH
HO
H
HO

≡

HO H H CH₂OH
OHC
O
H OH H

≡

Q. 3.3 (i) **A** is L-glucose.

B is a diastereoisomer of D-glucose, called D-mannose.

C is also a diastereoisomer of D-glucose called D-galactose.

(ii)

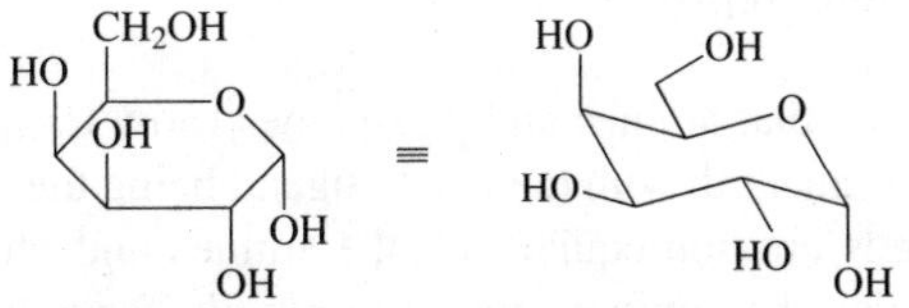

CH₂OH
HO
OH
O
OH
OH

≡

HO OH
HO
HO OH
O

Q. 3.4 The change of optical rotation with time is called *mutarotation*. It is caused by the equilibration of the α- and β-anomers through the open-chain form. The equilibration is catalysed by acid or base but even in the absence of catalyst the process is quite rapid.

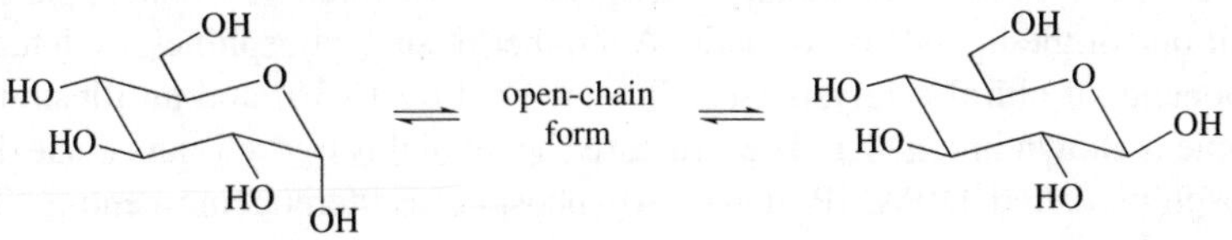

OH
HO
HO
O
HO OH

⇌ open-chain form ⇌

OH
HO
HO
O
OH
HO

$[\alpha]_D^{20} = +112°$

$[\alpha]_D^{20} = +18.7$

equilibrium mixture

$[\alpha]_D^{20} = +52.7$

4   Chirality in systems lacking a stereogenic carbon atom

Thus far we have considered chirality almost exclusively in terms of carbon stereogenic centres. However, chirality arises from other structural features and these are discussed in this chapter.

4.1 Point chirality

The stereogenic centres discussed in Chapter 2 are examples of point chirality and this phenomenon is also observed in compounds containing nitrogen, phosphorus and sulfur.

4.1.1 Tertiary amines and phosphines

You will recall from Chapter 1 that amines and phosphines (now termed phosphanes) are approximately tetrahedral, with the fourth ligand being the lone pair. If the remaining three ligands are non-equivalent, the amines and phosphines become chiral. Remember also that amines invert too rapidly for either of the enantiomeric forms to be isolated and so there is no observed optical activity for these compounds. (An exception to this is when the nitrogen of the amine forms part of a cyclic system which prevents inversion and locks one particular conformation – see Chapter 2.)

Asymmetrically substituted phosphines, on the other hand, have much higher energies of inversion and optically active isomers can often be isolated, particularly if one of the ligands is aromatic. A number of such phosphines are integral components of chiral catalysts (see Chapter 15, Fig. 15.36) and an illustrative example is shown in Fig. 4.1. The structure shown in this figure is an ethane-1,2-diphosphine called DIPAMP. It has two phosphorus stereogenic centres. The

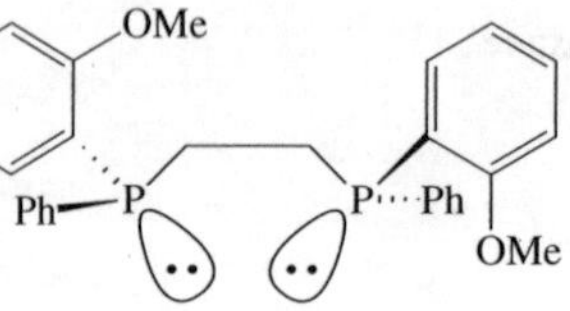

Fig. 4.1

stereoisomeric forms are enantiomeric, *RR* and *SS* (optically active), and diastereoisomeric, *RS* (*SR*). Because of the plane of symmetry the *RS* and *SR* forms are identical and constitute the *meso* isomer which is optically inactive (for a recap of *meso* isomerism, see Chapter 2). The *SS* enantiomer is shown; remember that the lone pair is the least-preferred ligand.

4.1.2 Phosphine oxides, amine oxides and sulfoxides

Phosphinates, phosphine oxides, amine oxides and sulfoxides are all configurationally stable, and, given a suitable diversity of substituents, are capable of optical activity (Fig. 4.2).

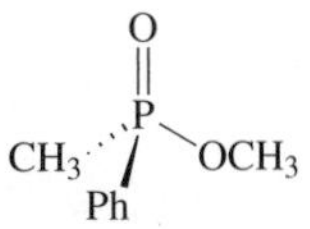

(*S*)-methyl methyl(phenyl)phosphinate

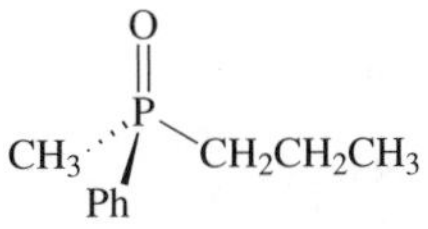

(*S*)-methyl(phenyl)propylphosphine oxide

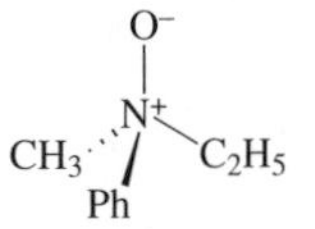

(*S*)-*N*-ethyl-*N*-methylaniline *N*-oxide

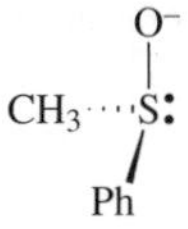

(*R*)-methyl phenylsulfoxide

Fig. 4.2

For pentavalent phosphorus compounds optical activity is observed only when the ligancy is four. For λ^5-phosphanes (PH_5 derivatives, with a ligancy of five) such as that shown in Fig. 4.3, the ligands are arranged so that they point to the vertices of a trigonal bipyramid. When the substituents are all different it would appear at first glance that enantiomerism is possible because the mirror images are apparently non-superimposable. However, this is illusory because the object and mirror image forms are able to interconvert, without bond breakage, through a process known as *pseudorotation*. Briefly, for a generalized trigonal bipyramidal λ^5-phosphane $PR^1R^2R^3R^4R^5$ (Fig. 4.3), the sequence of events comprises:

  (i) deformation of the axial bonds (R^2—P and R^5—P) to give a square-based pyramid;
 (ii) horizontal rotation about R^1—P;
(iii) migration of R^1—P and R^3—P to axial orientation.

Clearly there are several bond deformation/rotation combinations, so that after several iterations interconversion of object and mirror image is achieved.

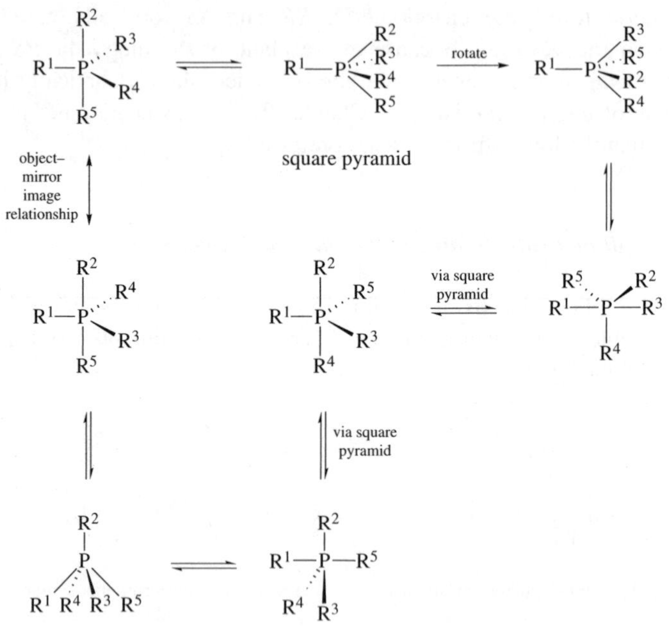

Fig. 4.3 Pseudorotation in λ^5 phosphanes.

4.2   Axial chirality

4.2.1   *Allenes*

We have already come across a class of compounds that show axial chirality and these are the allenes. (Bonding in allenes was introduced in Chapter 1.) An allene that is substituted at each end has no stereogenic centre yet can exist as enantiomers (non-superimposable mirror images, see 1,3-dichloroallene Fig. 4.4). The non-superimposability of the two forms may not be immediately obvious from the two-dimensional illustration, in which case you may find it helpful to construct a model of each enantiomer.

The procedure for assigning stereodescriptors to the two mirror images is as follows. The structure is regarded as an elongated tetrahedron and is viewed along the axis. To take the left-hand structure of Fig. 4.4 as an example, a tetrahedron is constructed with the substituents at the vertices [Fig. 4.5(a)].

The double bonds are removed for clarity, but as in Fig. 4.4, the top pair of atoms are in the plane of the paper and the bottom pair are perpendicular, with the chlorine at the back and the hydrogen at the front. The solid diagonal lines from H to H and H to Cl represent connections at the front of the tetrahedron, while the dotted line represents a connection at the rear. Next, the ligands are assigned an order of priority $a > b > c > d$ (> means 'is preferred to') [Fig. 4.5(b)]. The highest preferences, a and b, are given to the pair of atoms nearest

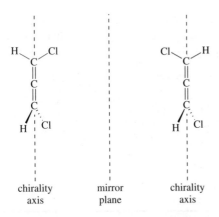

Fig. 4.4 Enantiomers of 1,3-dichloroallene (1,3-dichloropropa-1,2-diene). The top pair of ligands are in the plane of the paper, while the bottom pair are perpendicular.

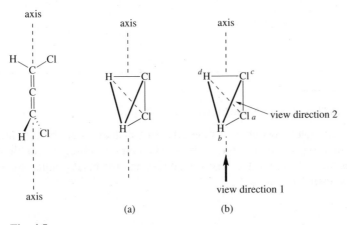

Fig. 4.5

the view direction 1 and in order of the CIP sequence rule. Therefore, the bottom rear Cl is *a* and bottom forward H is *b*. The remaining two ligands are simply assigned according to the sequence rule, so that top right Cl is *c* and top left H is *d*. To apply the sequence rule and determine the stereodescriptor, we must view the structure from opposite the least preferred ligand (*d*) (view direction 2). Now we are presented with a trigonal face with the remaining groups *a*, *b* and *c*. For the allene in Fig. 4.5 the order is clockwise and so the structure is (R_a)-1,3-dichloroallene (the subscript designates axial chirality).

In order to make this assignment we have viewed the structure along the chirality axis from the bottom. However, it makes no difference which end of the axis the structure is viewed from: the stereodescriptor will still be R_a. This can be deduced from Fig. 4.6 where the process is repeated. Thus the two ligands nearest the view direction are assigned *a* and *b*; this is now the top right Cl and top left H, respectively. Groups *c* and *d* now become bottom rear Cl and bottom forward H, respectively. To determine the order of *a*, *b* and *c*, we must view from the side opposite the least-preferred ligand, *d*, and we find that it is still clockwise, confirming the above *R* assignment.

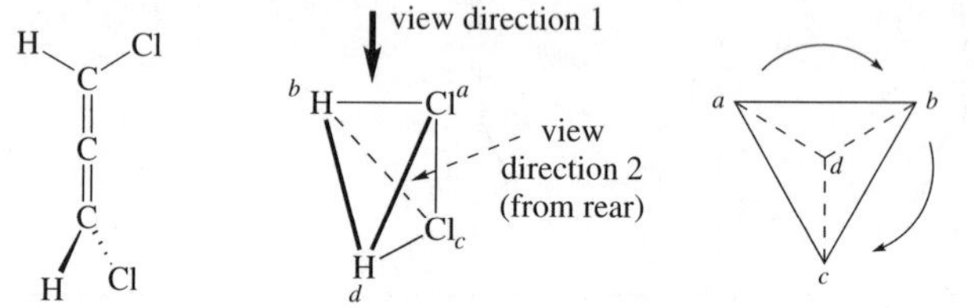

Fig. 4.6

Q. 4.1 Assign a stereodescriptor to 2-chlorobuta-2,3-diene shown below:

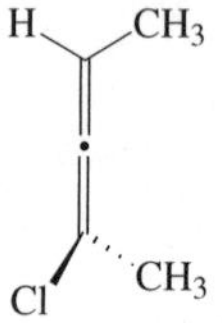

4.2.2 *Biphenyls and binaphthyls*

Other commonly encountered molecules that display axial chirality are *ortho*-substituted biphenyls and binaphthyls. The chirality observed for these compounds arises from the fact that rotation around the bond joining the aromatic units is restricted (Fig. 4.7).

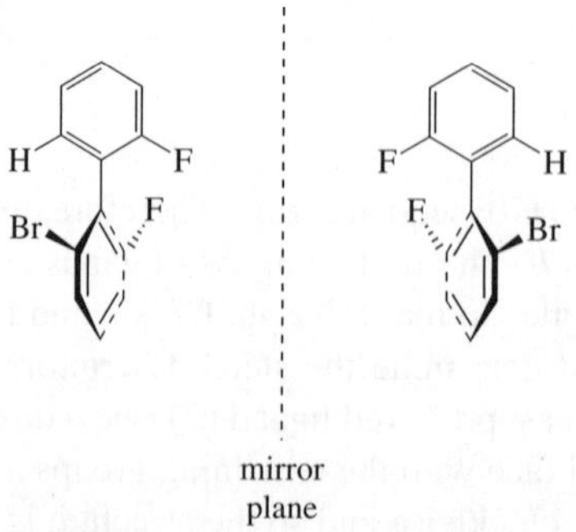

mirror
plane

Fig. 4.7

The substituents at the *ortho* positions restrict free rotation of the phenyl rings and prevent the rings from lying in the same plane. The result is that the rings remain orthogonal (perpendicular) to each other. When the substituents on either side of this inter-ring bond are different, as in Fig. 4.7, then the structure is able to exist in two enantiomeric forms. Such stereoisomers, resulting from restricted rotation about single bonds are called *atropisomers*.

2,2′-Binaphthol is an example of an asymmetric biaryl. It will be mentioned again in Chapter 15, where it is a constituent of a useful chiral reagent, BINAL-H (Fig. 4.8).

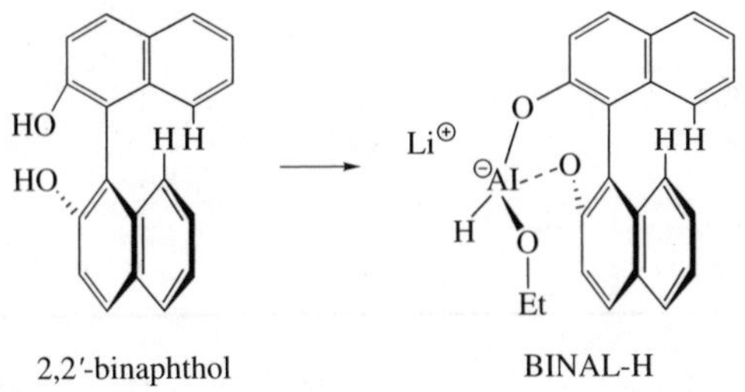

2,2′-binaphthol   BINAL-H

Fig. 4.8

The procedure for assigning stereodescriptors for biphenyls and binaphthyls is the same as for chiral allenes. The following is the assignment for the binaphthol in Fig. 4.8. The chiral axis is along the inter-ring bond and the four groups to be assigned in order of preference are shown as black dots in Fig. 4.9(a). The elongated tetrahedra are then constructed from these preference orders [Fig. 4.9(b)], and viewed from the face opposite the least-preferred ligand d [Fig. 4.9(c)]. The order is found to be anticlockwise, therefore the structure is (S_a)-2,2′-binaphthol.

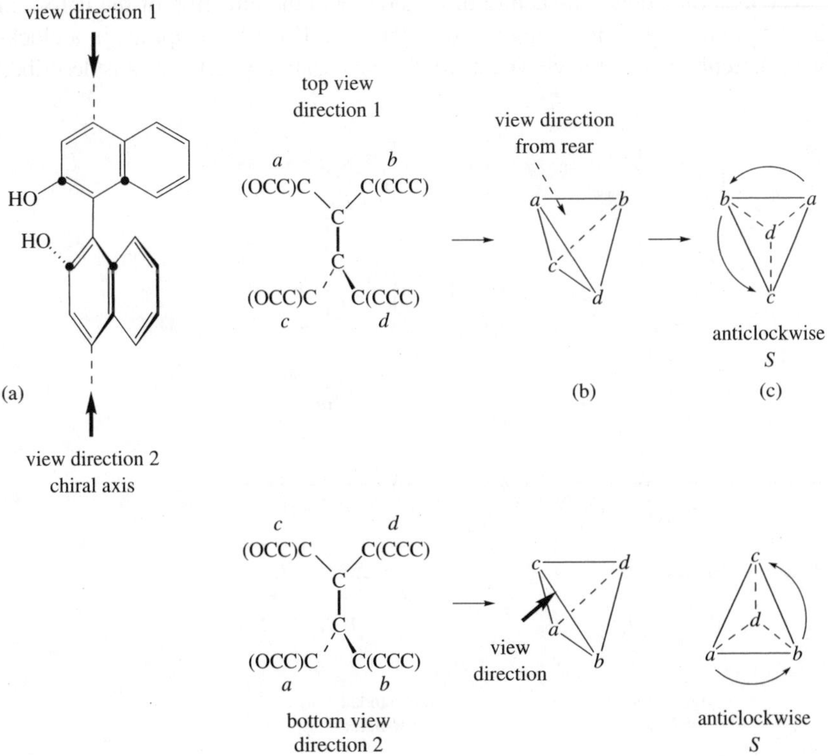

Fig. 4.9

Q. 4.2 Assign a stereodescriptor to the biaryl compound shown below:

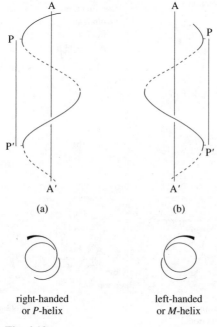

4.3 Helical structures

The helix is a chiral structure that is widespread in organic chemistry. Examples include DNA, some proteins, helicenes and, less obviously, biaryls, allenes (and other odd-numbered cumulenes), (*E*)-cyclooctene and some octahedral inorganic structures. In this section we look at the generalized helix and the different ways in which helicity and helical structures are considered.

Figure 4.10 shows a generalized helical segment (termed thus because, strictly speaking, a helix is of infinite length). The characteristics of a helix are its axis (line AA′), the pitch (line PP′) and the direction.

The axis runs down the centre of the helix and the direction of the helix can be determined when it is viewed down the axis. If the helix spirals in a clockwise direction when it is viewed down the axis, as in Fig. 4.10(a), it is described

(a)

(b)

right-handed
or *P*-helix

left-handed
or *M*-helix

Fig. 4.10

as a right-handed helix, to which the stereodescriptor *P* (plus) is applied. If the helix proceeds in an anticlockwise direction [Fig. 4.10(b)] then it is a left-handed helix and the associated stereodescriptor is *M* (minus). Note that it does not matter if we look down the axis A → A′ or A′ → A, the direction of helicity is the same.

An easily recognizable example of a helical structure is hexahelicene, shown in Fig. 4.11. The crowding of the six aromatic rings forces them out of planarity and causes one of the terminal rings to lie above or below the other so that the whole structure adopts a helical shape.

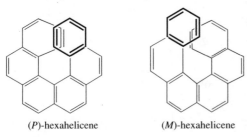

      (*P*)-hexahelicene   (*M*)-hexahelicene

Fig. 4.11

The other helix characteristic mentioned above is the pitch. This is the distance in which the helix makes one full turn.

4.3.1 Polynucleotides

Of the many helical structures in chemistry, the polynucleotide DNA (deoxyribonucleic acid) (Fig. 4.12) is probably one of the best known. The structure, as elucidated by Watson and Crick in 1953, is a double-stranded helix (a duplex) consisting of two antiparallel, right-handed (*P*) helices held together by hydrogen bonds. The backbone of each individual helix is made up of alternating deoxyribose and phosphate residues with any of the four bases thymine (T), cytidine (C), guanine (G) or adenine (A) projecting into the interior of the helix. The two strands are held together by hydrogen bonds between the base pairs A–T and G–C.

Polynucleotide helices are defined by the following characteristics: $n =$ number of residues per turn; $h =$ unit height translation per residue along the helix axis; $t = 360°/n =$ unit twist (angle of rotation per residue about the helix axis); $p =$ pitch height of the helix $= nh$. DNA has 10 base pairs per turn ($n = 10$) and a pitch of 34 Å ($p = 34$ Å) and therefore a unit twist $t = 36°$ and unit height translation $h = 3.4$ Å.

4.3.2 Poly(amino acids)

The set of helix characteristics above can be used to define the helical conformations adopted by many poly(amino acids) or polypeptides. However, helical polypeptide segments can also be defined in two other ways.

The first is in terms of the torsion angles of individual bonds in a residue. Figure 4.13 shows a perspective drawing of a residue in a polypeptide chain. The

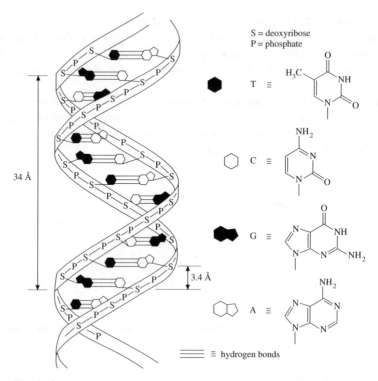

Fig. 4.12 Double helix structure of DNA.

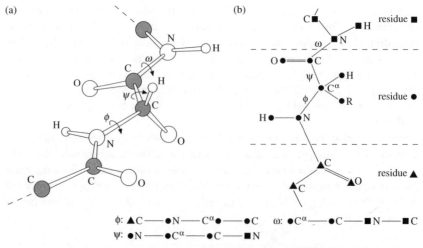

$\phi:$ ▲C —— ●N —— C$^\alpha$● —— ●C $\omega:$ ●C$^\alpha$ —— ●C —— ■N —— ■C

$\psi:$ ●N —— ●C$^\alpha$ —— ●C —— ■N

Fig. 4.13 (a) The atoms that make up the backbone of a polypeptide chain and the bonds with which the torsion angles ϕ, ψ and ω are associated. In (b) the backbone is drawn schematically to show the number of residues spanned by the defining torsion angles. The key gives the atoms between which the torsion angles are measured. A residue is defined as —NH—C$^\alpha$H—CO—. Any atoms attached to —NH or CO— belong to adjacent residues. C$^\alpha$ is the carbon bearing the amino acid side chain.

three repeating bonds in the backbone are N—CO, CO—CHR and CHR—N. In polypeptides that assume regular conformations, these bonds have fixed torsion angles associated with them labelled ω, ψ, ϕ, respectively. For example, poly(alanine) exists as a right-handed or (*P*)-α-helix. The torsion angles are

$$\phi = -57°, \psi = -47° \text{ and } \omega = 180°$$

In the mirror image (*M*)-α-helix (left-handed) the values are reversed:

$$\phi = +57°, \psi = +47° \text{ and } \omega = 180°$$

Other, non-helical conformations of polypeptides are discussed in Chapter 13.

The second way of denoting a helical structure of a polypeptide is used when hydrogen bonding serves to maintain the helical conformation. The torsion angle notation discussed above takes no account of hydrogen bonding. In this second designation, the helix is referred to as an n_r helix, where n is the number of residues per turn and r is the number of atoms in the ring that results from two points on the main chain being linked by a hydrogen bond (Fig. 4.14).

An α-helix with $n = 3.6$ and $r = 13$ can also be termed a 3.6_{13} helix. Another example is a 3_{10} helix, which is found, with 3.6_{13} helices, in the protein myoglobin (Fig. 4.14).

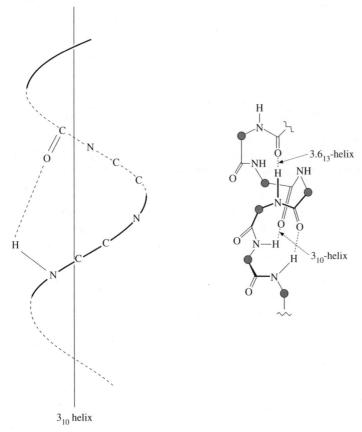

Fig. 4.14 3_{10} helix in part of the structure of myoglobin.

4.3.3   Biaryls and allenes

The perpendicular arrangement of the substituents about the chiral axis of allenes and biaryls means that they can also be considered as very short helical segments and as such can be described using the helix descriptors M and P. To take (R_a)-1,3-dimethylallene [(R_a)-penta-2,3-diene] as an example (Fig. 4.15),

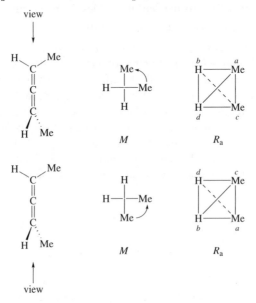

Fig. 4.15 (R_a)-1,3-dimethylallene.

the structure is viewed directly down the axis and the torsion angle between the group of highest CIP priority at the front and that at the back (the fiducial groups, the methyl groups in this case) defines the helicity. If the torsion angle

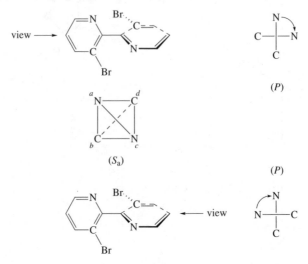

Fig. 4.16 (S_a)-3,3′-dibromo-2,2′-bipyridyl.

is anticlockwise going from front (nearest to view direction) to back, then the stereodescriptor is *M*; if the angle is clockwise, the descriptor is *P*.

Figure 4.16 shows the designation of (S_a)-3,3'-dibromo-2,2'-bipyridyl. From Figs. 4.15 and 4.16, you will see that there is a correlation between the two types of stereodescriptor: $R_a \Rightarrow M$; $S_a \Rightarrow P$.

Q. 4.3 (*E*)-Cyclooctene is a chiral molecule. Draw the two enantiomeric forms of this molecule.

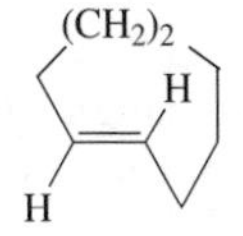

Answers to questions

Q. 4.1 (S_a)-2-chlorobut-2,3-diene: CH_3 has priority over H; Cl has priority over CH_3. Follow the format given in Fig. 4.5.

Q. 4.2 The stereodescriptor is S_a; note that F has priority over $N(O_2)$.

Q. 4.3

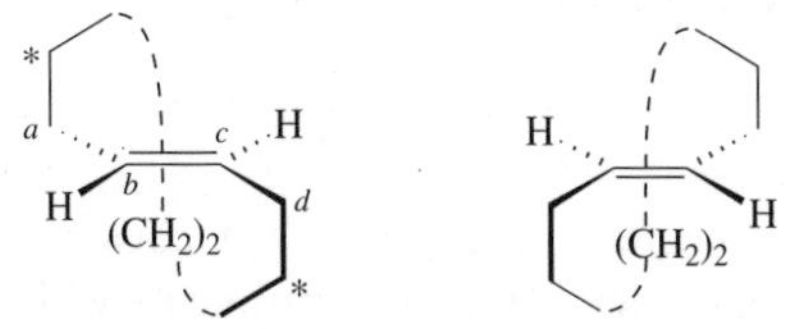

Note the barrier to racemization of (*E*)-cyclooctene is 150 kJ mol^{-1}. To assign a helical stereodescriptor to molecules such as (*E*)-cyclooctene the relationship of the carbon atoms (*) to the 'chiral plane' containing atoms C_a, C_b, C_c and C_d is considered.

For the left-hand molecule the two relevant projections are as follows:

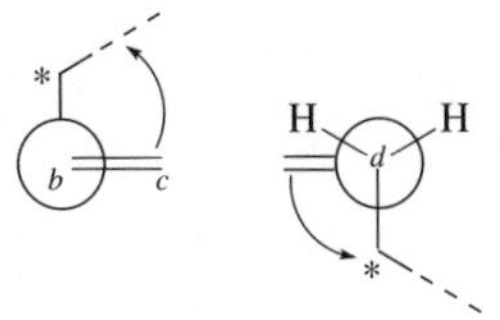

The motion in moving to the pilot atom (*) from the plane of chirality is anti-clockwise (*M*).

Similar considerations and manipulations result in the right-hand molecule being labelled (*P*).

5 Stereoisomerism about bonds of restricted rotation: *cis–trans* isomerism

We have already discussed the arrangement of groups around single atoms and axes giving rise to chiral entities. In this chapter we consider the arrangement of atoms or groups about bonds that have very restricted or zero rotation. This is generally referred to as *cis–trans* isomerism. You may come across the expression 'geometrical Isomerism' in older texts, but use of this term is now deprecated. *cis–trans* Isomerism describes the disposition of substituents about a reference plane: if the groups are on the same side of the reference plane they are described as *cis*, and if they are on opposite sides they are *trans*.

cis–trans Isomers sometimes, but not always, display chirality. Chiral cyclic compounds fall into this category and the *cis–trans* system provides a useful alternative way of describing these structures when only the relative stereochemistry is known.

5.1 Stereochemistry in cyclic systems

5.1.1 cis–trans *Nomenclature*

A simple example of *cis–trans* isomerism in cyclic compounds is dichlorocyclohexane (Fig. 5.1). Here we have two identical substituents which can be placed either on the same side of the ring plane (the reference plane) or on opposite sides. The relative ring positions of the substituents are immaterial and all three positional isomers in Fig. 5.1(a) are *cis*-dichlorocyclohexane and those in Fig. 5.1(b) are *trans*-dichlorocyclohexane.

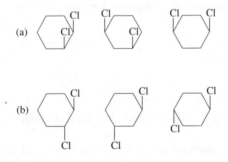

Fig. 5.1 (a) *cis*-Dichlorocyclohexanes; (b) *trans*-dichlorocyclohexanes.

Of course in reality the cyclohexane is not flat as shown for convenience in Fig. 5.1. In the chair form (see Chapter 1) the *cis* isomer of 1,2-dichlorocyclohexane has one chlorine in an axial and one in an equatorial position, Fig. 5.2(a).

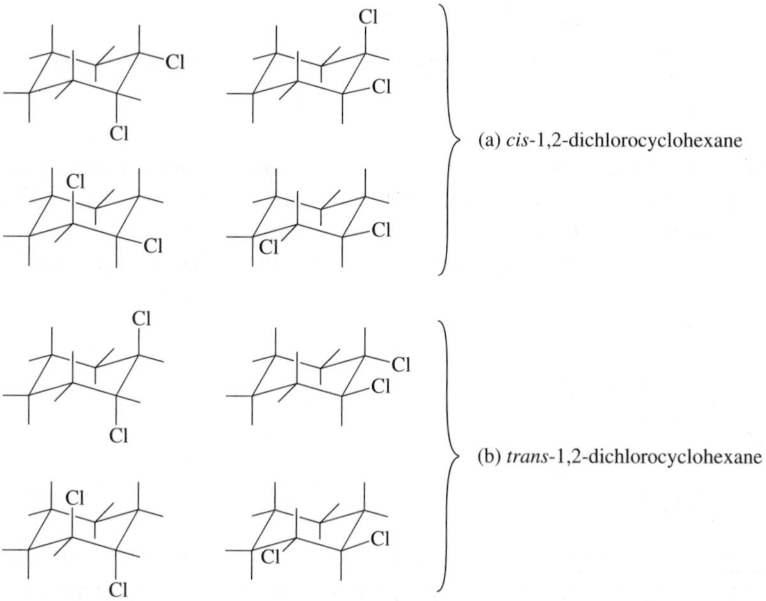

(a) *cis*-1,2-dichlorocyclohexane

(b) *trans*-1,2-dichlorocyclohexane

Fig. 5.2

The *trans* isomer, on the other hand, has both chlorines occupying either the axial or (preferentially) the equatorial positions. The situation is reversed for 1,3- and the same for the 1,4-dichlorocyclohexanes.

Q. 5.1 *cis*-1,2-Dichlorocyclohexane exists mainly in the chair form with one chlorine atom in the axial and one in the equatorial configuration. As such the molecule is chiral (**A** is not superimposable on **B**). Explain why *cis*-1,2-dichlorocyclohexane and like molecules do not show optical activity.

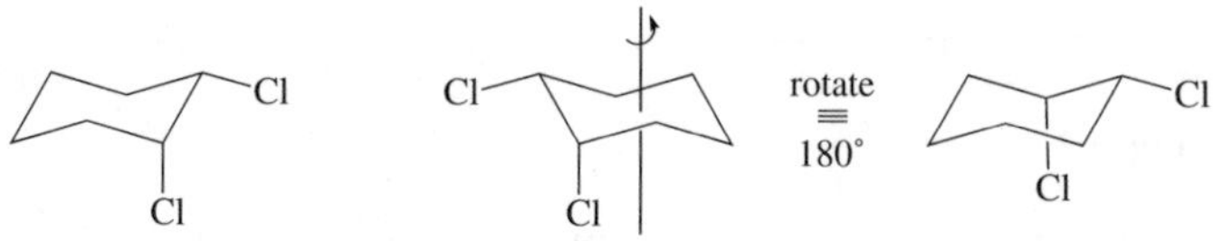

rotate
≡
180°

A   **B**

The two substituents need not be the same: 1-bromo-2-methylcyclopentane, for example, can exist as *cis* and *trans* isomers (Fig. 5.3).

trans-1-bromo-2-methylcyclopentane   cis-1-bromo-2-methylcyclopentane

Fig. 5.3

The rules for assigning *cis* or *trans* stereodescriptors are extended to allow for the presence of a third substituent. Taking again our example of 1,2-dichloro-cyclohexane, if another chlorine is introduced onto one of the already chlorine-bearing carbons then the *cis–trans* isomerism is lost and *RS* nomenclature for C* must be used to define the enantiomer under scrutiny (Fig. 5.4).

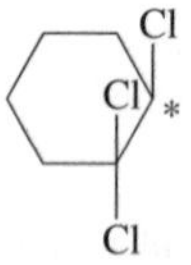

Fig. 5.4

However, when the third substituent is different and at a different position, *cis–trans* isomers still exist, but the notation is changed slightly. For a compound such as 5-chlorocyclohexane-1,3-dicarboxylate, where the three substituents are present on three different ring positions, one substituent is taken as a reference (*r*) (the one on the carbon atom with the lowest locant, in this case C1—COOH) and the other two are assigned as *cis* (*c*) or *trans* (*t*) to the reference. [The locant is the number given to a carbon atom to locate it relative to the principal functional group.] The notation used is *c*, *t* and *r*, and these are placed in front of the corresponding locant in the structure name, as illustrated in Fig. 5.5. The name in this case is *t*-5-chlorocyclohexane-*r*-1,*c*-3-dicarboxylic acid.

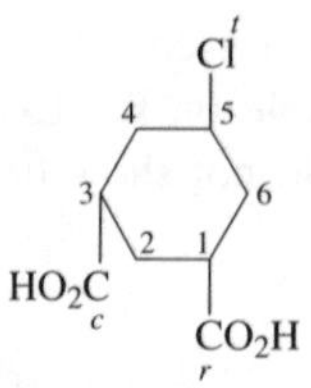

Fig. 5.5

Where there is a choice of substituent for the reference, as in Fig. 5.6 where the substituents are geminal pairs (that is, two substituents on the same carbon atom), that which is preferred by the CIP priority sequence rule is used. Only one of the other pair of substituents need then be designated to define the whole structure, and that again is the one of higher priority. The name therefore becomes *r*-1-bromo-1-chloro-*t*-3-ethyl-3-methylcyclohexane.

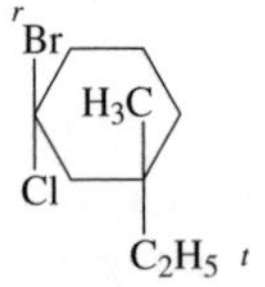

Fig. 5.6

The terms *cis* and *trans* are sometimes used to denote the relative stereo-chemistry of ring junctions in fused ring systems. Although *RS* nomenclature can usually be applied to the individual bridgehead positions, this does not necessarily give any indication of the geometrical shape of the molecule and the *cis–trans* alternative, which does, is sometimes more useful. Figure 5.7 shows the *cis* and *trans* forms of decalin.

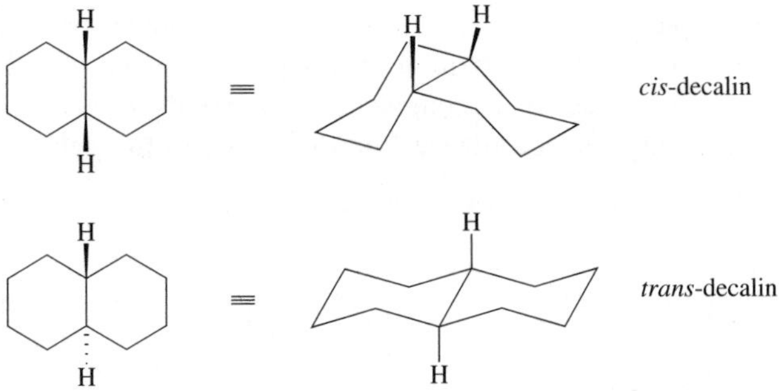

Fig. 5.7

Where there are two ring fusions in a structure, the stereochemistry of the two can be related by use of the descriptors *cisoid* and *transoid*. These define the dispositions of the nearest atoms of the two fusions or, if the fusions are the same distance apart at both ends, then between the end of the fusion with the lowest locant and its nearest neighbour. In Fig. 5.8 the perhydrophenanthrene illustrated is *cis-cisoid-trans* and the perhydroacridine is *cis*-4a-*cisoid*-4a, 10a-*trans*-10a.

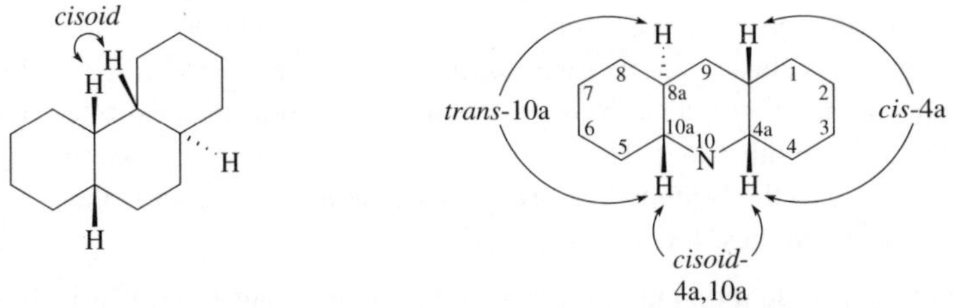

cis-cisoid-trans-perhydrophenanthrene   *cis*-4a-*cisoid*-4a,10a-*trans*-10a-perhydroacridine

Fig. 5.8

5.1.2   exo–endo, syn–anti *Nomenclature*

These often misused stereodescriptors provide a means of identifying the relative positions of non-bridgehead substituents on bicyclic systems of general formula bicyclo [*x.y.z*]alkane where $x \geq y \geq z > 0$, as shown in Fig. 5.9.

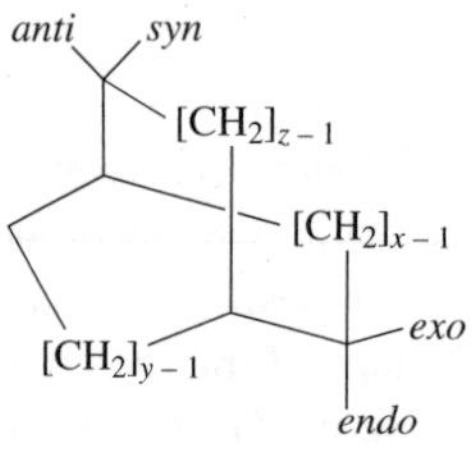

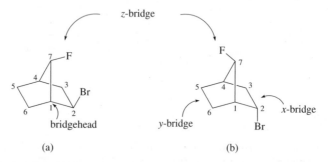

Fig. 5.9

A specific example, shown in Fig. 5.10, is a dihalonorbornane or, according to the von Baeyer nomenclature system, a bicyclo[2.2.1]heptane derivative. The two structures are substituted by a bromine next to the bridgehead and a fluorine on the methylene bridge. In accordance with the rules governing the numbering

Fig. 5.10 (a) 2-*exo*-bromo-7-*syn*-fluorobicyclo[2.2.1]heptane
(b) 2-*endo*-bromo-7-*anti*-fluorobicyclo[2.2.1]heptane

of bicyclic systems, the numbering starts at a bridgehead and follows the largest bridge first. In Fig. 5.10 there is a choice as two bridges are the same size. In this instance the numbering is fixed by the bromine, which requires the lowest possible locant, and so is assigned position 2, the numbering commencing at the adjacent bridgehead. The numbering proceeds to the next bridgehead (C4) and round the next ethylene bridge, leaving the fluorine-substituted methylene bridge to be numbered 7. The bromine atom concomitantly fixes the assignment of *x*, *y* and *z* to the bridges: the bridge with the lowest locant is the *x*-bridge, the one with the highest locant is the *z*-bridge.

- If the substituent on either the *x*- or *y*-bridge is orientated towards the *z*-bridge it is termed *exo*. If instead it points away from the *z*-bridge then it is *endo*.

- If a substituent on the *z*-bridge is directed towards the *x*- or lowest-numbered bridge it is *syn*, but if it points towards the *y*-bridge then it is *anti*.

Therefore, following the guidelines above, Fig. 5.10(a) is 2-*exo*-bromo-7-*syn*-fluorobicyclo[2.2.1]heptane and Fig. 5.10(b) is 2-*endo*-bromo-7-*anti*-fluoro-bicyclo[2.2.1]heptane.

Q. 5.2 Draw 2-*exo*-bromo-3-*endo*-hydroxybicyclo[3.2.1]octane.

5.2   Stereoisomerism about double bonds

5.2.1   E,Z nomenclature

The principles behind assigning the stereochemistry of double bonds is very similar to that discussed above for cyclic structures. We know from Chapter 1 that double-bond (sp²) structures are flat. The reference plane is therefore through the double bond, perpendicular to the substituents [Fig. 5.11(a)] which then occupy positions either above or below the plane. Usually, though, double bonds are written with the substituents in the plane of the paper [Fig. 5.11(b)] so that the reference plane is perpendicular to the page.

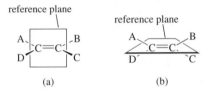

reference plane

reference plane

(a)   (b)

Fig. 5.11

The simplest example is when two adjacent substituents are the same. For example, when A = B = H and C = D = CO$_2$CH$_3$ [Fig. 5.12(a)], the two ester groups are on the same side of the reference plane and so the structure is *cis*. In Fig. 5.12(b) the ester groups are on opposite sides and so have a *trans* disposition.

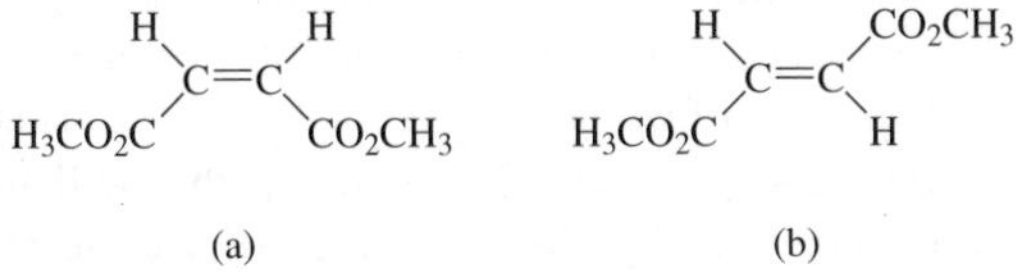

$$H_3CO_2C \quad H \quad H \qquad H_3CO_2C \quad H \quad CO_2CH_3$$

(a)   (b)

Fig. 5.12

When considering in general terms the relative orientation of the groups on a double bond it is fine to use the terms *cis* and *trans* to mean 'on the same side' and 'on opposite sides', respectively. However, when specifying a double bond in a compound name, the stereodescriptors Z and E [from the German *zusammen* (together) and *entgegen* (opposite)] are used. E and Z are generally placed in parentheses at the beginning of the name, with associated locants if there is more than one double bond. If the double bond is not part of the main chain E and/or Z is placed next to the substituent containing the double bond. Some more examples are given in Fig. 5.13.

(2E,4E)-hexa-2,4-diene

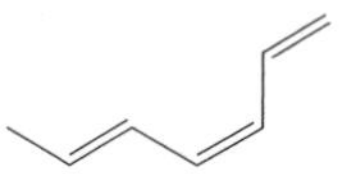

(3Z,5E)-hepta-1,3,5-triene

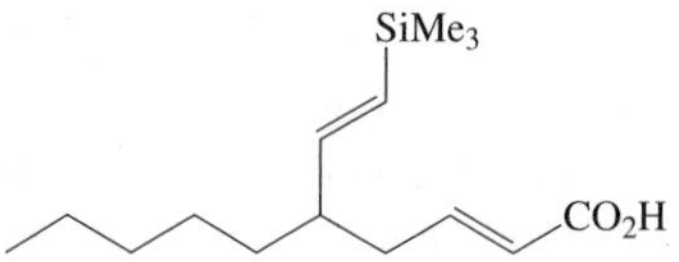

(2E)-5-[(E)-2-trimethylsilylvinyl]dec-2-enoic acid

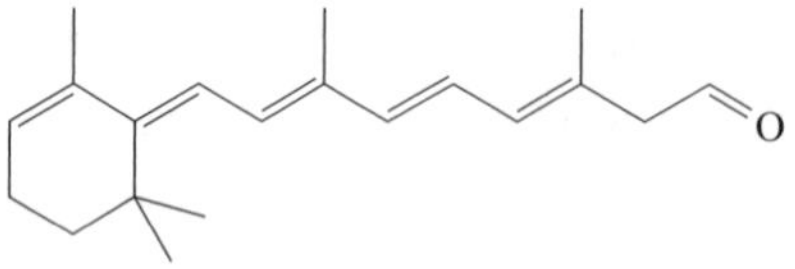

(3E,5E,7E)-3,7-dimethyl-9-[(E)-2,6,6-trimethylcyclohex-2-enylidene]nona-3,5,7-trienal
[(6E,8E,10E,12E)-4,14-*retro*-retinal]

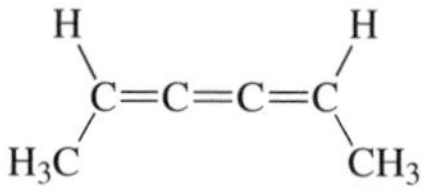

(Z)-hexa-2,3,4-triene

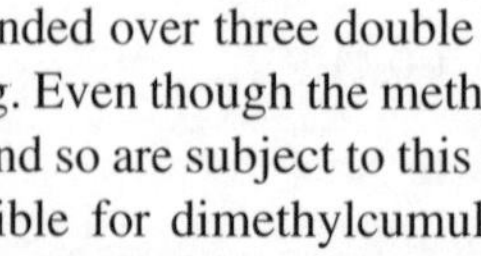

(E)-3,6-diethylidenecyclohexa-1,4-diene

Fig. 5.13

The last two examples in Fig. 5.13 illustrate how the *cis/trans* dispositions can be extended over three double bonds and over two double bonds separated by a flat ring. Even though the methyl groups are so far apart, they remain in the same plane and so are subject to this type of isomerism. Although *cis–trans* isomerism is possible for dimethylcumulene [(Z)-hexa-2,3,4-triene], this will not be the case for the lower homologue, dimethylallene, as the pairs of terminal substituents are in perpendicular planes and the molecule overall is an elongated tetrahedron, rather than flat (see section 4.2).

Returning now to the generalized structure in Fig. 5.11, when three of the substituents are the same, A = B = C = Br, say, then the stereoisomerism is lost. If, however, two substituents are the same and a third is different, *cis–trans* isomerism can still exist, but not for all positional isomers. For example, for the isomers of bromodichloroethene, Fig. 5.14, (a) displays no stereoisomerism

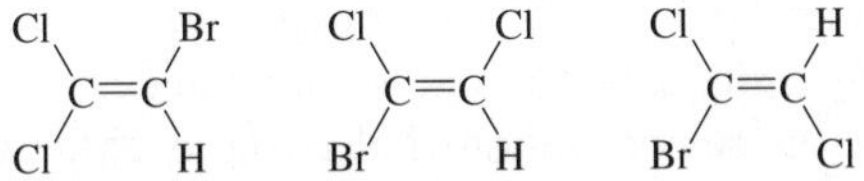

Fig. 5.14 Bromodichloroethene.

whereas (b) and (c) do. Isomer (b) is (*E*)- and (c) is (*Z*)-1-bromo-1,2-dichloro-ethene. At first glance it may appear in (b) and (c) that the stereodescriptors are misassigned, since the chlorines are on the same side of the reference plane in the *E*-isomer and on opposite sides in the *Z*-isomer. However, the groups defining *cis–trans* isomerism in this case are the chlorine on C2 and the bromine. As you may have guessed from criteria governing the selection of stereodescriptors discussed in other chapters, this is because the bromine has higher priority in the CIP sequence rule. It is possible to say, in general terms, that the chlorines are *cis* disposed in (b), but in terms of a specific designator of the stereochemistry of the whole molecule, the descriptor *E* must be used. Similar arguments apply to the Z-isomer, in which the chlorines are *trans* disposed. While this system of nomenclature may not seem very sensible in the above structure, it is a necessary condition when the three substituent groups (and therefore all groups) are different, that is A ≠ B ≠ C ≠ D (Fig. 5.15).

(a)
(*Z*)-3-chloro-2-ethylbut-2-enoic acid

(b)
(*E*)-3-chloro-2-ethylbut-2-enoic acid

Fig. 5.15

The groups defining the *EZ* descriptors are the adjacent pair of higher priority according to the CIP sequence rule. Therefore (a) in Fig. 5.15 is (*Z*)-3-chloro-2-ethylbut-2-enoic acid, the COOH group being preferred to the ethyl group and Cl being preferred to Me. Figure 5.15(b) is the corresponding *E*-isomer. Further examples of this type are given in Fig. 5.16.

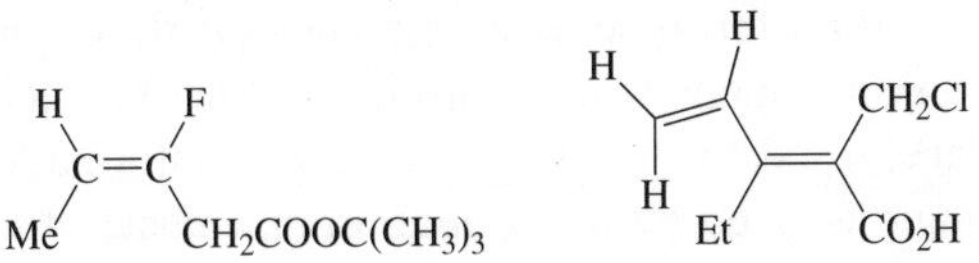

E-isomer

Z-isomer

Fig. 5.16

5.2.2 Structures of partial bond order

Structures containing adjacent single and double bonds can be capable of displaying partial bond order and consequently *cis–trans* isomerism. Two common examples of this phenomenon are buta-1,3-dienes (or other structures containing two double bonds separated by a single bond) and *N*-alkyl amides.

To deal with 1,3-dienes first, the delocalization of electrons over the π-systems raises the bond order of the central bond, compared with an isolated single bond, Fig. 5.17.

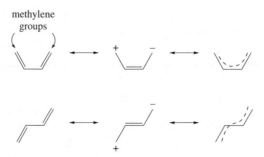

Fig. 5.17

Although the central bond is not raised to the status of double bond by this operation, it does have conferred upon it a degree of rigidity which increases the population of two possible conformations over and above that expected when the bond is able to rotate freely. The two conformations are shown in Fig. 5.18, together with their respective Newman projections.

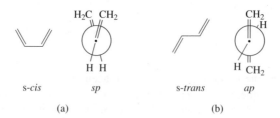

Fig. 5.18

The stereodescriptors used for this type of system are s-*cis* for (a) and s-*trans* for (b) because the two terminal methylene groups are on the same and opposite sides of the reference plane, respectively. This is entirely a conformational effect in which the dihedral angle is fixed by the spatial arrangements of the double bonds and is not affected by the presence of groups or atoms of higher priority. Alternative descriptors are the conformational ones *sp* (synperiplanar) and *ap* (antiperiplanar), respectively (Chapter 1).

In contrast, for *N*-alkylamides, the terms *E* and *Z* are used, as illustrated in Fig. 5.19 for *N*-methylbenzamide. The substituents of highest priority defining the isomerism are the oxygen and the *N*-methyl groups. In (a) these are on the same side of the reference plane and so the descriptor is *Z*, and in (b) they are on opposite sides and the structure is *E*.

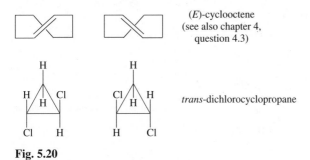

(a) 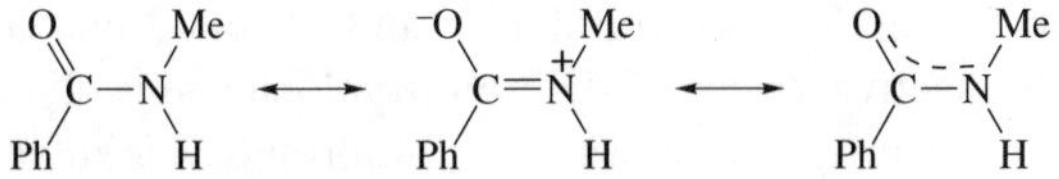 *(Z)*-*N*-methylbenzamide

(b) 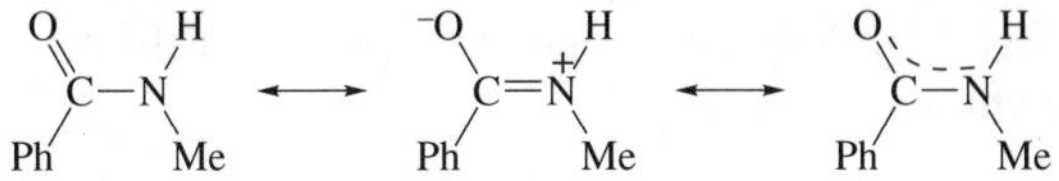 *(E)*-*N*-methylbenzamide

Fig. 5.19

5.3 *cis–trans* **Isomerism, enantiomerism and diastereoisomerism**

Bearing in mind the definition of enantiomerism given in earlier chapters, the structural pairs shown in Fig. 5.20 are enantiomers.

(E)-cyclooctene
(see also chapter 4,
   question 4.3)

trans-dichlorocyclopropane

Fig. 5.20

Remember that enantiomers are stereoisomers with non-superimposable mirror images. The pairs in Fig. 5.21 are diastereoisomers, that is, stereoisomers *not* related as object to mirror image.

(E)-but-2-ene

(Z)-but-2-ene

trans-1,2-
dibromocyclobutane

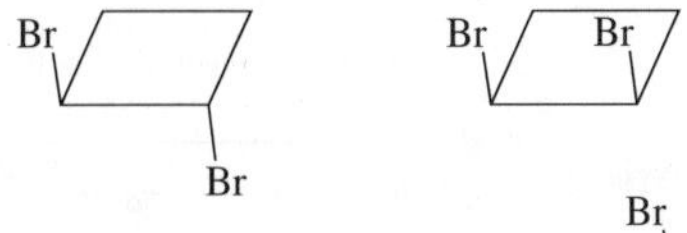

cis-1,2-
dibromocyclobutane

trans-1,3-
dibromocyclobutane

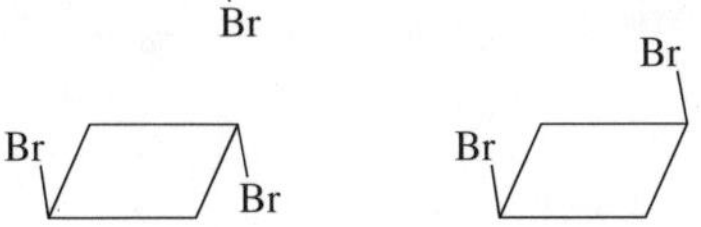

cis-1,3-
dibromocyclobutane

trans-1-chloro-3-
methoxycyclopentane

cis-1-chloro-3-
methoxycyclopentane

Fig. 5.21

Not all the pairs in Fig. 5.21 are chiral. (*E*)- and (*Z*)-but-2-ene are achiral and therefore optically inactive. *trans*-1,2-Dibromocyclobutane is chiral (has a non-superimposable mirror image) whereas *cis*-1,2-dibromocyclobutane is optically inactive, being a *meso* compound (having a plane of symmetry). Neither *cis*- nor *trans*-1,3-dibromocyclobutane is chiral. The *cis* compound is a *meso* isomer and the *trans* isomer has a centre of symmetry and is thus optically inactive (see Chapter 2, Question 2.5). Both the *cis* and *trans* isomers of 1-bromo-3-methoxy-cyclopentane are chiral.

5.4 *cis–trans* **Isomerism in nitrogen-containing compounds**

The nitrogen-containing compounds featured in Fig. 5.22 display *cis–trans* isomerism.

Oximes

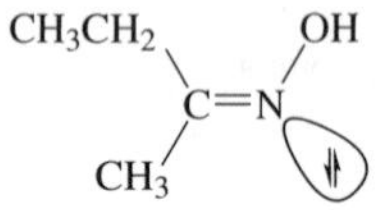

(*Z*)-butanone oxime

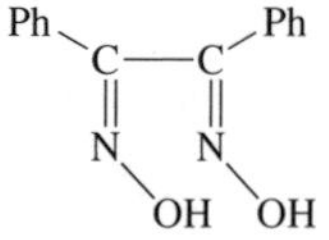

(*Z,E*)-benzil dioxime
[(*Z,E*)-1,2-diphenylethanedione dioxime]

Q. 5.3 In (*Z,E*)-benzil dioxime which carbon–nitrogen double bond is *Z* and which one is *E*?

Hydrazones

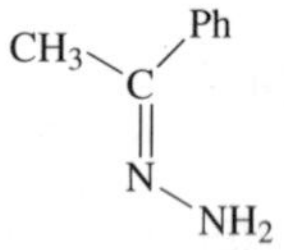

(*Z*)-acetophenone hydrazone

Semicarbazones

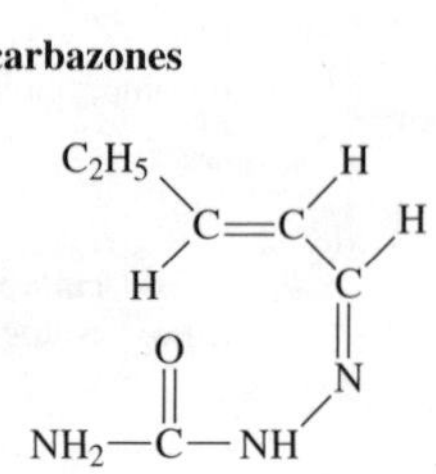

(*Z*)-[(*E*)-pent-2-enal semicarbazone]

Imines

H   Me
 \   /
  C=N   (*E*)-benzaldehyde methylimine
 /
Ph

Azo compounds

         Ph
        /
   N=N   (*E*)-azobenzene
  /
Ph

N≡C(CH₃)₂C   C(CH₃)₂C≡N   (*Z*)-azoisobutyronitrile (AIBN)
            \   /
             N=N

Fig. 5.22

Answers to questions

Q. 5.1   The enantiomer **A** is rapidly converted into its mirror image **B** through the 'boat' conformation.

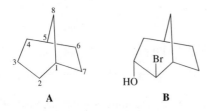

Q. 5.2

Formula **A** shows the basic structure of a bicyclo[3.2.1]octane with the numbering system. Formula **B** shows the bromine and hydroxy units in the appropriate positions.

Q. 5.3

Note that C(NNC) takes preference to C(CCC).

6 Methods for estimating ratios of stereoisomers in a mixture and the separation and identification of the individual components

The general reaction A + B → C is a rare and wonderful reaction. More often than not, A + B are more likely to give a mixture of C + D + E. When the components of a reaction mixture are stereoisomers (enantiomers, diastereoisomers or *cis–trans* isomers) the practical chemist requires methods to determine both the identity of the individual components and the proportions in which they are present, in order to judge the usefulness of a particular reaction.

There are many ways in which these evaluations may be achieved and some of the most widely used techniques are outlined in this chapter. The discussion is divided into three sections: the estimation of ratios of stereoisomers in a mixture, separation of isomers and the identification of individual components isolated from a mixture. As you can appreciate, this is a very wide subject area and a detailed analysis of all the techniques used to isolate and then determine the structural identity of reaction products is quite beyond the scope of this text. What follows is, however, a passing reference to those techniques that are particularly useful to the estimation of isomer ratios and the separation and identification of stereoisomers.

6.1 Estimating ratios of stereoisomers

6.1.1 NMR spectroscopy

NMR spectroscopy is one of the most useful techniques at the chemist's disposal and much information can be gleaned from the chemical shifts (δ), spin–spin coupling constants (J) and techniques such as nuclear Overhauser enhancement. We must stress that the following represents a cursory glance at the use of NMR spectroscopy in relation to stereochemical problems. Other texts should be consulted to appreciate the full power of the methodology. What follows here is simply an *aide-mémoire* regarding the origins of chemical shifts and coupling constants.

Chemical shift

In very qualitative terms, the NMR phenomenon arises because certain nuclei such as ^{1}H, ^{13}C, ^{19}F and ^{31}P, all of which have nuclear spin values $I = \frac{1}{2}$, are able to behave like tiny magnets. In common with such magnets, these nuclei display directionality. When they are subjected to a larger uniform external magnetic field (as when a sample is placed in an NMR spectrometer) the nuclei can adopt one of two orientations: they can line up so that their directionality is the same as that of the applied magnetic field (this requires relatively little energy) or they can oppose the direction of the magnetic field (a higher-energy orientation) (Fig. 6.1).

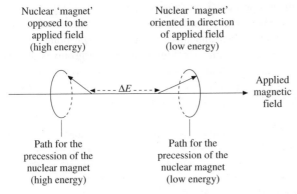

Nuclear 'magnet' opposed to the applied field (high energy)

Nuclear 'magnet' oriented in direction of applied field (low energy)

ΔE

Applied magnetic field

Path for the precession of the nuclear magnet (high energy)

Path for the precession of the nuclear magnet (low energy)

Fig. 6.1

More graphically, the behaviour of nuclei can be likened to that of a spinning top: they precess (the axis of their spin 'wobbles') around the direction of the magnetic field rather like a spinning top precesses in the field of gravity. The frequency of this precession (the resonance frequency) depends on the type of nuclei (whether ^{1}H, ^{13}C or other), on the magnetic field strength and, most importantly, there is a slight but precisely measurable dependence on the chemical environment, that is, on the position of the nuclei in the molecule.

Resonance is observed when energy of the resonance frequency is supplied to make the nuclei flip from one orientation to the other. You can do this either by keeping the magnetic field constant and sweeping a range of frequencies (nuclei in different environments will require energy of different frequency to resonate) or by applying a constant frequency and sweeping a range of field strengths.

When, for example, a ^{1}H NMR spectrum is run on a continuous wave NMR spectrometer, the latter condition is operative, with the frequency at, say, 60 MHz. The spectrometer provides the external magnetic field with which the nuclei in the sample orientate themselves, and also supplies the energy which enables them to resonate. The signals recorded reflect the different field strengths at which the different types of ^{1}H nucleus in a structure (that is, the nuclei in different environments), resonate when the frequency is fixed.

In a modern higher field strength spectrometer the sample is placed inside a strong magnetic field and irradiated for a very short time (pulse) by a radio-frequency transmitter at, say, 250 MHz. This causes the nuclei to precess in phase (in step) with each other. After the pulse the nuclei continue to precess in step for a few seconds, each type of nucleus with its own resonance frequency. These frequencies are then detected with what is essentially a radio receiver, analysed by a computer, and recorded as a spectrum. The spectrum consists of peaks (signals) of various intensities appearing in various places on the horizontal axis of frequencies, which is calibrated in hertz (Hz) or in parts per million (ppm).

The position of a signal in an NMR spectrum is usually called the chemical shift (δ), and is measured relative to the signal from a reference compound, usually tetramethylsilane (δ_0). The chemical shift gives us information on the type of environment a particular nucleus is in and Table 6.1 gives some typical δ values or ranges. The chemical shift is affected by a number of things, notably (a) the inductive effect of electronegative atoms (a through-bond effect) in which the δ values of, say, CH_2 nuclei are higher when the group is attached to an oxygen, halogen or nitrogen than when it is attached to another carbon atom, and (b) the anisotropy of neighbouring bonds or functional groups (a through-space effect). Figure 6.2 illustrates an example of effect (b).

Table 6.1 Chemical shifts of some protons in common environments.

Proton	Chemical shift (δ) (approximate values)
CH_3—C(alkyl)$_3$	0.9
CH_3—C—O—	1.3
CH_3CO—OR	2.3
CH_3—aryl	2.3
CH_3—OR	3.3
—CH_2—OR	3.4
—CH_2—Cl	3.6
—CH_2—Br	4.3
—CH_2—NO$_2$	4.7
CH_2=CH_2	5.3
H—Ph	7.3
H—OR	range 0.5–4.5
H$_2$NR	range 1.0–5.0
HOCR	range 9.0–13.0

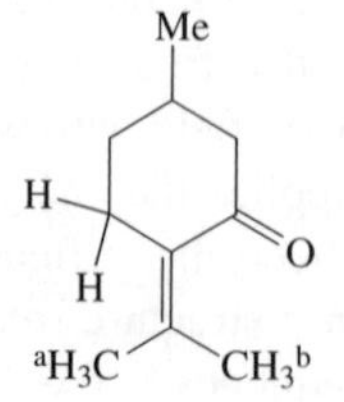

$\delta H^a = 1.77$ (affected by neighbouring methylene group)

$\delta H^b = 1.95$ (affected by neighbouring carbonyl group)

Fig. 6.2

Spin–spin coupling

The NMR signal of a proton in a given chemical (and magnetic) environment often appears not as a single line, but as a group of closely spaced lines. These splitting patterns arise because of so-called spin–spin coupling. For example, the signal of a methyl group may appear as a triplet because of an adjacent methylene group in which the two hydrogens (or more precisely their nuclei) can be oriented (a) both with the field, (b) one with and one against the field or (c) both against the field. Each of these situations results in a different peak in the signal of the methyl group (it is said to be coupled to the methylene group). Since there is only one possible arrangement for (a) and (c) and two possible arrangements for (b) the signals appear as a 1:2:1 triplet. A typical picture of a methyl group split into a triplet by an adjacent methylene group is shown in Fig. 6.5. In this simple example (coupling patterns can be complicated), the distance between two adjacent lines of the triplet is called the coupling constant (J), and is, after the chemical shift, the second most important piece of information about the structure of the molecule. Some typical J-values are given in Table 6.2. When the proton under scrutiny is coupled to more than one type of proton-partner, the different coupling constants may be distinguished in the composite (often quite complicated) signal.

Table 6.2 Some typical J-values for proton–proton coupling.

Coupling partners	Typical J values
H⧵C⧸H (geminal)	12–15
CH—CH (freely rotating)	6–8
CH—CH (in locked system)	0–12 (depending on dihedral angle [Karplus correlation])
Z-alkene	7–11
E-alkene	12–18
=CH—CH=	9–11

6.1.2 NMR and isomer ratios

Since different compounds (for example, different isomers) can be distinguished and identified by the chemical shifts of hydrogen atoms and spin–spin coupling constants, the use of NMR for the elucidation of structure and the estimation of isomer ratios is particularly widespread. Undoubtedly the estimation of the ratio of isomers in a mixture can be accomplished by a wide variety of physical methods but NMR spectroscopy is particularly useful since it is non-invasive and the isomer ratio can be estimated without recourse to the separation of the components.

For instance the ratio of *E* and *Z* isomers in a mixture of disubstituted alkenes can be assessed by NMR spectroscopy. Almost invariably the signals due to the alkene protons (normally in the range δ5–7) will be distinct and may be separately integrated to give the isomer ratio. Notice that the *E* alkene will show the larger *J*-value for the coupled alkene protons (Table 6.2).

The ratio of two diastereoisomers present as a mixture can also be monitored directly by NMR spectroscopy. Most, if not all, of the ¹H and ¹³C nuclei in the two diastereoisomers will be magnetically (and chemically) non-equivalent, the NMR signals will be distinct, more or less, and the ratio of isomers can be measured by integration of these signals. The ratio of two diastereoisomers A and B is formulated as the **diastereoisomeric excess** (d.e.), given by the expression:

$$\text{d.e.} = \frac{\% \text{ diastereoisomer A} - \% \text{ diastereoisomer B}}{\% \text{ diastereoisomer A} + \% \text{ diastereoisomer B}} \%$$

If more than two diastereoisomers are present in the mixture then the estimation of the ratios by NMR becomes increasingly difficult, particularly for the minor components.

NMR spectroscopy can be used to estimate ratios of enantiomers, thus allowing the calculation of enantiomeric excesses, using the following equation:

$$\text{enantiomeric excess (e.e.)} = \frac{\% \text{ enantiomer A} - \% \text{ enantiomer B}}{\% \text{ enantiomer A} + \% \text{ enantiomer B}} \%$$

For example a 90:10 ratio of two enantiomers gives an enantiomeric excess of 80%. Measurement of enantiomeric excess can be accomplished by converting the enantiomers into distinguishable diastereomeric entities by treatment with an appropriate reagent. A shift reagent, which is a paramagnetic metal [such as a

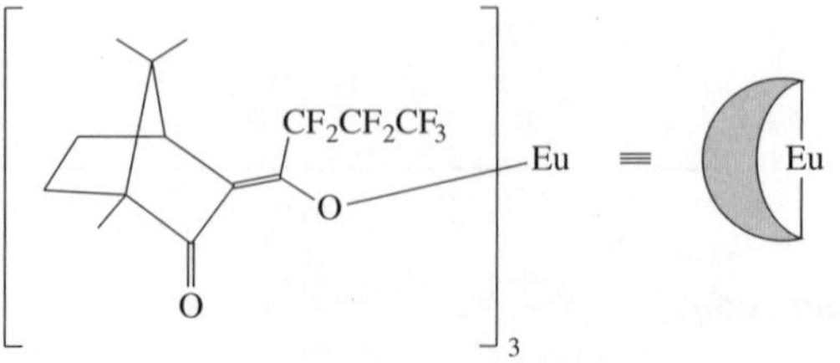

Fig. 6.3 Tris[3-(heptafluoropropylhydroxymethylene)-(+)-camphorato]-europium(III) [Eu(hfc)₃].

europium(III) salt] derivatized to allow solubility in an organic solvent, will associate with polar functional groups in a molecule, causing a downfield shift of the resonance frequency of protons in the locality. If the ligand around the transition metal is chiral, for example Eu(hfc)$_3$ (Fig. 6.3), then the two enantiomers of the chiral compound will form diastereoisomeric complexes with the organometallic agent and some protons may then resonate at slightly different chemical shifts. In Fig. 6.4 the interaction of the europium shift reagent with the two enantiomers

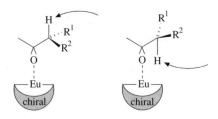

Fig. 6.4

of a chiral carbonyl compound is shown. The signals of the arrowed hydrogen atoms may well be shifted downfield to different extents by the paramagnetic transition metal. Increasing the amount of shift reagent will increase the amount of complex present at any given time, leading to a greater difference in the chemical shift values ($\Delta\delta$). The effect of adding Eu(hfc)$_3$ to a solution of racemic ester **1** is shown in Fig. 6.5: in this case even the methyl group at the end of the chain monitors the differing magnetic field on complexing the two enantiomers of the ester with Eu(hfc)$_3$. Care must be taken, however, not to add too much shift reagent as this leads to broadening of the signals in the spectrum and makes the estimate of enantiomer ratios more difficult.

A large variety of compounds, particularly those that have a heteroatom relatively close to a stereogenic centre, can be analysed in this way. Alternatively, the mixture of enantiomers may be reacted with an optically pure substance to give a new covalent bond and two different diastereoisomers which will have different spectral characteristics. The optically pure substances used for this purpose should contain easily identifiable features in the NMR spectrum. For example, secondary alcohols can be reacted with Mosher's acid to give the corresponding esters (Fig. 6.6). The signals of the methoxy group in the ^{1}H NMR spectrum and/or the signals of the CF$_3$ group in the ^{19}F NMR spectrum give the enantiomer ratio. Part of the NMR spectra of the Mosher's ester derivatives of (*R*)- and (*S*)-1-phenylbutanol are shown in Fig. 6.6b, c and d.

In a second example a chiral lactone may be reacted with an enantiomerically pure diol to give different *ortho*-esters from the two enantiomers. Many of the NMR signals of the diastereoisomeric products will be distinguishable and the ratio can be assessed (Fig. 6.7).

The employment of NMR spectroscopy to identify individual stereoisomers is dealt with later in this chapter.

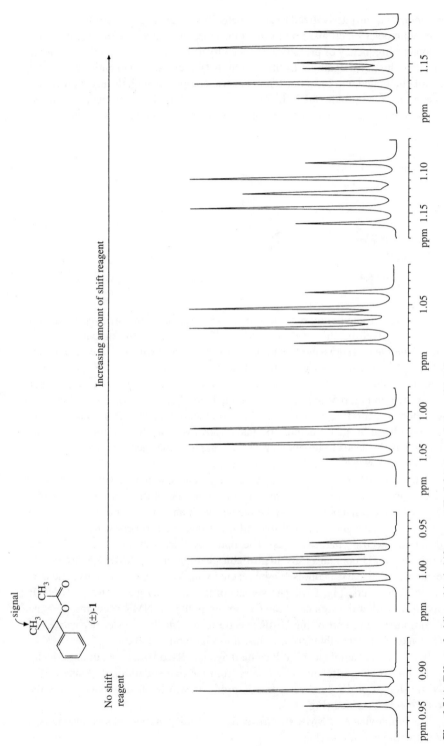

Fig. 6.5(a) Effect of adding chiral shift reagent Eu(hfc)₃ to a solution of (±)-**1**.

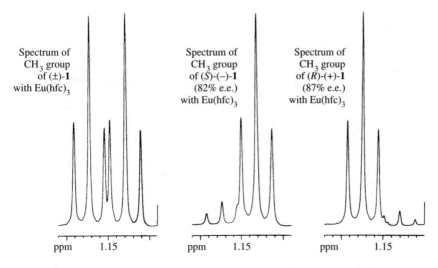

Fig. 6.5(b) Effect of adding Eu(hfc)$_3$ to solutions containing (±)-**1** and samples of **1** enriched with the (+)- or (−)-enantiomer.

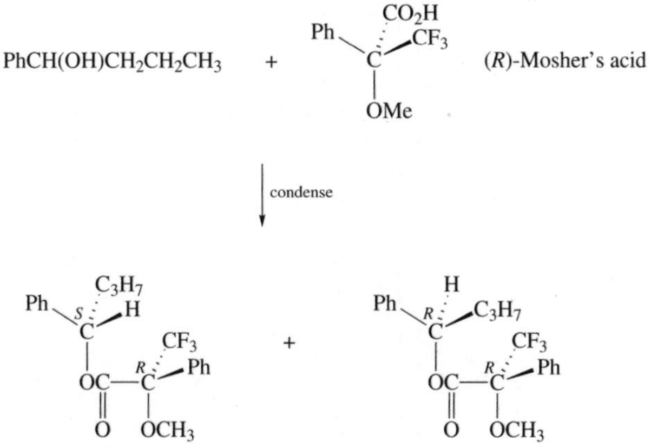

PhCH(OH)CH$_2$CH$_2$CH$_3$ + [Ph—C(CO$_2$H)(CF$_3$)(OMe)] (*R*)-Mosher's acid

condense

Fig. 6.6(a) Phenylbutanol reacts with Mosher's acid to give two diastereoisomeric esters.

6.2 Separating isomers

cis–trans Isomers (e.g. *E* and *Z* alkenes) and diastereoisomers (e.g. *RS* and *SS* isomers) can often be separated by chromatographic methods such as gas chromatography (GC), high pressure liquid chromatography (HPLC) or column chromatography. For GC and HPLC the isomer ratios can be assessed by integration of the signals given by the different components of the mixture, using a suitable detection system and a recorder.

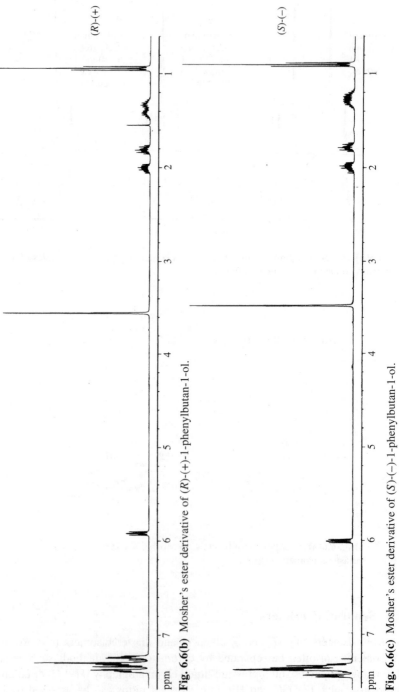

Fig. 6.6(b) Mosher's ester derivative of (R)-(+)-1-phenylbutan-1-ol.

Fig. 6.6(c) Mosher's ester derivative of (S)-(−)-1-phenylbutan-1-ol.

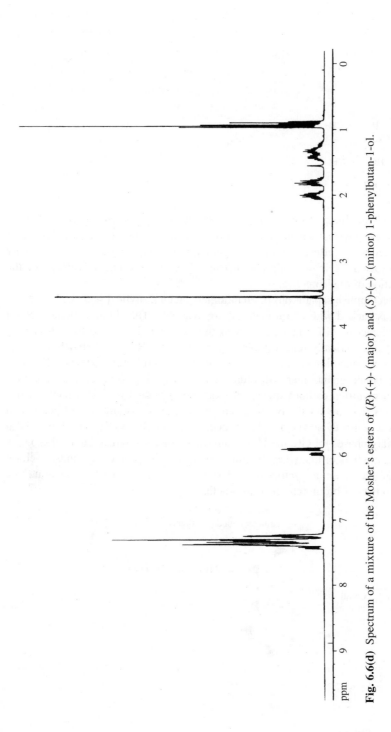

Fig. 6.6(d) Spectrum of a mixture of the Mosher's esters of (*R*)-(+)- (major) and (*S*)-(–)- (minor) 1-phenylbutan-1-ol.

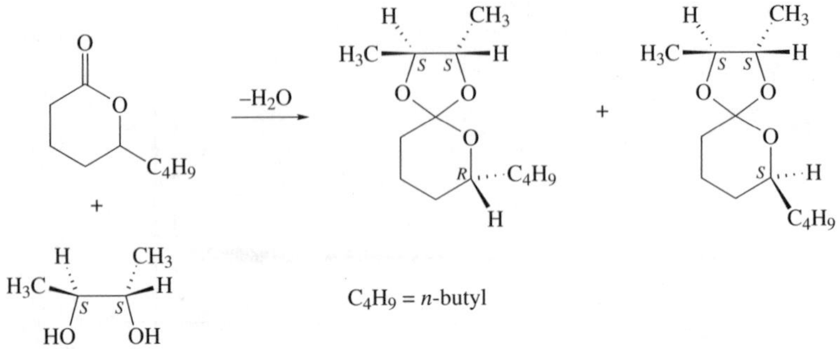

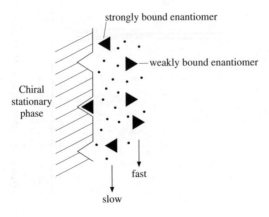

Fig. 6.7

HPLC or column chromatography over a stationary phase of silica or alumina allows the physical separation of substantial quantities of material. By using such methods to effect a complete separation of materials the presence of many diastereoisomers in the original mixture does not present a problem in the estimation of the isomer ratios.

To accomplish separations of enantiomers on an analytical scale GC or HPLC columns packed with chiral material are available. The chiral stationary phase will interact to different extents with the two enantiomers to form transient, diastereoisomerically related molecular complexes. The chiral stationary phase will adsorb enantiomers dissolved in the eluent to different extents. The more tightly bound enantiomer will elute more slowly (Fig. 6.8). Popular columns contain modified carbohydrates, for example cyclodextrins (see Chapter 13) – materials that are relatively cheap and easy to manufacture. Per-*O*-pentylated derivatives of cyclodextrins have been particularly widely used as chiral stationary phases for GC and HPLC analyses, since such materials will separate the enantiomers of many acyclic, monocyclic and bicyclic compounds. A chiral column based on a cyclodextrin derivative has been used to separate the enantiomers of bromo(chloro)fluoromethane.

strongly bound enantiomer

—weakly bound enantiomer

Chiral
stationary
phase

fast

slow

Fig. 6.8

In some rare cases a non-racemic mixture of two enantiomers has been separated on a non-chiral column. In this case the eluent, containing unequal numbers of the two enantiomers, provides the chiral environment and one of the enantiomers has a stronger preference for remaining in solution (due to having more favourable interactions with the prevalent enantiomer) and is eluted more rapidly, as illustrated in Fig. 6.9.

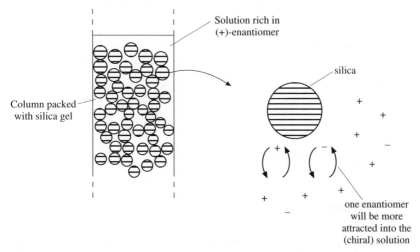

Fig. 6.9

To perform a separation of enantiomers it is often advantageous to react the mixture with an optically pure compound to produce a mixture of diastereo-isomers. (In practice it is always advisable to make the diastereoisomeric mixture using the appropriate racemate so that the detection of low levels of one or other enantiomer can be ensured.) It is important to note that two enantiomers may not react at the same rate with a chiral compound and hence the deriviti-zation should be taken to completion to avoid a mistake in the analysis due to the more rapid formation of one of the two possible products.

Having formed diastereoisomers from the different enantiomers (for example, in the racemate), physical separation can be attempted by crystallization, careful distillation, chromatography, etc., since the diastereoisomers will have different solubilities, boiling points, affinity for silica and so on. After effecting separation the different enantiomers can often be regenerated, for example by hydrolysis (Fig. 6.10). If a clean separation of diastereoisomers is achieved, the enantiomers obtained after regeneration should have equal and opposite rotations when measured using a polarimeter. The separation of the enantiomers in a racemate (resolution of the racemate) can be very important (commercially) in asymmetric synthesis (Chapter 14).

Another method for the separation of mixtures of diastereoisomers and enantiomers is capillary electrophoresis. This technique separates molecules according to their different electrophoretic mobilities in solution. Until recently, capillary electrophoresis was limited to charged molecules but this has changed

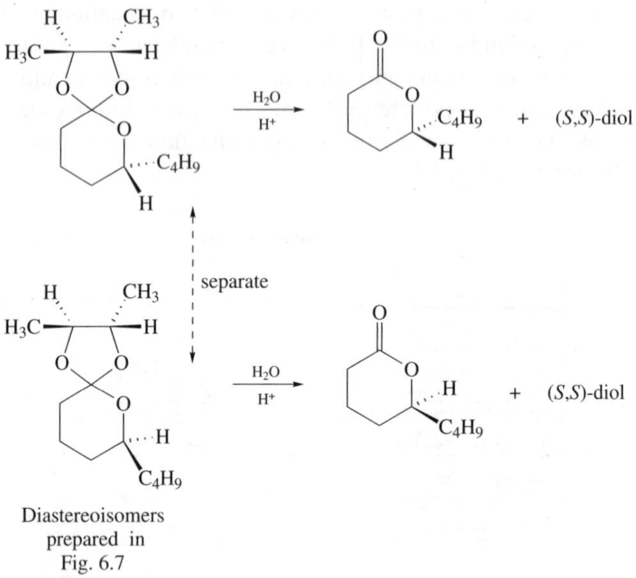

Diastereoisomers
prepared in
Fig. 6.7

Fig. 6.10

with the introduction of micellar electrophoretic separation. Using this technique neutral solutes (such as diastereoisomers) can be separated according to their different partitioning between the aqueous and hydrophobic phases of micelles. The micelles form a second phase (corresponding to the stationary phase in conventional chromatography).

To effect separation of enantiomers a chiral additive, for example a cyclo-dextrin derivative, is added to a micellar solution. The enantiomers form transient associations with the chiral additive and this temporary non-covalent binding affects differently the mobility of the two enantiomers. The separated compounds are generally detected by UV-absorption or fluorescence techniques.

6.3   Identifying individual stereoisomers

6.3.1   NMR spectroscopy

The identity of a geometrical isomer or a particular diastereoisomer can be ascertained by comparison with authentic material, if the compound in question has been made previously by an unambiguous method. If the compound is novel then, to elucidate the structure, recourse is made to spectroscopy and other physical techniques. For disubstituted alkenes a distinction can often be made between the *cis* and *trans* isomers by NMR spectroscopy because the coupling between the vinylic protons is generally greater for the *E* isomer ($J_{ab} \approx 15 \pm 3$ Hz) than for the *Z* isomer ($J_{ab} \approx 9 \pm 2$ Hz) (Table 6.2 and Fig. 6.11). For more heavily substituted alkenes, nuclear Overhauser experiments will often distinguish the isomers. The nuclear Overhauser effect (nOe) is a through-space

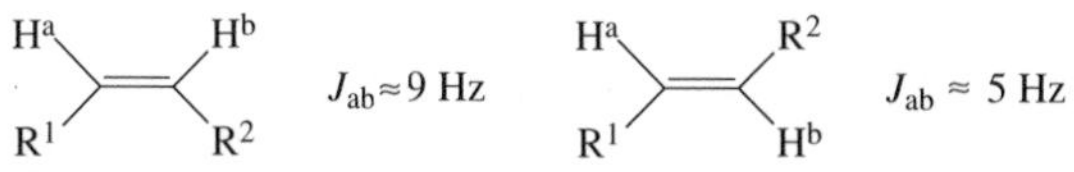

Fig. 6.11

phenomenon diminishing as the distance (r) between the two relevant nuclei increases ($\propto \frac{1}{r^6}$). Using such a technique the spatial proximity of two groups about a double bond or across a ring (such as a cyclobutane ring) can be assessed (see Fig. 6.12).

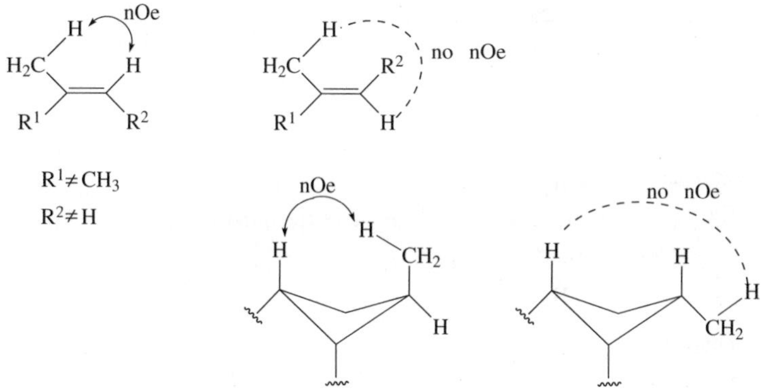

Fig. 6.12 The nuclear Overhauser effect (nOe).

Q. 6.1 (*E*)-Citral and (*Z*)-citral occur naturally. They have the molecular formula $(CH_3)_2C=CHCH_2CH_2C(CH_3)=CHCHO$. The carbonyl carbon atom is numbered one. On irradiation of the C3 methyl group of one isomer (**A**) an 18% enhancement of the signal due to H2 was observed. A similar experiment on the other isomer (**B**) led to no enhancement in the signal due to H2. Assign *Z* and *E* structures to the isomers **A** and **B**.

NMR spectroscopy is also widely employed for the identification of individual diastereoisomers. One particularly useful parameter for deciding the disposition of substituents along a non-freely rotating carbon backbone is the Karplus correlation, which allows us to estimate the dihedral angle between protons on vicinal carbon atoms. The Karplus equation, relating dihedral angle θ and coupling constant J_{ab} for two protons H_a, H_b, gives rise to a typical Karplus curve, illustrated in Fig. 6.13. Some examples of coupling constants from a decalin system are shown in Fig. 6.14.

By assimilating all such NMR data, including coupling constants and nOes, a distinction between two or more possible diastereoisomeric structures can be made. Of course, other spectroscopic data (infrared, ultraviolet, mass

Karplus curve

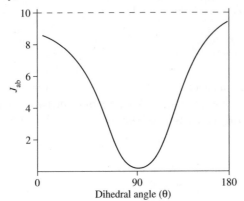

Fig. 6.13

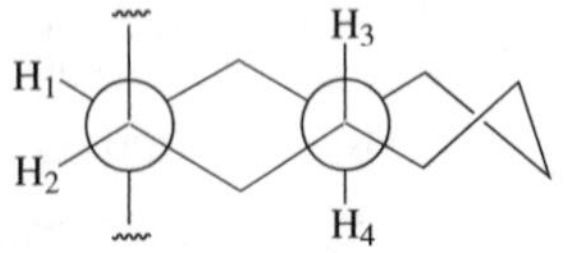

$J_{1,2} = 3.5$ Hz ($\theta \approx 60°$)

$J_{3,4} = 10.0$ Hz ($\theta \approx 180°$)

Fig. 6.14

spectrometry) and different methodology, especially X-ray crystallography, can help enormously in confirming the validity of the proposed structure.

Q. 6.2   Compounds **A** and **B** display different coupling constants for protons H1, H2 and H3 as shown below. Explain why this difference is observed (it is necessary to make a model of the skeleton of the bicyclic ketone, with relevant Hs attached, in order to answer this question).

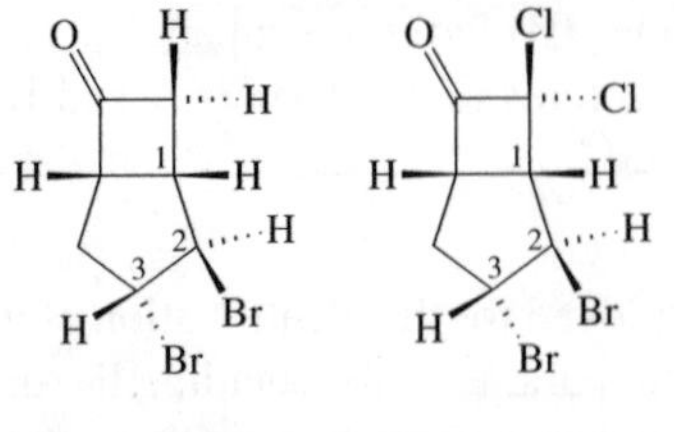

$J_{1,2} = 0.5$ Hz

$J_{2,3} = 0.1$ Hz

A

$J_{1,2} = 3.5$ Hz

$J_{2,3} = 4.0$ Hz

B

6.3.2 Optical activity

The determination of the absolute configuration of an optically active substance can involve more painstaking work. If the compound has been prepared previously a comparison of the reported and the measured optical rotations will provide the required answer. It is important that the rotation is measured at the same concentration in the same solvent, at about the same temperature, to ensure that a sound comparison is made (see Chapter 2). The direction of the rotation (laevorotatory or dextrorotatory) will designate the absolute configuration of the product. The value of the rotation should give a measure of optical purity. Thus if the measured rotation is +50 and the literature reports a rotation of +100 for the same compound, then the newly made sample would seem to be about 75% optically pure. However, many chemists are suspicious about quoting values of optical purity based on optical rotations alone, and most would prefer to base such estimations on NMR, GC and/or HPLC data. In short, optical rotation is generally used to identify the absolute configuration of a new sample of a known compound but is not recommended for an accurate estimation of optical purity.

If the optically active compound has not been prepared previously then more benchwork is necessary. In order to be sure of the new compound's structure, it should be converted into a substance of known absolute configuration, making sure that the stereochemistry at the stereogenic centre is unaffected (or is affected in a predictable, non-random fashion) during the transformation of the new substance into the established compound. In a simple example, optically active bromopropanoic acid (initially of unknown configuration) can be correlated to (+)-(*S*)-lactic acid or (+)-(*S*)-alanine (Fig. 6.15).

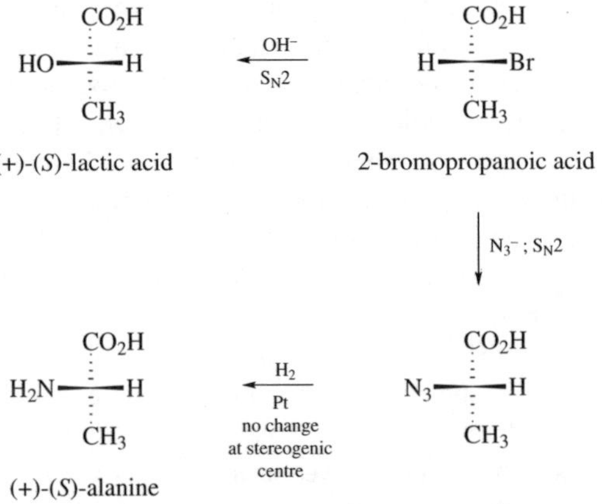

Fig. 6.15 Correlation of 2-bromopropanoic acid with (+)-(*S*)-lactic acid and (+)-(*S*)-alanine. Since we know the configuration of the reference compounds and that S$_N$2 reactions lead to inversion of configuration, we can work back and deduce the configuration of the 2-bromopropanoic acid. In the case illustrated the unknown stereogenic centre turned out to be *R*.

6.3.3   X-ray crystallography

Another safe method for the determination of the absolute configuration of a compound is to derivatize the material with a compound containing one (or more) stereogenic centre(s) of known configuration to give a crystalline product which can be subjected to analysis by X-ray crystallography. For example, the

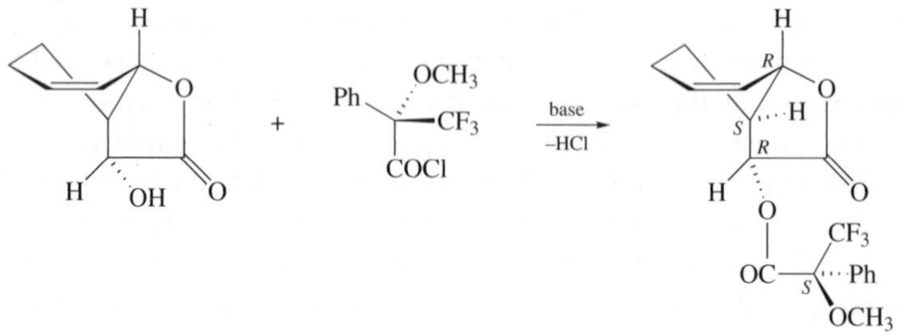

Fig. 6.16

hydroxy lactone pictured in Fig. 6.16 (initially unknown absolute configuration) was converted into the corresponding Mosher's ester using the requisite *R*-acid chloride. The ester is crystalline and, based on the *S*-configuration known to exist in the side chain, the stereochemistry of the bicyclic system was assigned by X-ray crystallography on this pure diastereoisomer. (Note that the absolute stereochemistry of the exocyclic stereogenic centre does not change in the course of the coupling reaction; the change from *R*-configuration for the acid chloride to the *S*-configuration in the product is a consequence of the sequence rule and the replacement of the chlorine atom by an alkoxy substituent: Cl > F > O).

The direct determination of the absolute configuration of a molecule is impossible by ordinary methods of X-ray diffraction. However, success can be achieved using the method of anomolous X-ray scattering. This method uses the fact that, when the wavelength of the incident beam is slightly shorter than the absorption edge of an atom in the molecule, the inner electrons of that atom are excited and fluorescence occurs in anomolous scattering. In contrast with ordinary diffraction, the phase lag between the waves scattered from successive layers in the crystal differs according to whether the layer containing the fluorescing atom is in front of or behind other reference atoms, all relative to the source of radiation. Under these circumstances crystal structures of enantiomerically pure compounds give different diffraction patterns which can be interpreted in terms of the spatial disposition of the nuclei in the crystal relative to the source of radiation, that is, in terms of the absolute configuration.

The conditions under which anomolous scattering occurs are critical, since certain relationships between the frequency of the incident beam and the atomic number of the scattering atom must be satisfied. Suitable anomolous scattering centres are the elements iodine, bromine and rubidium (as in sodium rubidium tartrate, *vide infra*). Thus the job is made easier by inclusion of a heavy atom (I, Br, Rb) in the structure but for compounds containing only C, H, N and O, interpretation of the X-ray data is not straightforward and literature data dealing with the designation of the absolute configuration of simple compounds, without the appropriate derivatization, should be treated with caution.

Historically, the most important X-ray experiment of this type was performed by Bijvoet in 1951. In this experiment the sodium rubidium salt of (+)-tartrate anion was found to have the *RR* configuration. Thus, for the first time, the absolute configuration of a chiral compound was established unequivocally.

6.3.4   The Cotton effect

Other more specialized spectroscopic techniques can be employed that help in the elucidation of the absolute configuration of certain structural types, for instance, compounds containing a nitro group or a carbonyl group. To understand the origin of the important phenomena in these more specialized techniques we need to reconsider the rotation of plane-polarized light by asymmetric compounds.

The value for $[\alpha]_{\lambda}^{\theta}$ depends on the wavelength of the incident light (the Cotton effect). For comparison of $[\alpha]$ values of different compounds and for different samples of the same compound, measurement is made at a fixed wavelength, the sodium D-line at 589 nm hence $[\alpha]_D$ (Chapter 2). The variation of $[\alpha]$ with wavelength of the incident polarized light (λ) is called optical rotary dispersion (ORD) and typical curves are shown in Fig. 6.17. The slope of the ORD curve changes sign at two wavelengths: these positions are called the *extrema*. If the rotation at the extremum on the longer wavelength side (first extremum) is more positive than the rotation at the extremum on the shorter wavelength side (second extremum) the ORD curve is termed positive (a positive Cotton effect). At the inflection point of the ORD curve the optical rotation is zero (at λ_0). When λ_0 is at short wavelengths it and the first extremum might not be recorded by the instrument and the observed tail of the curve is called a *plain curve*. Enantiomers have oppositely signed ORD curves (Fig. 6.17). The amplitude of the ORD curve is a measure of the magnitude of the optical rotatory power.

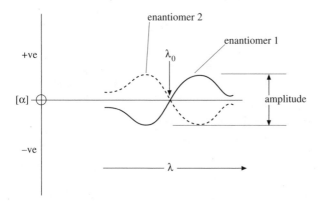

Fig. 6.17 An ORD curve.

If a molecule containing a symmetric chromophore (such as a carbonyl group*) has an adjacent asymmetric unit (such as a stereogenic centre) then the sign of the Cotton effect of such a chromophore is determined by the chirality of the adjacent, perturbing environment. Since the sign of the Cotton effect reflects the stereochemistry of the environment of the chromophore, it follows that,

---

* The carbonyl group acts as a chromophore as a result of excitation of an electron in a non-bonding orbital (n) (oxygen lone pair) into a low-lying antibonding orbital (π bond).

when two similar substances exhibit curves of the same sign and shape, the configuration(s) of the asymmetric grouping(s) near the chromophore are alike. On the other hand, Cotton effects of opposite sign signal mirror image environments and therefore enantiomeric types. Thus a chromophore such as a carbonyl group acts as a probe of the chirality of its environment.

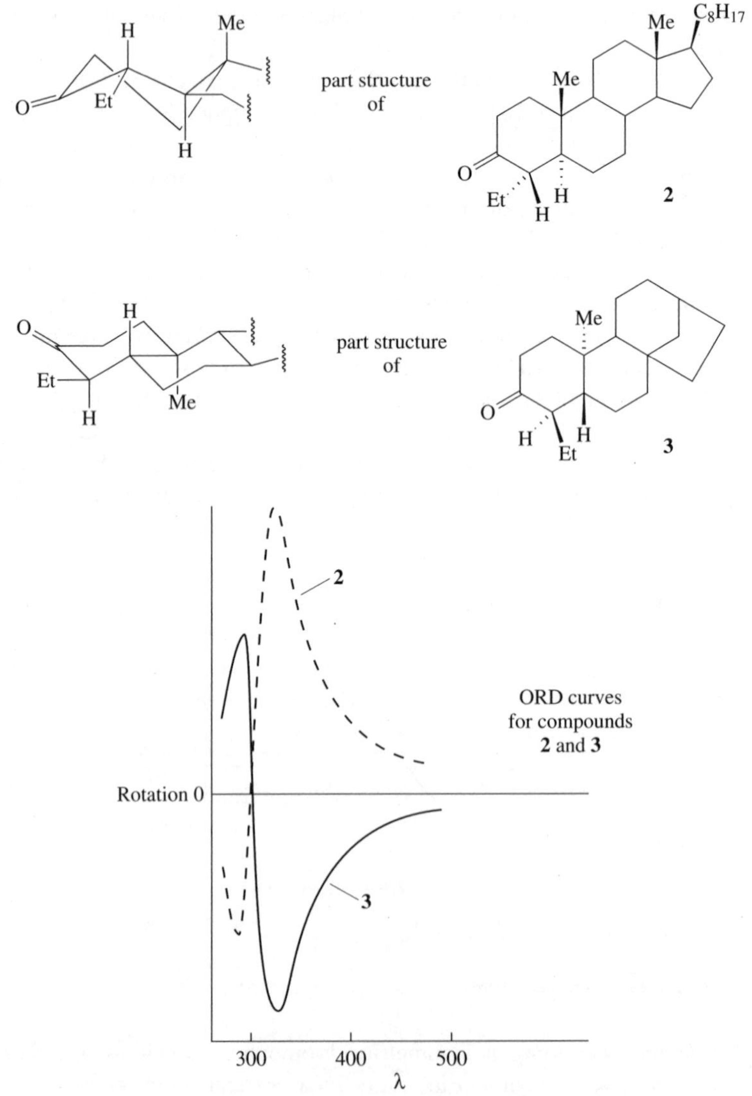

Fig. 6.18

The compounds **2** and **3** shown in Fig. 6.18 are not enantiomers. However, the groups in the immediate vicinity of the carbonyl group have an object:mirror image relationship and this is reflected in the ORD curves, where oppositely signed Cotton effects, attributable to the n → π* absorption of the carbonyl group, are observed.

Observations such as these have led to the formulation of the octant rule, which can be used to determine the absolute configuration for compounds of unknown stereochemistry. In the octant rule, the environment of the carbonyl group is dissected into eight octants, each of which is associated with a sign. The overall sign of the n → π* Cotton effect is estimated from the sum of the contributions made by the perturbing groups in each of the eight quadrants.

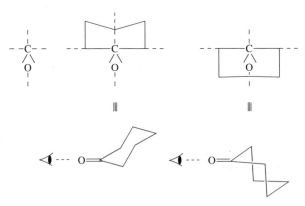

Fig. 6.19 The octant rule.

The necessary division of space is best understood by looking along the oxygen–carbon bond of the carbonyl group (Fig. 6.19). There are four sectors which converge on the carbonyl carbon atom. Behind the plane of the paper are four areas (blocks of space) where substituents can lie. The other four areas that make up the octet are again delineated by the lines shown in Fig. 6.19 but extend towards the reader in front of the plane of the carbonyl carbon atom. (Only in rare cases are substituents positioned in the front four segments.)

For cyclohexanone, two situations are shown in Fig. 6.19. In one, carbons 3–5 occupy the space to the top of, and back from, the carbonyl group; in the other (the mirror image) the same carbon atoms occupy the 'bottom-back' segments. These two conformers of cyclohexanone are present to an equal extent, being rapidly interconverted by rotation about the C—C bonds (cf. cyclohexane, Chapter 1).

For 2- and 3-substituted cyclohexanones the quadrants will not be occupied by atoms or groups to the same extent. The octant rule states that atoms lying in the far lower right and far upper left quadrants make positive contributions to the n → π* Cotton effect, while atoms lying in the far lower left and far upper right octants make negative contributions. Though required for consideration more rarely, positive contributions are also made by the near lower left and near upper right octants while negative contributions accrue from the near upper left and near lower right quadrants.

Consideration of molecule **3** (Fig. 6.20) shows the methyl group to be equally shared between the far upper left and far upper right segments (thus making no contribution) while most other functionality is based in the far upper right segment leading to a negative Cotton effect, as observed. On the other hand, molecule **2** has the contributing substituents in the far lower right segment, leading to the positive Cotton effect.

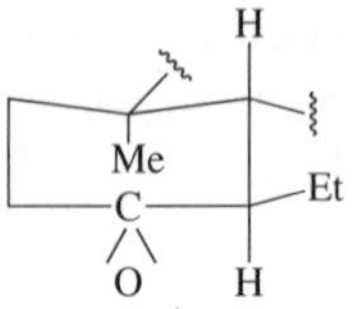

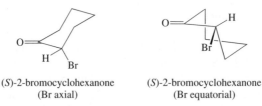

3

Fig. 6.20

The Cotton effect can also give useful information about the preferred conformation of a molecule. For instance (2*S*)-bromocyclohexanone has a strong positive Cotton effect: this indicates that the bromine substitutent occupies an axial position since the equatorial substituent is, by definition, situated close to the plane bisecting the carbonyl group, C2 and C6, and this would make a relatively small contribution to the Cotton effect (Fig. 6.21).

(*S*)-2-bromocyclohexanone
(Br axial)

(*S*)-2-bromocyclohexanone
(Br equatorial)

Fig. 6.21

Substituents more remote from the carbonyl group (for example, axial and equatorial substituents at C3 and/or C5) make contributions to the Cotton effect but these are less marked than groups in the axial situation at C2 and C6.

> **Q. 6.3** Explain why (2*R*,5*R*)-2-chloro-5-methylcyclohexane shows a negative Cotton effect when dissolved in octane and a positive Cotton effect when dissolved in methanol.

If the molecule in question contains an asymmetric chromophore, the sign and magnitude of the Cotton effect is, to all intents and purposes, solely associated with such a chromophore. For example, the biphenyl compound shown in Fig. 6.22 has a negative Cotton effect. When a π-system such as a benzene ring is suitably oriented in one of the far quadrants prescribed for the octant rule, the presence of that π-system is sufficient to determine the sign of the Cotton effect

and overrides all other considerations that might arise with regard to contributions by other weakly perturbing groups, including those with asymmetric
atoms. Hence absolute configuration can be assigned to a molecule having a
composite twisted chromophore.

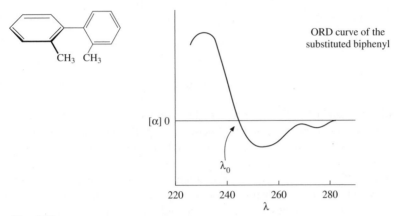

ORD curve of the
substituted biphenyl

Fig. 6.22

Answers to questions

Q. 6.1

From the above structures it can be seen that the Z isomer is compound **A** and
the E isomer is compound **B**.

Q. 6.2 There is a change in conformation of the five-membered ring on going
from compound **A** to compound **B**. Thus compound **A** has the conformation
shown below. This conformation gives dihedral angles of about 90° between H1
and H2 and between H2 and H3. Thus very small coupling constants are
observed. In order to minimize transannular interactions between the chlorine
and bromine atoms, compound **B** prefers an alternative conformation as shown.

The dihedral angle between H1 and H2 is about 150° and the dihedral angle between H2 and H3 is about 160°, accounting for the larger coupling constants in this case. The four-membered ring is essentially planar in both cases.

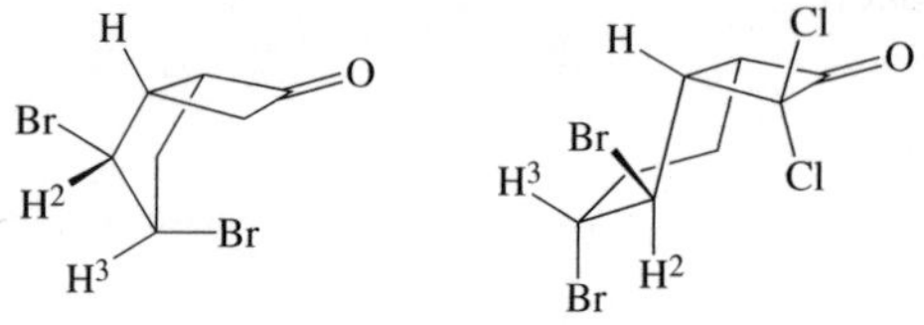

A   B

Q. 6.3 In octane (2*R*,5*R*)-2-chloro-5-methylcyclohexane exists in a conformation with the chlorine atom and the methyl group in axial situations (**A**). The more highly influential halogen atom fits into the bottom left-hand octant behind the carbonyl group. In methanol the compound exists with both substituents equatorial (**B**). The equatorial chlorine atom does not have much influence in this case but the other substituent in the system, the methyl group, clearly occupies the bottom right-hand segment behind the carbonyl group, leading to the positive Cotton effect.

In octane the dipoles of the bonds C=O and C—Cl are kept apart. Solvation effects due to methanol mean that the dipoles can lie parallel in this solvent.

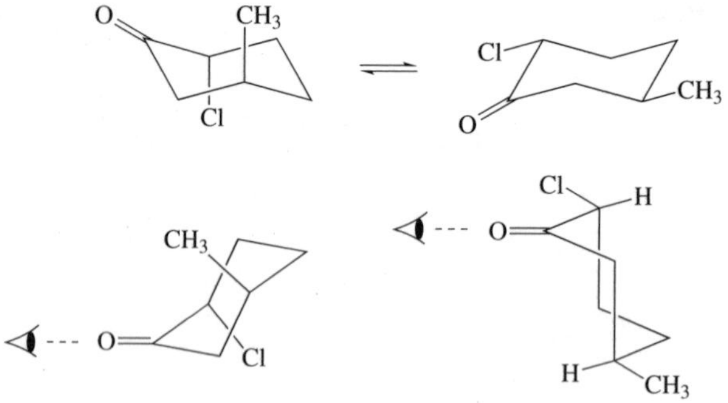

A   B

7 Some transformations at or near a stereogenic centre: racemization and resolution

Much of organic chemistry is concerned with reactions involving functional groups, either directly or at adjacent positions. Many such transformations take place at or near stereogenic centres. In this chapter, two processes that directly involve stereogenic centres, namely racemization and resolution, are described for simple chiral compounds.

7.1 Racemization

Racemization is the process by which an enantiomer is converted into its mirror image until a mixture is obtained that contains equal quantities of each enantiomer. Such a mixture is optically inactive and is called a racemate, Fig. 7.1.

$$(+)\text{-enantiomer} \xrightarrow{\text{racemization}} (\pm)\text{-racemate} \xleftarrow{\text{racemization}} (-)\text{-enantiomer}$$

$$\text{\textbar\textbar\textbar}$$

50:50 mixture
of (+) and (−)
forms

Fig. 7.1

Racemization in some ways has a large nuisance value, in that while a reaction may initially give a stereochemically defined product, the prevailing reaction conditions may be favourable to subsequent racemization and the final outcome may be an undesired mixture of isomers. Therefore, careful planning in the design of synthetic strategies to produce enantiomeric structures is often crucial. There are cases, however, when the propensity of a structure to racemize can be used positively, and an example is outlined later in this chapter (Fig. 7.11).

There are a number of ways in which optically active compounds undergo racemization and three typical mechanisms are described below.

(a)   Via an S_N2 reaction

In an S_N2 reaction a nucleophile Nu⁻ approaches a tetrahedral substrate R_3C—Lg, where Lg is a leaving group, from the back (that is, from the side opposite Lg). As a bond begins to form between Nu and C, the existing bond between C and Lg begins to dissolve and the three groups R are pushed outwards until a transition state is reached in which the R groups and the C are coplanar (Fig. 7.2). As the reaction proceeds, the bond between Nu and C strengthens and eventually displaces Lg altogether; at the same time the R groups flip into the opposite configuration.

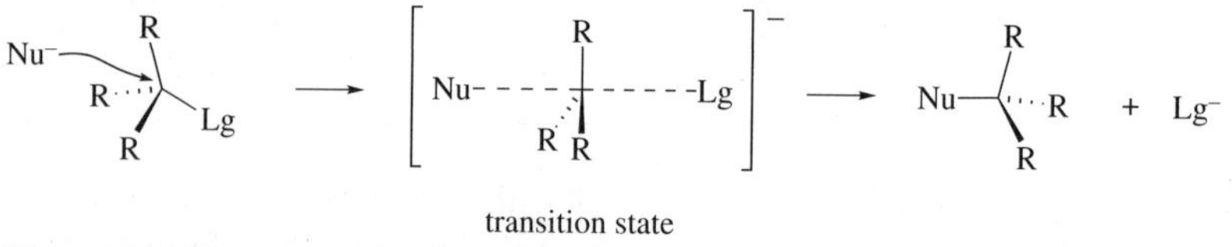

transition state

Fig. 7.2 S_N2 reaction.

An example of S_N2 racemization is (*S*)-2-iodooctane, which, when treated with sodium iodide and monitored polarimetrically, is seen to undergo a gradual fall in its optical activity until eventually the value reaches and remains at zero. Examination of the reaction mixture finds the products to be constitutionally identical, so the inference must be that by some process an optically pure starting material is converted into a 50:50 mixture of enantiomers, namely a racemate.

That the mechanism of this racemization is S_N2, as described above, was ascertained by Hughes *et al* in the mid-1930s, when they carried out the reaction using radiolabelled iodine (I*). The sequence is exactly as in Fig. 7.2, with, in this case, I*$^-$ as the nucleophile and —I as the leaving group. I*$^-$ approaches a molecule of 2-iodooctane from the back, displaces I$^-$ and causes the configuration to invert, Fig. 7.3. There is, of course, nothing to prevent the reverse reaction from occurring and when the optical activity reaches zero, the reaction is in equilibrium.

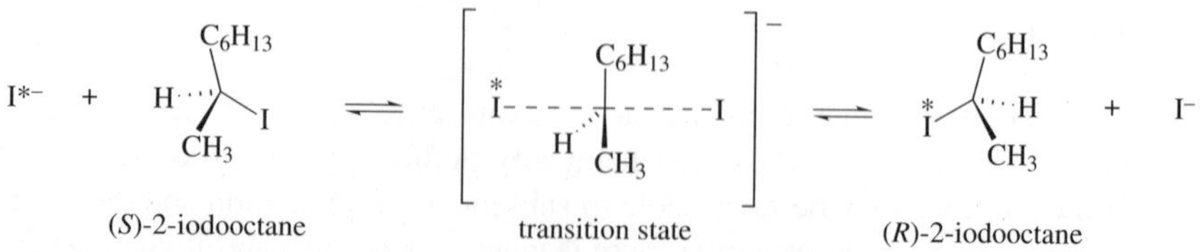

(*S*)-2-iodooctane   transition state   (*R*)-2-iodooctane

Fig. 7.3 Racemization of (*S*)-2-iodoctane via an S_N2 reaction.

This experiment was one of the significant reactions that led to the elucidation of displacement mechanisms. Polarimetric monitoring enabled the initial rate of racemization to be measured, and thus the rate of inversion [the inversion rate is half the rate of racemization, since for every pair of *S* isomers, only one is inverted to produce the racemate: $S + S \rightarrow S + R$ (racemate)]. The rate of iodine label incorporation could also be determined from the product and in fact was found to be the same as the rate of inversion. In other words, it was possible to say that each time a radiolabelled iodide displaced a leaving-group iodide, an inversion took place: the S_N2 mechanism was confirmed.

(b) Via an S_N1 reaction

Molecules that contain good leaving groups (such as iodide) and adjacent substituents that are capable of stabilizing a positive charge are prone to undergo S_N1 substitutions. The difference between this mechanism and the S_N2 mechanism described in (a) is that the departure of the leaving group and the approach of the nucleophile are not linked. Here there are two distinct steps, the first of which is the loss of the leaving group, the second being the introduction of the nucleophile.

(S)-1-Iodo-1-phenylethane is such a molecule and is also chiral. Under conditions where a weak nucleophile is present in a polar solvent, the iodide leaving group will be displaced to give a racemic product. Figure 7.4 illustrates what happens when the nucleophile is water. The positive charge formed on the carbon upon departure of the iodide is stabilized mesomerically by the phenyl group and to a lesser extent inductively by the methyl group. As disclosed in the answer to Question 1.1, the carbenium ion can be considered as sp^2 hybridized in structure and therefore flat. Consequently the nucleophile, in this case water, can approach from either face with equal ease; attack from side (a) leads to inversion of configuration whereas approach from side (b) results in retention of configuration and so a 50:50 mixture of R and S enantiomers is the result.

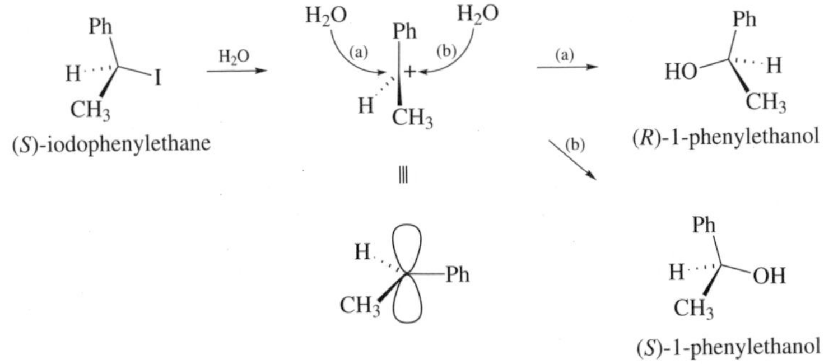

Fig. 7.4 S_N1 reaction.

> **Q. 7.1** (S)-Pentan-2-ol reacts with thionyl chloride ($SOCl_2$) to give 2-chloropentane with retention of configuration at the chiral centre. Propose a mechanism to explain this result.

(c) Via an enol

Careful consideration must be given to the reaction conditions when preparing compounds in which an H-containing chiral centre is present next to a carbonyl group. Such compounds often undergo enolization in the presence of aqueous acid or base, because of the labile nature of the hydrogen adjacent to the carbonyl group. Figure 7.5 shows (S)-3-phenylbutanone and the two possible

mechanisms by which it can enolize. The enol structure is achiral and susceptible to attack by aqueous acid or base and, because the enol is flat, the approach
of the electrophile (H_3O^+ or H_2O) can be towards either face, leading to a 1:1
mixture of enantiomers – a racemate.

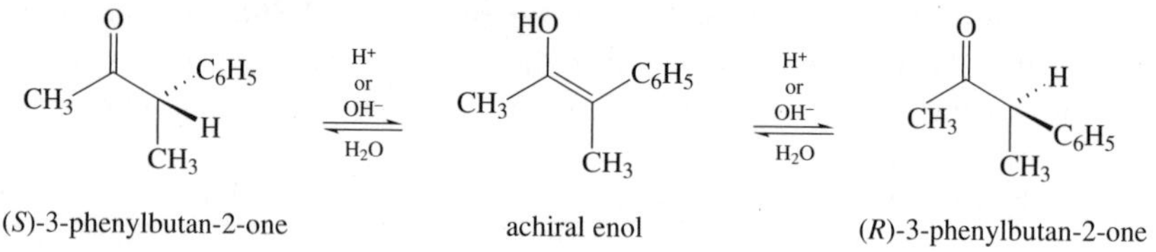

(S)-3-phenylbutan-2-one   achiral enol   (R)-3-phenylbutan-2-one

Fig. 7.5

Q. 7.2 Groups other than the carbonyl group activate adjacent positions
such that the protons are easily removed by base. Examples include nitro-
compounds, sulfones and nitriles. Thus the optically active nitrile **A** would,
at first sight, be expected to racemize when treated with base. However, **A**
undergoes deuterium exchange 4000 times faster than it racemizes when
treated with sodium methoxide in deuteriomethanol. Explain why **A** is an
exception to the usual rule.

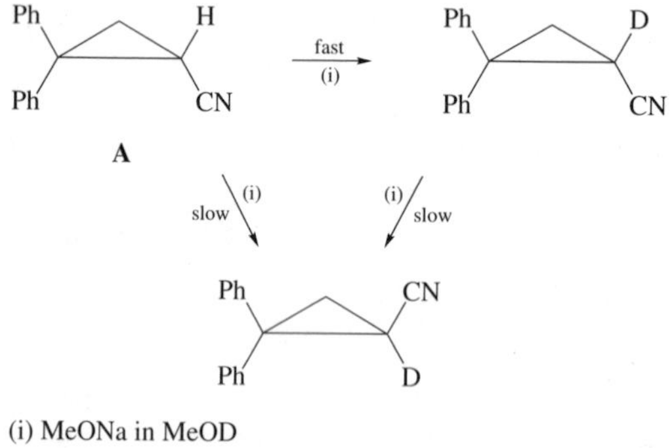

(i) MeONa in MeOD

Racemization of other chiral species can often be accomplished by heating the
sample to a high temperature. For example, helicenes such as hexahelicene (Fig.
4.11) are completely racemized in 10 minutes at 380°C. Bond deformations over
the whole molecule allow the planar structure to form as a transient species.
Similarly, the biphenyl compounds featured in Fig. 7.6, if resolvable, can often
be racemized by heating.

7.2 Resolution of a racemate

Non-natural compounds produced by synthesis are often obtained as racemates
from which the individual enantiomers may have to be separated. This separation process is termed *resolution* (Fig. 7.7).

(a) **Non-resolvable**

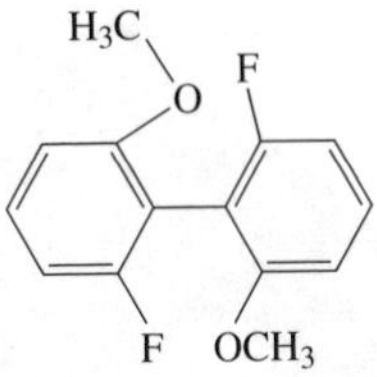

F, OCH₃ (also H) in *ortho*
positions do not prevent rotation

(b) **Resolvable but easily racemized**

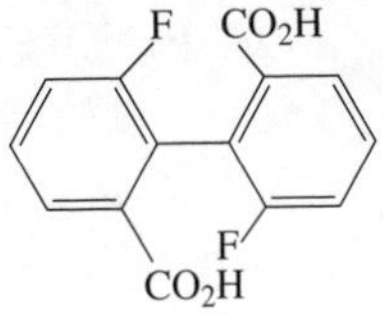

Molecules contain two amino
groups, or two carboxy groups,
or one amino and one carboxy
group; the remaining groups in
the *ortho* position must be F,
OCH₃ or similar moieties

(c) **Not easily racemized**

Molecules contain at least
two nitro groups; the remaining
groups in the *ortho* position
must be F, OCH₃ or similar

Fig. 7.6 Some examples of biaryls differing in optical stability.

(±)-racemate $\xrightarrow{\text{resolution}}$ (+)-enantiomer + (−)-enantiomer

Fig. 7.7

In very rare cases the two enantiomers of a racemate will crystallize separately, to give one set of crystals containing only the (+)-enantiomer and another containing only the (−)-enantiomer. This phenomenon was observed by Pasteur in his work on the different forms of tartaric acid. Pasteur's seminal work was sponsored and supported by the French wine-growing industry: Pasteur and the vintners were intrigued that the tartaric acid present in wine early in the fermentation process was different from that which was investigated earlier by Biot in 1815. The difference was due to the fact that the form derived from wine was the racemate, while Biot had worked on the dextrorotatory enantiomer (Fig. 7.8).

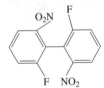

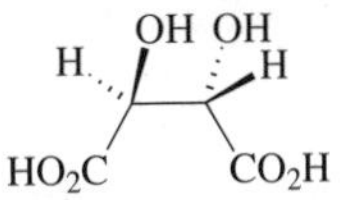

(−)-(*S*,*S*)-tartaric acid

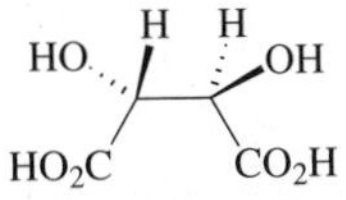

(+)-(*R*,*R*)-tartaric acid

Fig. 7.8

(a)

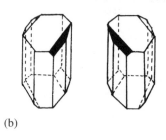

(b)

(c)

Fig. 7.9 (a) Models used by Pasteur.
(b) Diagrammatic representation of
the crystals of (+)- and (−)-forms of
sodium ammonium tartrate.
(c) Extracts from Pasteur's notes.
*(By kind permission of the Musée
Pasteur, Paris.)*

Pasteur prepared the sodium ammonium salt of tartaric acid and crystallized
this material by slow evaporation of an aqueous solution. The crystals of the
sodium ammonium salts of (+)- and (−)-tartaric acid that formed are themselves
related as object and mirror image. Pasteur was able to separate the two forms

using a lens and tweezers. It is noteworthy that only rarely will a racemic compound crystallize in this way [the (+)-enantiomer packing with other (+)-molecules]. It has been estimated that heterochiral packing (the preferred formation of crystals of the racemate) occurs in more than 90% of cases.

In order to resolve a racemate, it is more often the case that the racemate has to be reacted with a second chiral compound to give a mixture of diastereo-isomers which are relatively easy to separate. You should recall from Chapter 2 that the two enantiomers of butan-2-ol will both form esters with a single enantiomer of 2-chloropropanoic acid (Fig. 7.10). Two diastereoisomers are formed, the *SR* and the *RR* compounds, and these two molecules will have different physical properties (melting point, boiling point, adsorption charac-teristics, solubilities, etc.), allowing them to be separated, for example by column chromatography. Having effected a separation the two diastereoisomers can then be hydrolysed separately to regenerate the *R*-acid and the *S*-alcohol in one case, and the *R*-acid and the *R*-alcohol in the other.

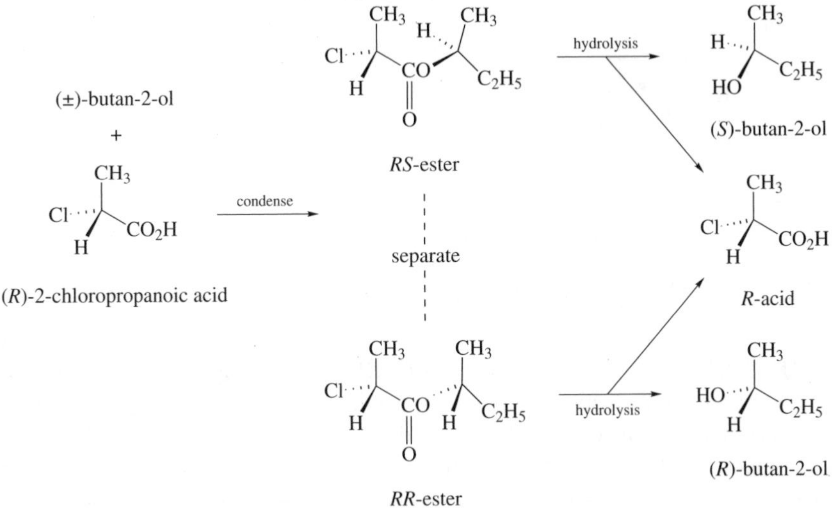

Fig. 7.10 Resolution of (±)-butan-2-ol.

This type of procedure is known as classical resolution and is very important from a commercial standpoint in a number of different areas, for example, in the preparation of semisynthetic penicillins. Figure 7.11 illustrates an example of this. Ampicillin is a clinically useful antibacterial agent and is made from 6-aminopenicillanic acid (6-APA) (in turn derived from penicillin G, made in large quantities by fermentation) and the amino acid phenylglycine. The phenyl-glycine component generally comes in the form of a racemate and must first be resolved before it is reacted with 6-APA, since ampicillin requires only the side-chain configuration provided by the *R* form of phenylglycine. Resolution is achieved with (+)-camphorsulfonic acid (CSA) as outlined in Fig. 7.11. The unwanted *S* form is not wasted, however; it can be actively racemized with strong base and recycled.

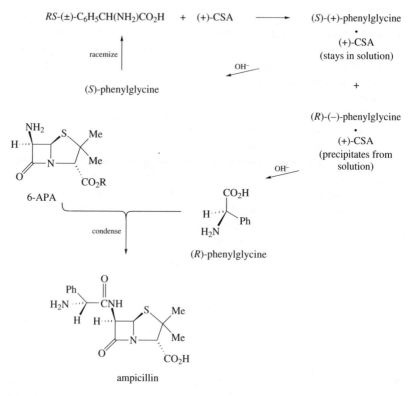

Fig. 7.11 Synthesis of ampicillin.

Q. 7.3 In Fig. 7.11 a mixture of (*RS*)-(±)-phenylglycine and (+)-camphorsulfonic acid produced diastereoisomeric salts; the (−) (+)-salt precipitated from solution. How could (+)-phenylglycine be isolated as a component of the less soluble compound?

It is equally easy to separate racemic carboxylic acids through the formation of diastereoisomer salts that, once again, can be separated by fractional crystallization. (1*R*,2*S*)-Ephedrine [Fig. 7.12(a)] has been used in this connection. Alternatively, a racemic acid can be reacted with an optically pure alcohol such as (−)-menthol [Fig. 7.12(b)] to give a mixture of diastereoisomeric esters that can be separated by chromatography (cf. Fig. 7.10).

Racemates can also be resolved using an enzyme (enzymes are naturally occurring chiral catalysts, proteinaceous in nature, and made up of series of condensed L-amino acids). Enzymes offer the chemist the considerable advantage that they are often highly enantioselective: when presented with a racemic substrate they will often react exclusively with one enantiomer, leaving the other untouched. This kind of process is called a kinetic resolution.

Hog renal acylase (HRA) is an enzyme whose natural function is to effect deacylation in pigs' kidneys. *In vitro*, however, the chemist can use it to resolve,

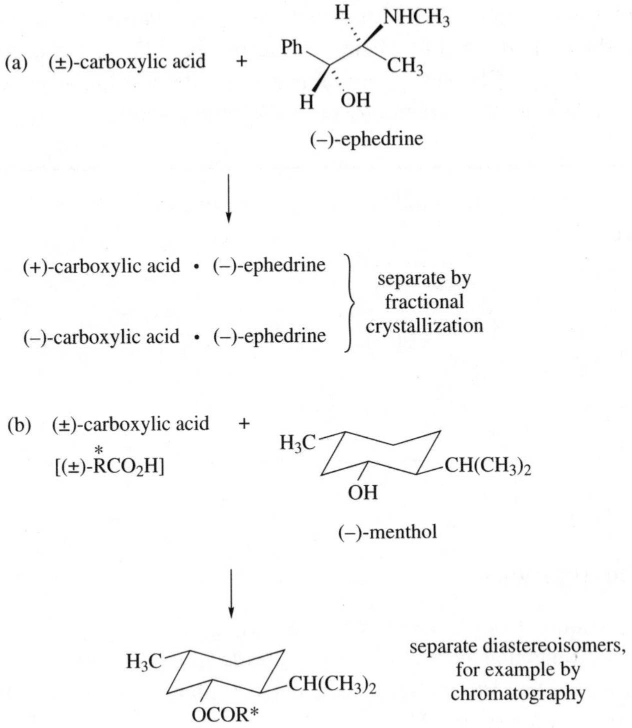

Fig. 7.12

for example, (*RS*)-leucine. The first step is to acylate the racemic leucine, thus converting it into a substrate (still racemic) which is appropriate for the enzyme. HRA, when added to this mixture, catalyses the deacylation only of (*S*)-*N*-acetylleucine and not the *R* form, liberating (*S*)-leucine. The product mixture of (*S*)-leucine and the *R*-acetylated derivative can then be easily separated, since

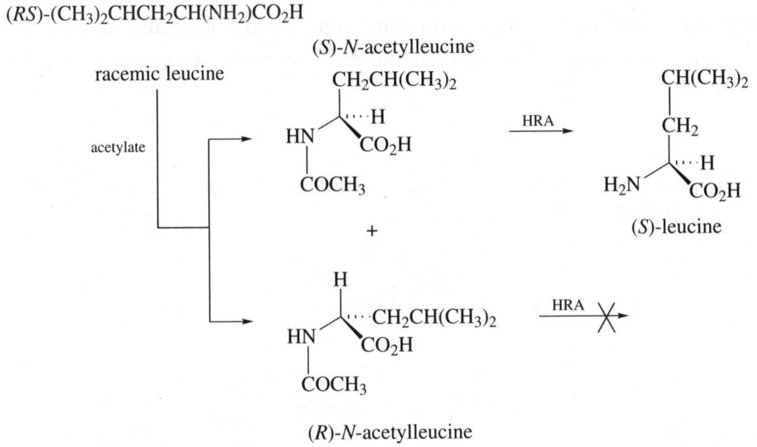

Fig. 7.13 Kinetic resolution of leucine.

they are different compounds, with different physicochemical properties. (*R*)-leucine can then be obtained by chemical hydrolysis of the (*R*)-*N*-acetylleucine (Fig. 7.13). Note that Chapter 15 contains a fuller description of enzymes and has a further example of a process involving kinetic resolution.

> **Q. 7.4** How many compounds are formed on treatment of menthone (**A**) with base? 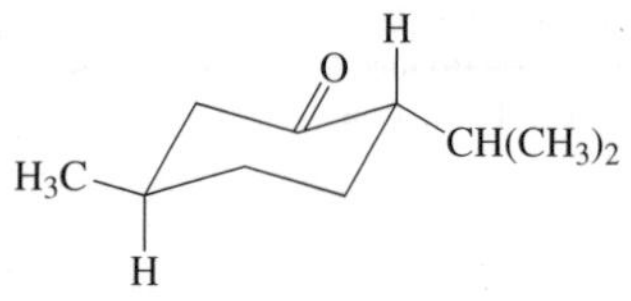
>
>
> **A**

Answers to questions

Q. 7.1 (*S*)-Pentan-2-ol (**A**) reacts with thionyl chloride to give a chlorosulfite ester (**B**). Intramolecular delivery of the chloride unit probably via a tight ion pair (with concomitant loss of SO$_2$) takes place, with retention of configuration, to give the observed product (**C**).

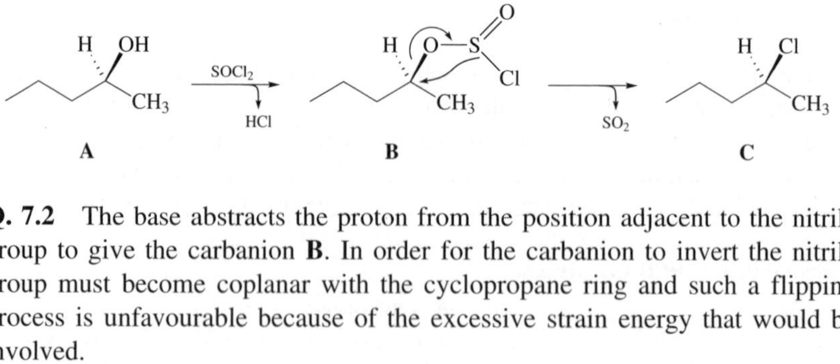

Q. 7.2   The base abstracts the proton from the position adjacent to the nitrile group to give the carbanion **B**. In order for the carbanion to invert the nitrile group must become coplanar with the cyclopropane ring and such a flipping process is unfavourable because of the excessive strain energy that would be involved.

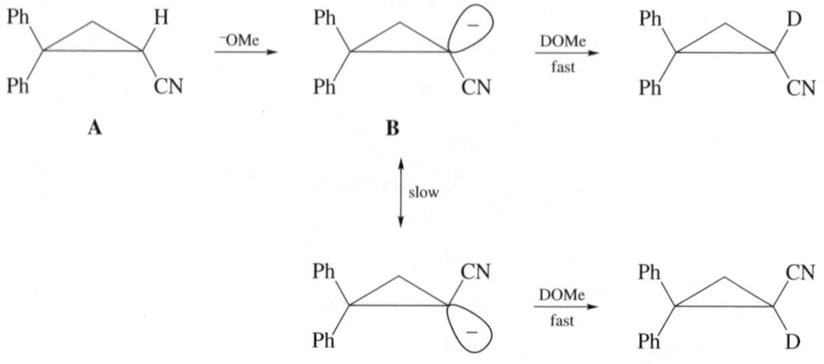

Q. 7.3   The answer is to use (−) camphorsulfonic acid as the resolving agent. This will produce (+)-phenylglycine·(−)-CSA and (−)-phenylglycine·(−)-CSA. Since (−)-phenylglycine·(+)-CSA is less soluble than (+)-phenylglycine·(+)-CSA it follows that the mirror images will have the same difference in solubility in water. Thus (+)-phenylglycine·(−)-CSA will be less soluble than (−)-phenylglycine·(−)-CSA and hence the (+)-phenylglycine moiety will be present in the precipitated salt.

Q. 7.4   Treatment of menthone **A** with base will produce the enolate **B** which will react with water to generate starting material and a small amount of iso-menthone (**C**). Menthone and isomenthone are referred to as epimers: they differ in the configuration at just one chiral centre in a molecule containing more than one chiral centre (N.B. mannose and galactose are epimers of glucose, see Chapter 3).

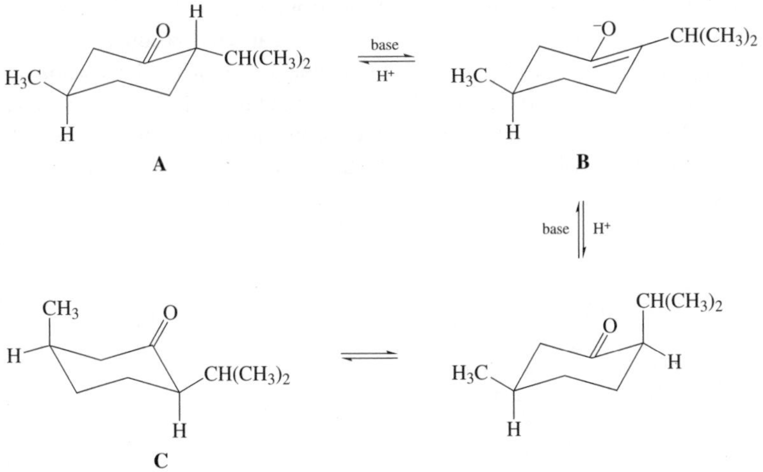

8 Some simple reactions of carbonyl compounds

8.1 The carbonyl group

The carbonyl group is one of the most versatile functional groups and its characteristics and range of transformations have been extensively exploited by the synthetic chemist. The difference in electronegativity between carbon and oxygen is the main reason for the reactivity of the carbonyl group as it gives rise to a charge differential along the C=O double bond, with the carbon atom having a partial positive charge and the oxygen having a partial negative charge (Fig. 8.1). This charge separation makes the carbon atom susceptible to attack by nucleophiles and much of the synthetic utility of the C=O group revolves around this fact.

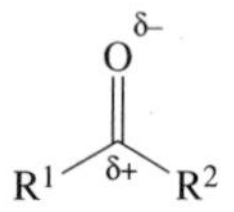

$R^1 = R^2 = H$ . . . formaldehyde
$R^1 = H; R^2 =$ alkyl or aryl . . . aldehyde
$R^1 = R^2 =$ alkyl or aryl . . . ketone

Fig. 8.1 The carbonyl group.

In order to make effective use of these nucleophilic reactions an understanding of the accompanying stereochemical changes is essential. As the carbon atom is sp^2 hybridized (see Chapter 1), the carbonyl group is planar, with the =O and the two substituents displaced trigonally (OCR angle = 120°). After reaction with a nucleophile, however, the carbon atom becomes tetrahedral and therefore potentially chiral. If the product is chiral then it can exist as stereoisomers which may not all be required, and, if this is so, the yield of the required enantiomer or diastereoisomer will be lowered and the reaction as a whole will be wasteful. However, knowledge of the stereochemical principles involved means that, with appropriate tuning of the reaction conditions, formation of unwanted stereoisomers can be minimized or suppressed and the reaction rendered more efficient overall.

Figure 8.2 shows a generalized carbonyl compound R^1—CO—R^2 with the molecular plane at right angles to the plane of the paper. As it is flat, it has two faces, one as viewed from underneath and the other as viewed from above.

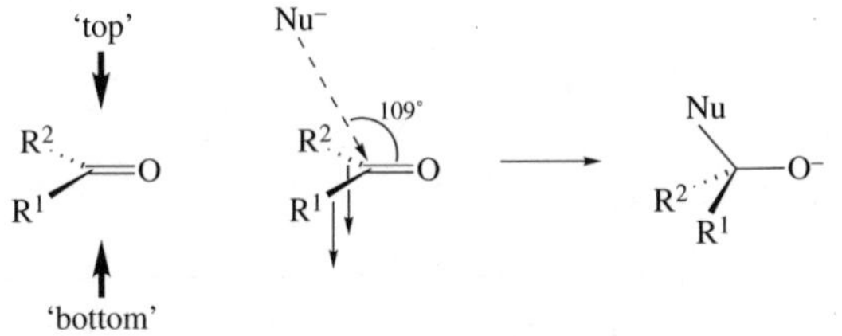

Fig. 8.2 Nucleophilic attack on a carbonyl compound from the top face.

Nucleophilic attack takes place at one or other of these faces at an angle of 109° to the carbonyl group. When $R^1 = R^2$, as in acetone or formaldehyde, the top face is identical with the bottom face and the molecules are described as homotopic. Such molecules can be identified as having two planes of symmetry and a two-fold axis of symmetry (Fig. 8.3). Attack of a nucleophile (Nu⁻) on a homotopic carbonyl group (R—CO—R) gives a product (RRNuCO⁻) which is achiral since it has two identical groups and is superimposable on its mirror image; in stereochemical terms it is immaterial from which side the nucleophilic attack takes place.

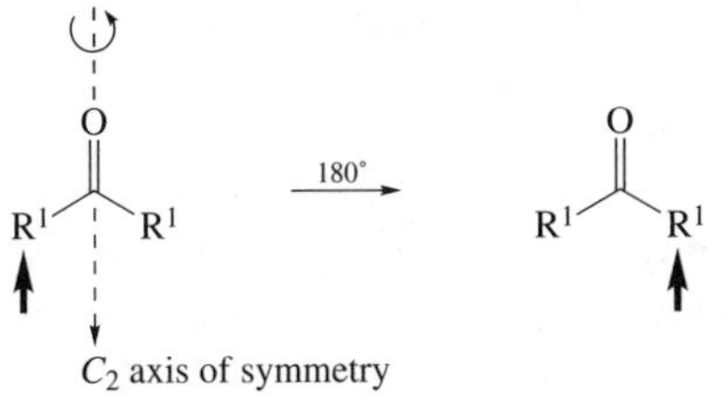

Fig. 8.3

When R^1 does not equal R^2, as in aldehydes (Fig. 8.4) and unsymmetrical ketones, the situation is different. Nucleophilic attack from the top face gives a

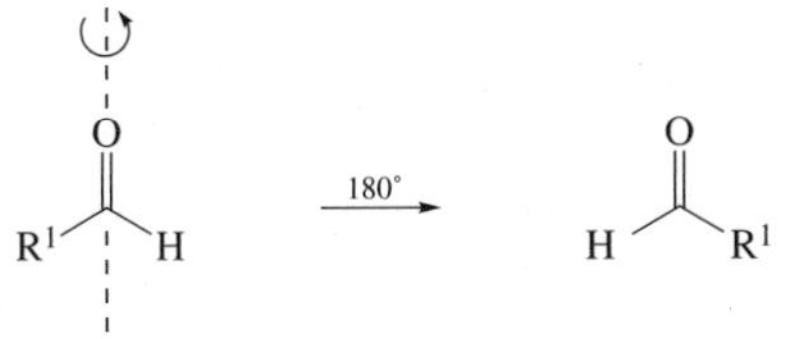

Fig. 8.4

product that is the mirror image of that produced by attack of the same nucleophile on the bottom face. Provided that the nucleophile is not R^1 or R^2, the products formed are enantiomers and the faces of the starting carbonyl compounds are therefore known as enantiotopic. Figure 8.5 gives a generalized example of nucleophilic attack on butan-2-one generating alkoxides that are related as object to mirror image. This is another example of prochirality, addition to one face in an achiral precursor giving rise to chiral products (see Chapter 2, p. 30).

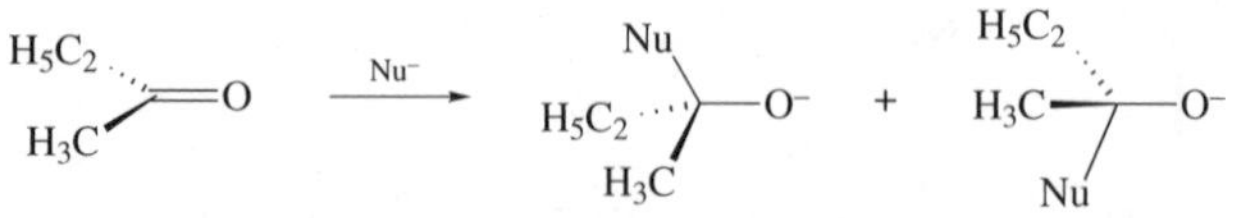

H₅C₂ and H₃C groups on C=O, Nu⁻ arrow, products

Fig. 8.5 Nucleophilic attack on butan-2-one.

Q. 8.1 (S)-3-Phenylbutan-2-one reacts with HCN to give two diastereo-isomers. Represent the two products by means of Fischer projections.

A different situation arises when the groups R¹, R² contain one or more stereogenic centres. The two faces of the carbonyl group are then described as diastereotopic. Nucleophilic attack on a diastereotopic carbonyl then produces diastereoisomers, that is to say stereoisomers that are not mirror image related. A simple example of this is the reduction of (2S,3S)-2,3-dimethylcyclobutanone shown in Fig. 8.6. The nucleophile (H⁻) can approach the ketone group from either face and two products are formed: (1R,2S,3S)-2,3-dimethylcyclobutanol and (1S,2S,3S)-2,3-dimethylcyclobutanol. It is important to note, however, that these two products are not formed to an equal extent and that the all-*cis* product is by far the major one. A glance at a three-dimensional representation of the starting ketone will tell us why this should be. The ring is puckered across the diagonal in an attempt to relieve its angle strain (see Chapter 1) and in either of the two possible conformations approach of the hydride ion to one of the faces is severely hindered by the two methyl groups. The other face, however, is completely open to approach by H⁻, which forces the newly forming hydroxy group onto the same face as the methyl groups (Fig. 8.6). [It is instructive in this case to make the relevant molecular models in order to check the approach paths available to the attacking nucleophile.]

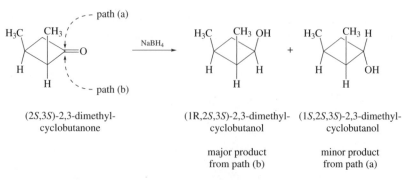

(2S,3S)-2,3-dimethyl-cyclobutanone

(1R,2S,3S)-2,3-dimethyl-cyclobutanol

major product
from path (b)

(1S,2S,3S)-2,3-dimethyl-cyclobutanol

minor product
from path (a)

Fig. 8.6 Path (a) is hindered by the methyl groups whereas path (b) allows easy approach of H⁻.

Groups or faces in molecules which are enantiotopic or diastereotopic are collectively known as heterotopic. In carbonyl compounds this simply means that R¹ and R² (Fig. 8.1) are different and that the two faces of the trigonal centre are different. A system has been devised that differentiates between the two faces, based on the Cahn–Ingold–Prelog priority rules already discussed. Taking again the example of (2*S*,3*S*)-2,3-dimethylcyclobutanone, one face has the sustituents in a clockwise order of priority (the side with the methyl groups pointing up) and this is designated the *Re* face: O > C(CCH) > C(CHH) (Fig. 8.7). The other face has the substituents in an anticlockwise order of priority and is designated the *Si* face. In the reduction of (2*S*,3*S*)-2,3-dimethylcyclobutanone the hydride ion approaches from the *Si* face and gives a new *R* stereogenic centre. It should perhaps be stressed, however, that because of the way the CIP priority rules operate, there is no relationship between the face attacked by the nucleophile and the stereodesignator of the product: it depends entirely on the nature of the groups R^1, R^2 and on the nucleophile.

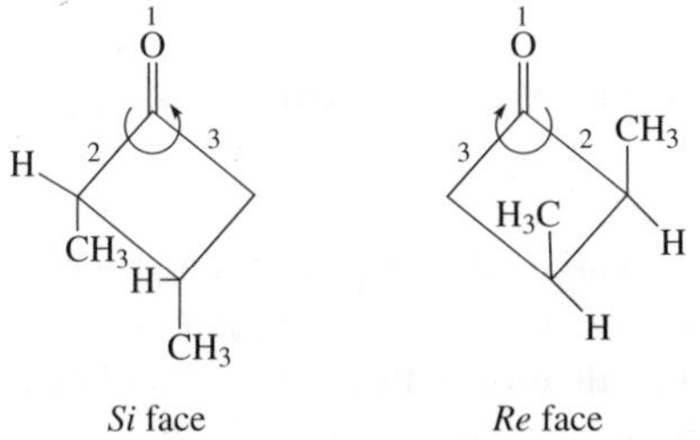

Si face   *Re* face

Fig. 8.7 1, 2 and 3 indicate the priority ratings in the Cahn–Ingold–Prelog system.

This process of taking an achiral position in a molecule and turning it into a stereogenic centre is known as chiral or asymmetric induction, which, when it can be carried out with a high degree of selectivity, is one of the most useful techniques used in synthesis. In some instances an existing stereogenic centre, which is not required in the final product, can be used solely to control chiral induction in a specific step and then be removed in a subsequent transformation.

As a simple stylized illustration, the reduction of a (2*R*)-2-substituted cyclopentanone with sodium borohydride followed by *O*-alkylation and elimination of HX leads to a methylenecyclopentyl ether with *R* stereochemistry at position 1 (Fig. 8.8). The H$^-$ ion is unable to approach the carbonyl carbon atom at the *Re* face because its 109° angle of approach is prevented by the large and bulky substituent on that face. Therefore, it can only approach the *Si* face and the result

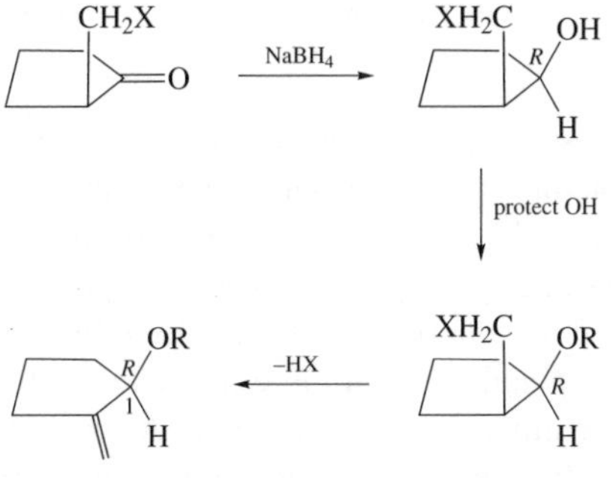

Fig. 8.8 Nucleophilic attack on a 2-substituted cyclopentanone.

is a *cis*-cyclopentanol. The alcohol functionality is protected by alkylation and the original chirality at position 2 is removed by elimination of HX to give an exocyclic double bond, the ring end of which, incidentally, is diastereotopic, it being a trigonal centre with an adjacent stereogenic centre.

Q. 8.2 Given that $=CH_2$ is of lower priority than $-CH_2-$ (for reasons explained in the answer to this question), determine the *Re* and the *Si* face of the alkene featured in Fig. 8.8.

8.2 Nucleophilic attack on acyclic carbonyl compounds

8.2.1 Cram's rule

Chiral induction is not restricted to carbonyl groups contained within a cyclic framework. It can also be applied to acyclic compounds, although their greater conformational freedom makes the situation rather more complicated. A rule was formulated by Donald Cram which enables the chiral induction to be predicted to a certain extent and this is illustrated in Fig. 8.9 for a Grignard reaction with a diastereotopic ketone. The four groups bonded to the stereogenic

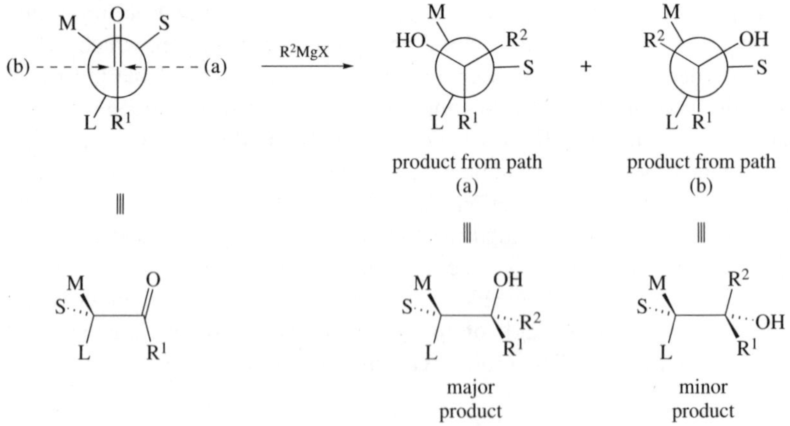

Fig. 8.9

centre are COR, S, M and L, the last three standing for small, medium and large. Cram's rule is based on the principle that the group L, being the largest, will prefer to be orientated as far as possible from the carbonyl group. As you will see from the Newman projection in Fig. 8.9, by adopting this position it must eclipse the group R^1. The Grignard reagent then has the choice of approach from the two faces of the carbonyl group and is much more likely to opt for the less hindered face, that is the face presenting the smallest group. A specific example that follows Cram's rule (with hydride ion as the nucleophile on this occasion) is given in Fig. 8.10.

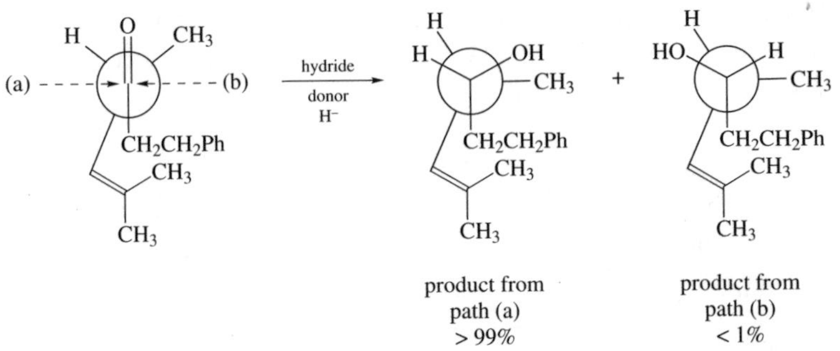

product from
path (a)
> 99%

product from
path (b)
< 1%

Fig. 8.10

8.2.2 The Felkin–Ahn rule

Whilst Cram's rule is useful up to a point, its predictive capability is not extensive. Felkin, Ahn and coworkers subsequently developed the theme and proposed a variation which accounts more fully for the chiral induction observed in many highly stereoselective transformations. It can also be used more effectively to predict the stereochemical outcome of planned reactions.

The difference between the Felkin–Ahn (FA) model and Cram's rule is in the conformation of the carbonyl compound. In the FA model group L is assumed not to be antiperiplanar to the CO group, but to be positioned at right angles. This has the advantage of removing unfavourable eclipsing interactions (or Pitzer strain, see Chapter 1) between L and R. There are two possible orientations of this sort, depending on the direction of the rotation about the single bond, and these are illustrated in the Newman projections in Fig. 8.11. The

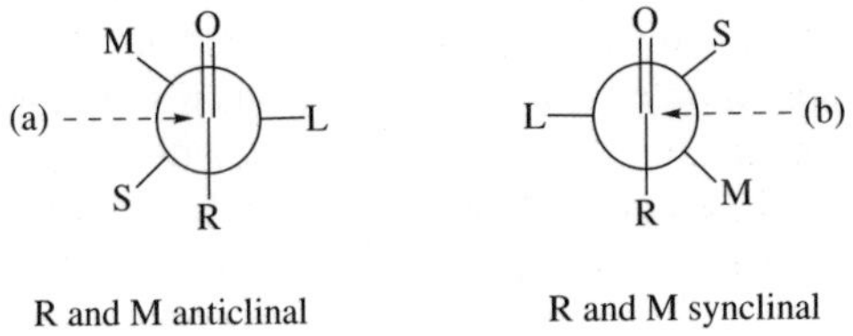

R and M anticlinal

R and M synclinal

Fig. 8.11

stereoselectivity of the ensuing nucleophilic reaction arises from the relative energies of the two transition states which in turn depend on the interactions between M and R. If the interactions between R and M are very unfavourable, the reaction will tend to pass through the transition state in which R and M are anticlinal (Fig. 8.11). The stereochemistry of the major product alcohol will be that resulting from approach of the nucleophile from the side remote from L [approach path (a)]. A minor proportion of the reaction is likely to pass through the transition state in which R and M are synclinal, and formation of a small amount of the other stereoisomer is probable [approach (b) in Fig. 8.11].

Figure 8.12 shows an actual example of this model: $LiAlH_4$ reduction of (*S*)-3-phenylbutanone gives (2*R*,3*S*)- and (2*S*,3*S*)-3-phenylbutan-2-ol in the ratio 2:1.

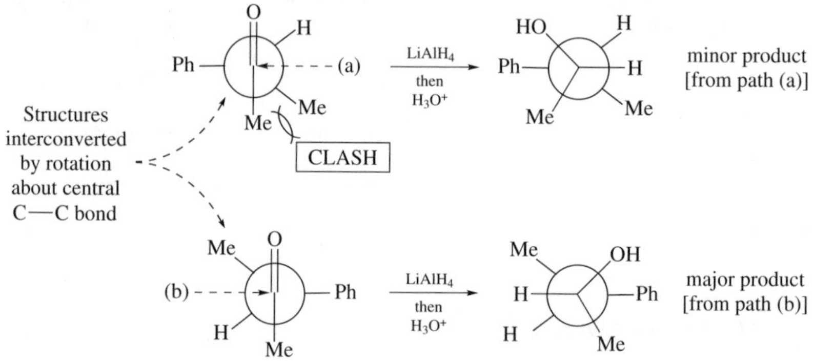

Structures
interconverted
by rotation
about central
C—C bond

CLASH

minor product
[from path (a)]

major product
[from path (b)]

Fig. 8.12

The preference for one product stereoisomer over the other will obviously become more pronounced as the difference in size between S and M increases. When S and M are very similar, there is likely to be little difference in their interactions with R and the diastereoselectivity of nucleophilic reactions involving such molecules is likely to be poor.

Electronic effects that disturb the Felkin–Ahn model

Steric effects are a major factor in determining the conformations of the diastereotopic ketones on reaction with a nucleophile, and consequently the stereochemistry of the products. However, electronic effects must also be taken into consideration where appropriate. For example, the methyl group and the chlorine atom are of approximately the same size but they differ markedly in their electronic interactions. The conformation adopted by the chlorine-containing structure in Fig. 8.13 places the chlorine at right-angles to the carbonyl

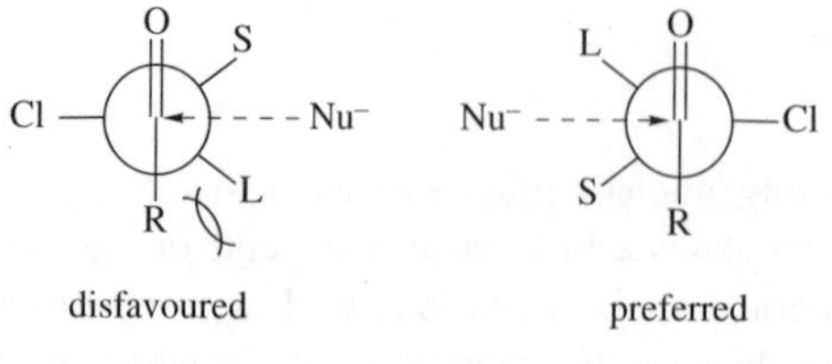

disfavoured   preferred

Fig. 8.13

group. Therefore, because of electronic interactions, the electron-rich chlorine with its greater repulsive power dictates the course of the reaction. The large group (Fig. 8.13) prefers to distance itself from the group R and the approach path of the nucleophile bisects L and S. A specific example is shown in Fig. 8.14, involving the reduction of (*S*)-2-chloro-1-phenylpropanone, which affords the *S,S*-diastereoisomer as the major product.

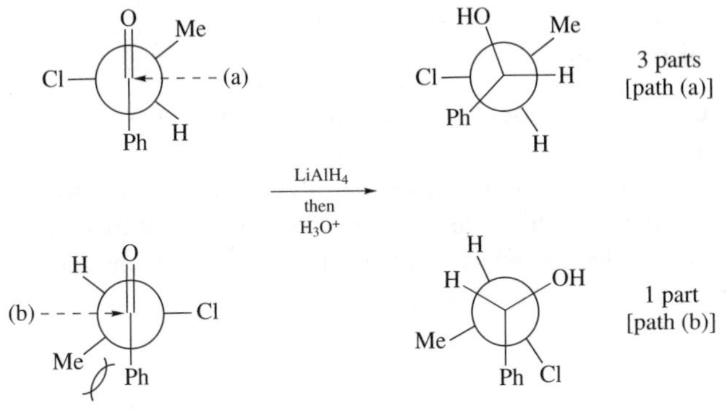

Fig. 8.14

8.3   Locking effects in nucleophilic reactions at carbonyl groups

It is possible in some cases to control the stereoselectivity of nucleophilic reactions of carbonyl compounds by appropriate choice of reagent. Lithium cations show a great propensity to coordinate with oxygen atoms and this property is put to good effect in the hydride reduction of 2-methoxy-2-phenyl-1-(*p*-tolyl)-ethanone. When this compound is treated with lithium aluminium hydride, the reagent acts not only as a reducing agent but also as a conformational lock, by coordinating both to the methoxy oxygen and to the ketone oxygen (Fig. 8.15).

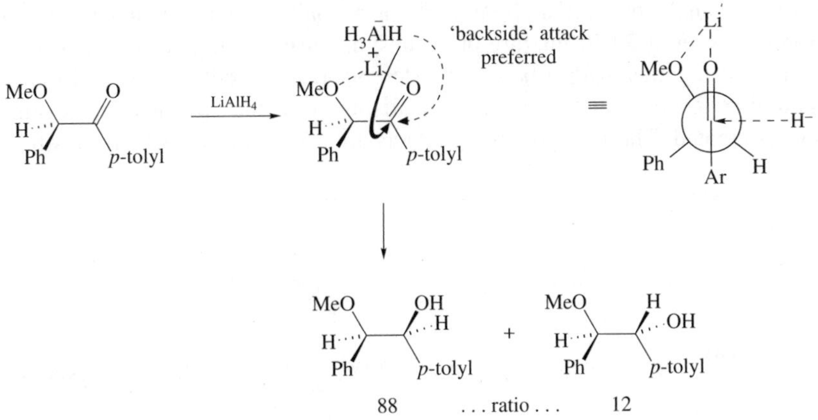

Fig. 8.15

The hydride ion is then delivered predominantly to the *Re* face (the side away from the phenyl group), this face offering the least resistance to the approach. The measured ratio of the product alcohols is 1*S*,2*R*:1*R*,2*R* = 88:12. This particular type of pre-reaction coordination has been found for hydroxy groups and small alkoxy groups but does not appear to operate for bulky moieties such as trialkylsilyloxy groups (R_3SiO).

8.4   The aldol reaction

Reaction between acetaldehyde and a base with an external supply of heat is known as the aldol reaction. The sequence of events is as shown in Fig. 8.16. The base abstracts a proton from a molecule of acetaldehyde and the resulting anion attacks a further acetaldehyde molecule. The product of this nucleophilic attack is a secondary alcohol, which can, under more forcing conditions, undergo elimination of water to give an achiral enal.

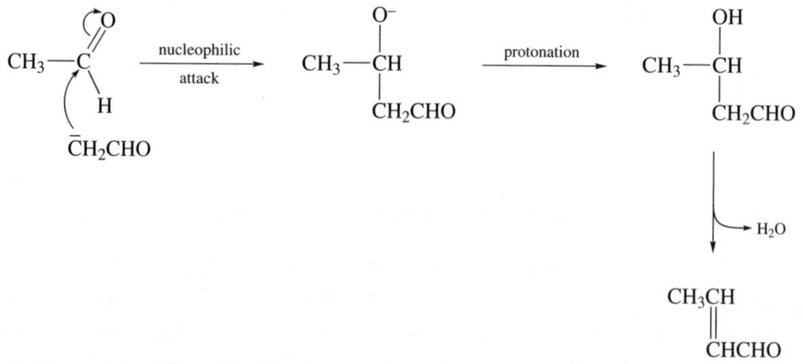

Fig. 8.16 The aldol reaction.

From a stereochemical viewpoint, the classical aldol reaction is not particularly exciting. However, when a combination of different carbonyl compounds is used the analogous initial-phase reactions become much more interesting. For example, when a 50:50 mixture of an unsymmetrical ketone and an aldehyde are heated together with a base, four stereoisomeric products can be isolated (assuming R^1, R^2 and R^3 are achiral) arising from the two new stereogenic centres created. The four possible products are shown in Fig. 8.17 from which

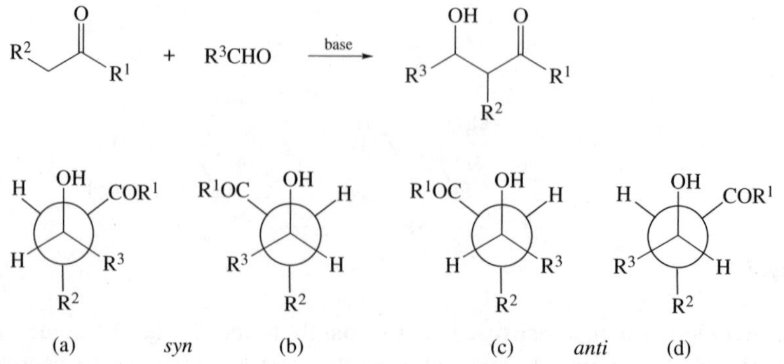

Fig. 8.17

you will see that (a) and (b), and (c) and (d) are enantiomeric pairs and that (a) and (c), and (b) and (d) are diastereomerically related. From the Newman projections you should also notice that one enantiomeric pair is shown in a synclinal conformation whilst the other is antiperiplanar.

In order for this reaction to be useful to the synthetic chemist, the reaction conditions must be carefully considered, so that an acceptable degree of stereoselectivity is accomplished. Good diastereoselection is achieved using lithium and boron enolates and a detailed consideration of aldol reactions of the boron species follows. In fact the configuration of the enolate often determines the stereochemistry of the product. First, observe the difference in the disposition of the boron moiety and the alkene hydrogen in the enolates described in Fig. 8.18.

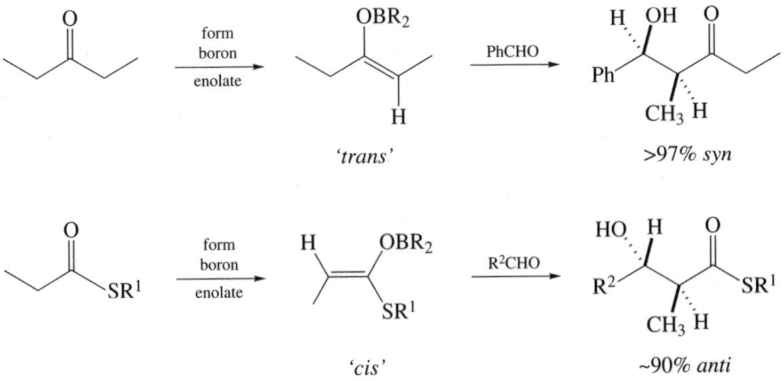

Fig. 8.18

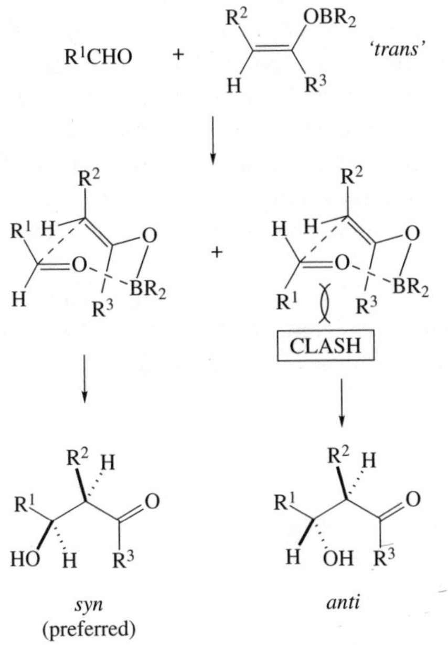

N.B. Enantiomers are formed by backside attack of R¹CHO giving four compounds in total.

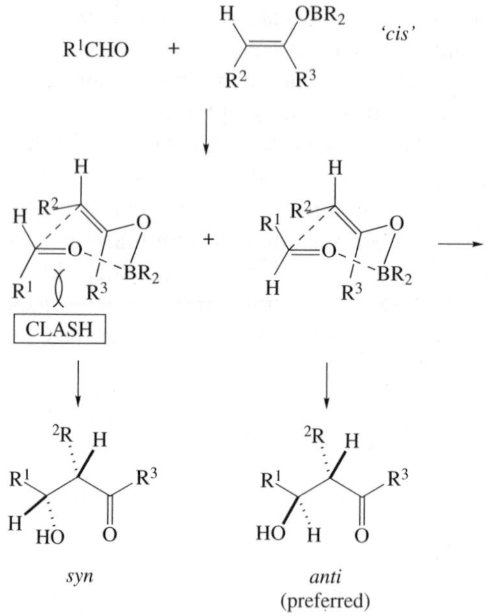

Fig. 8.19

Aldol reactions involving enolates having the *'trans'* arrangement give *syn*-products, while enolates having the *'cis'*-arrangement give mainly the *anti*-diastereoisomers. Why this should be so is explained in terms of the different interactions in the chair-like transition states (Fig. 8.19).

Thus the orientation of the aldehyde to the enolate is influenced by the interaction of the electron-rich oxygen atom of the aldehydic carbonyl group and the electron-poor boron atom contained in the enolate (boron has only six electrons in the outermost orbitals). The electron-rich enol double bond and the electrophilic carbonyl carbon atom are brought into close proximity. The substituents R^1 and R^3 get uncomfortably close in space in one of the two possible cases described in Fig. 8.19 (unfavourable interactions earmarked by the word CLASH). Hence the *'trans'*-enolate leads to *syn* products and the *'cis'*-enolate leads to *anti*-products.

In this particular scenario a detailed consideration of the stereochemistry of the possible transition states leads to a clear picture of why one set of products predominates over another. This story is amplified in Chapter 15.

Answers to questions

Q. 8.1

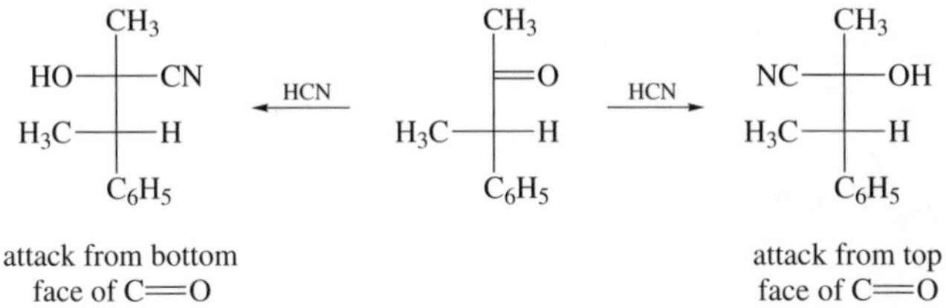

attack from bottom
face of C=O

attack from top
face of C=O

Q. 8.2

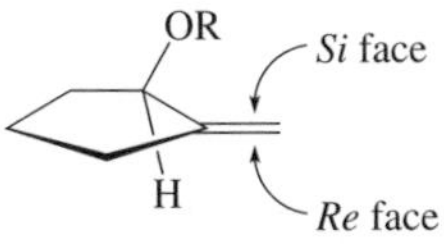

Si face

Re face

When, in the general structure R^1—$\overset{\overset{O}{\|}}{C}$—$R^2$, one of the R groups contains an oxygen, a choice has to be made as to which oxygen (the carbonyl oxygen or the R-group oxygen) has priority in order to determine the *Re* and *Si* faces. Esters are a case in point and the procedure for making this choice can be illustrated with methyl acetate. Here R^1 is CH_3 and R^2 is OCH_3.

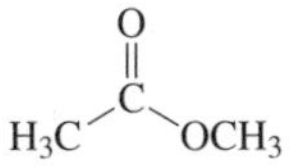

Clearly, in terms of the CIP sequence rule the methyl group is least preferred but how do we distinguish between =O and —O—? In Chapter 2 (p. 26) we state that multiple bonds are considered as the equivalent number of single bonds (that is, double bonds are given as two single bonds, triple bonds as three). The carbonyl oxygen can therefore be represented as:

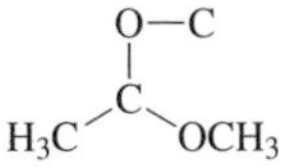

The new carbon atom is deemed not to be further substituted but is nominally attached to a 'phantom atom' of atomic number 0.

Returning to the choice we have to make, the pictorial representations are now:

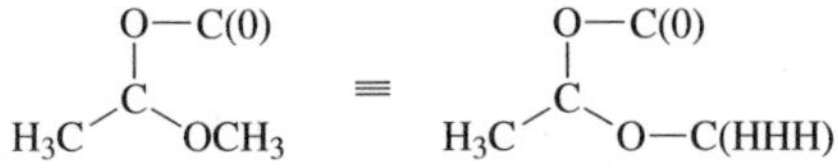

The first point of difference is that the carbon of the methoxy group has three substituents of atomic number 1 (hydrogens), whereas the carbonyl oxygen atom is attached to a carbon atom carrying a nominal substituent of atomic number 0. Therefore the methoxy oxygen has priority over the carbonyl oxygen and the face presented above is the *Si* face.

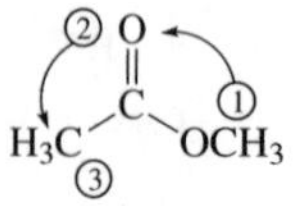

Applying this principle to the alkene featured in Fig. 8.8 we have (looking from the top face):

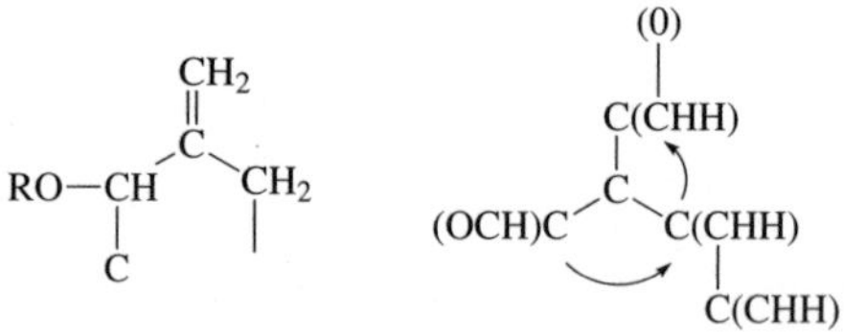

hence the stereodescriptor *Si* designates the top face.

9 Stereochemistry of some important reactions leading to the formation of alkenes

There is a plethora of different ways of forming selected alkenes, many of them involving oxidation and rearrangement reactions. However, there are three major types of olefin-forming reactions that have the most profound stereochemical implications, namely elimination reactions, Wittig reactions (and related procedures) and reactions of esters derived from β-hydroxy sulfones; these transformations will feature in this chapter.

9.1 Elimination reactions

One of the most commonly used methods for forming carbon–carbon double bonds utilizes β-elimination reactions (Fig. 9.1), for example acid-catalysed dehydrations of alcohols, base-induced eliminations from alkyl halides (or sulfonates) and Hofmann eliminations from quaternary ammonium salts.

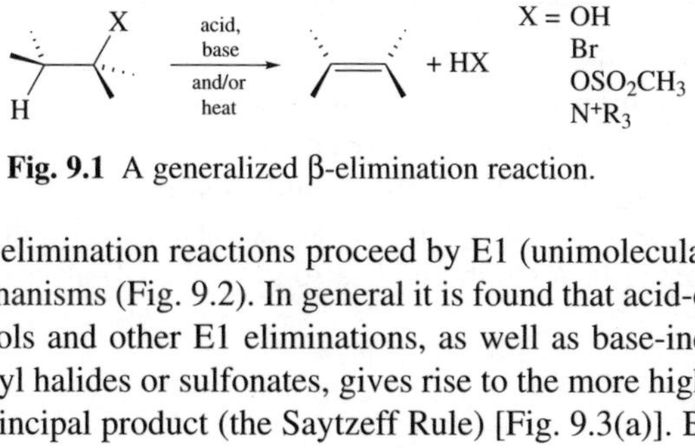

Fig. 9.1 A generalized β-elimination reaction.

These elimination reactions proceed by E1 (unimolecular) and E2 (bimolecular) mechanisms (Fig. 9.2). In general it is found that acid-catalysed dehydration of alcohols and other E1 eliminations, as well as base-induced E2 elimination from alkyl halides or sulfonates, gives rise to the more highly substituted alkene as the principal product (the Saytzeff Rule) [Fig. 9.3(a)]. Base-induced eliminations from quaternary ammonium salts give predominantly the less-substituted

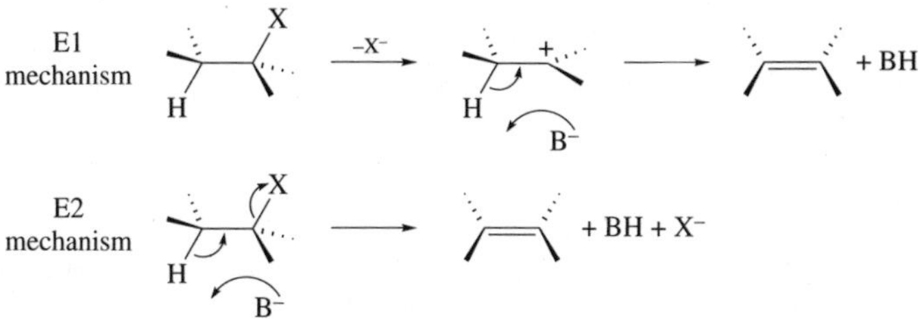

Fig. 9.2

alkene (the Hofmann Rule) [Fig. 9.3(b)]. Both reactions featured in Fig. 9.3 are regioselective but neither is regiospecific (two products are formed in both cases).

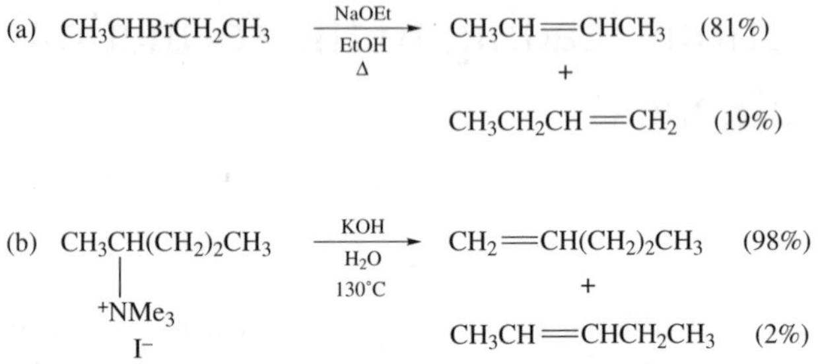

(a) $CH_3CHBrCH_2CH_3$ $\xrightarrow[\substack{EtOH \\ \Delta}]{NaOEt}$ $CH_3CH{=}CHCH_3$ (81%)

$+$

$CH_3CH_2CH{=}CH_2$ (19%)

(b) $CH_3CH(CH_2)_2CH_3$ $\xrightarrow[\substack{H_2O \\ 130°C}]{KOH}$ $CH_2{=}CH(CH_2)_2CH_3$ (98%)

$\underset{I^-}{\overset{|}{\underset{}{+NMe_3}}}$

$+$

$CH_3CH{=}CHCH_2CH_3$ (2%)

Fig. 9.3

The Hofmann reactions involving quaternary ammonium salts and base-induced eliminations from alkyl halides and sulfonates are generally E2 *anti*-elimination processes. That is to say, the hydrogen atom and the leaving group depart from opposite sides of the incipient double bond.

Q. 9.1 Given that a trimethylammonium substituent is very much larger than a bromine atom, present a rationale for the difference in pathways for the reactions (a) and (b) in Fig. 9.3. [Construction of Newman projections is advised.]

For example (Fig. 9.4), treatment of *meso*-(R,S)-1,2-dibromo-1,2-diphenylethane with base furnishes (E)-bromostilbene while racemic dibromide [i.e. (R,R)- and (S,S)-1,2-dibromo-1,2-diphenylethane] gives (Z)-bromostilbene under similar reaction conditions. These two elimination reactions are stereo-specific.* That is, the R,S-dibromide gives only (E)-stilbene, the S,S-dibromide yields only the (Z)-alkene [the R,R-dibromide also gives the (Z)-alkene].

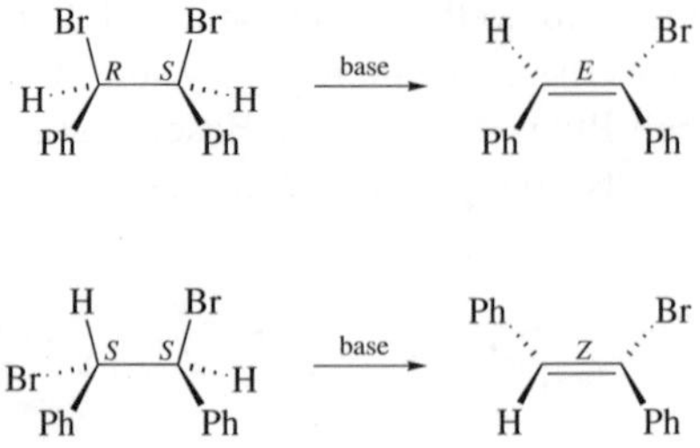

Fig. 9.4

---

*The word stereospecific is used here in the proper sense. A reaction is truly stereospecific if two (or more) starting materials, differing only in configuration, are converted into stereo-chemically different products. In modern day parlance 'stereospecific' is also used to mean 'very highly stereoselective'.

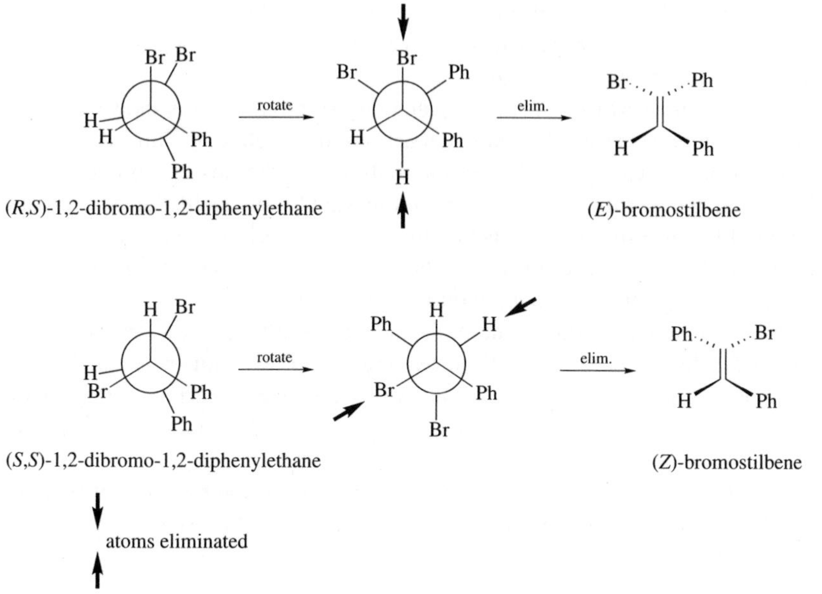

(*R,S*)-1,2-dibromo-1,2-diphenylethane   (*E*)-bromostilbene

(*S,S*)-1,2-dibromo-1,2-diphenylethane   (*Z*)-bromostilbene

↓ atoms eliminated ↑

Fig. 9.5

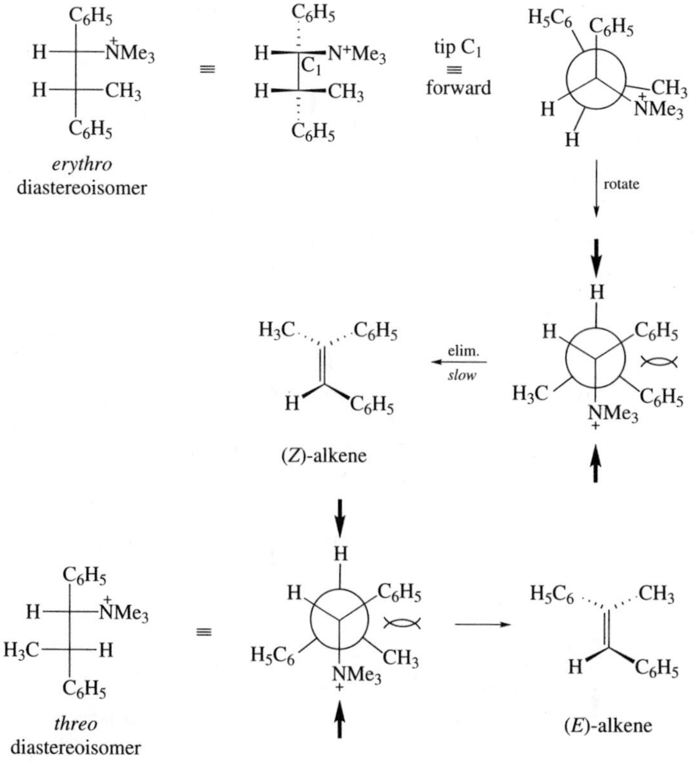

erythro
diastereoisomer

(*Z*)-alkene

threo
diastereoisomer

(*E*)-alkene

Fig. 9.6

Restating what was said above in another way, the elimination takes place most readily when the hydrogen atom and the leaving group are in the anti-periplanar arrangement (Fig. 9.5).

Similarly, the *erythro-* and *threo*-quaternary ammonium salts shown in Fig. 9.6 undergo stereospecific elimination reactions on treatment with base. Neither reaction occurs easily for the reasons outlined in the answer to question 9.1. However, the *erythro*-isomer reacts more slowly since, in the conformation required for the elimination reaction to proceed, clearly the phenyl groups have to approach each other closely (the phenyl groups take up a *gauche* or synclinal relationship), giving rise to unfavourable interactions.

In open-chain compounds the molecule can usually adopt a conformation in which the hydrogen atom and the leaving group are antiperiplanar. In cyclic systems this is not always the case and this may have a bearing on the outcome of the reaction. In six-membered ring systems the two leaving groups must be *trans*-diaxial so as to assume the requisite arrangement prior to a concerted (E2) elimination. Thus menthyl chloride gives only 2-menthene on treatment with base, while neomenthyl chloride gives a mixture rich in 3-menthene (Fig. 9.7).

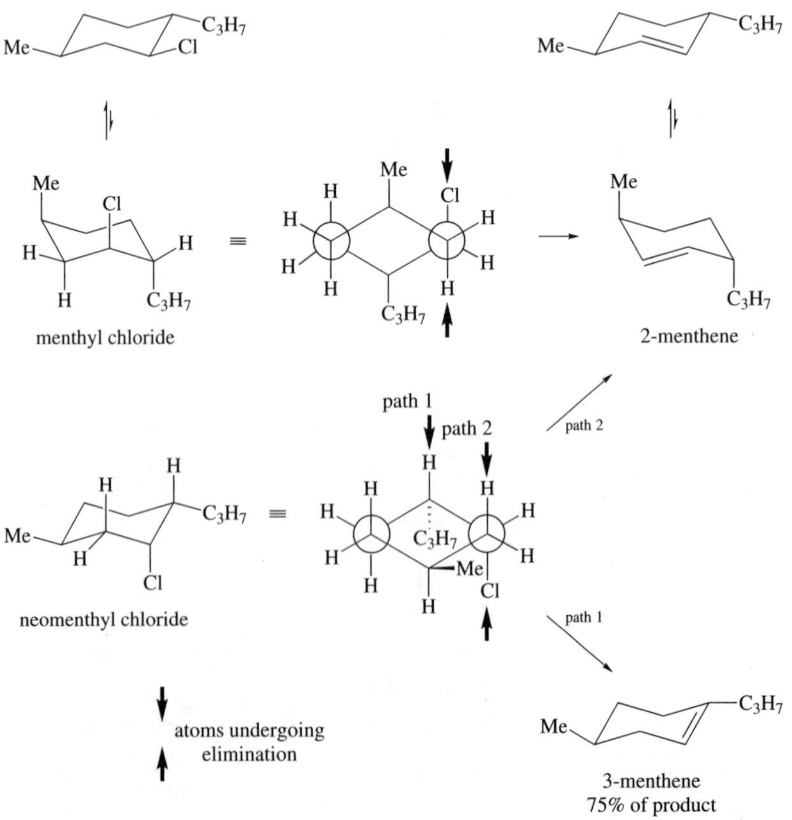

Fig. 9.7

Pyrolytic eliminations, for example involving carboxylic esters and xanthates, are another important group of alkene-forming reactions. These elimination reactions are generally *syn*-eliminations and the reactions proceed by a concerted pathway (Fig. 9.8). As suggested in this figure, *N*-oxides and selenoxides

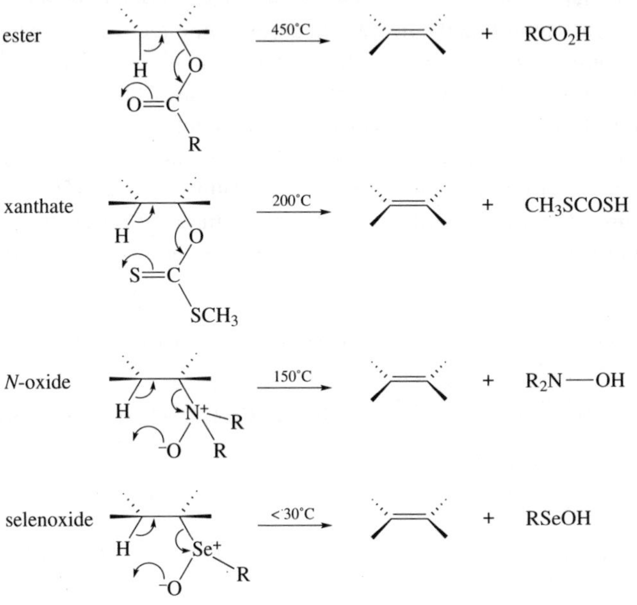

Fig. 9.8

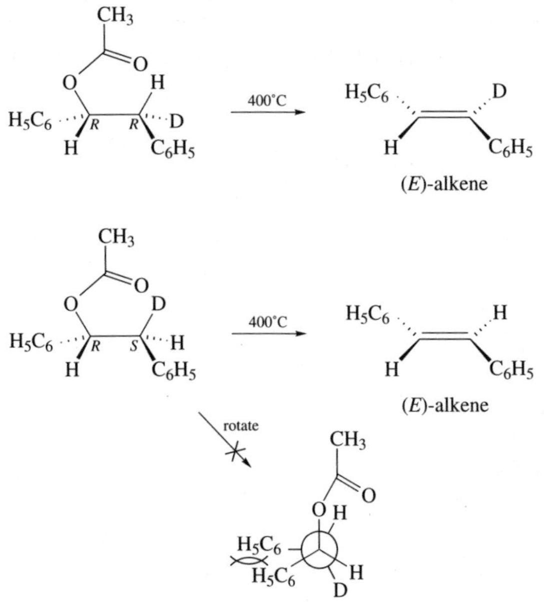

Fig. 9.9

undergo elimination reactions by similar mechanisms, the latter at room temperature or below.

The predilection of esters to eliminate in a *syn*-mode is well illustrated by the reactions of the deuterium-labelled compounds featured in Fig. 9.9. Thus pyrolysis of the (*R*,*R*)-isomer of 1-acetoxy-2-deuterio-1,2-diphenylethane gave (*E*)-deuteriostilbene (retaining the deuterium atom) while the (*R*,*S*)-diastereoisomer led to deuterium-free stilbene. In each case the alternative product could only be formed if the phenyl groups became eclipsed (as illustrated for the *R*,*S*-compound in Fig. 9.9), a most unsatisfactory situation.

Pyrolysis of amine oxides (the Cope reaction) and methyl xanthates (the Chugaev reaction) takes place at lower temperatures (about 150–200 °C) than those required for esters, while selenoxides eliminate at ambient temperature. (Some specific examples of the latter processes are shown in Fig. 9.10.)

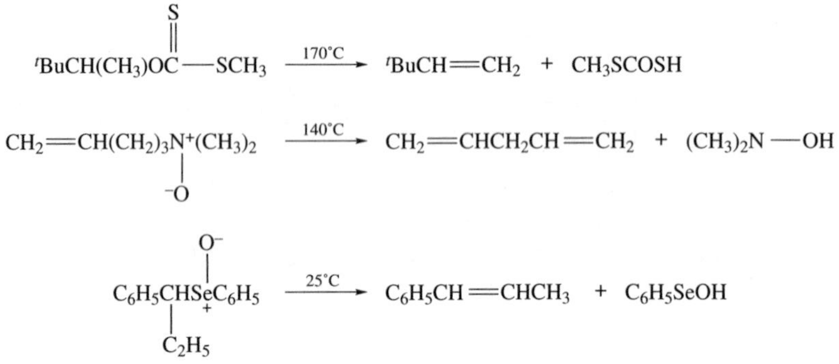
ᵗBuCH(CH3)OC(=S)—SCH3
170°C
ᵗBuCH=CH2 + CH3SCOSH
CH2=CH(CH2)3N+(CH3)2
-O
140°C
CH2=CHCH2CH=CH2 + (CH3)2N—OH
O−
C6H5CHSeC6H5
C2H5
25°C
C6H5CH=CHCH3 + C6H5SeOH

Fig. 9.10

Once again, with alicyclic compounds some restrictions are imposed by the conformation of the leaving groups and the necessity to form the cyclic intermediate. Thus the cyclohexyl acetate shown in Fig. 9.11 eliminates ethanoic acid

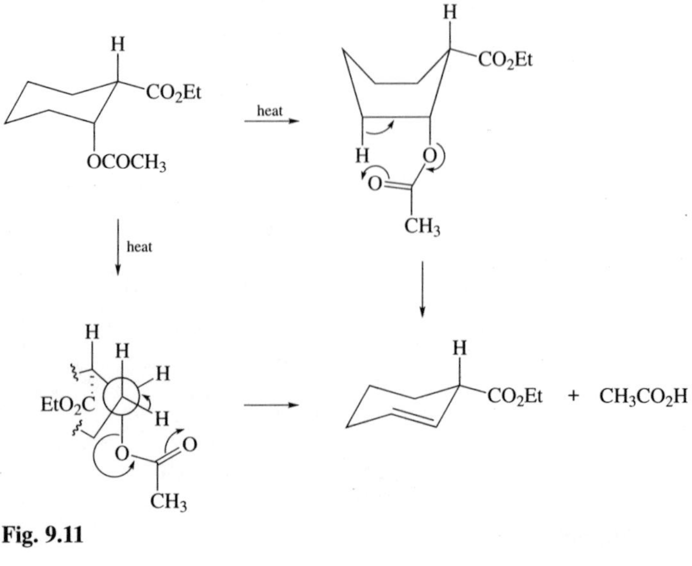
H
CO2Et
OCOCH3
heat
H
CO2Et
H
O
O
CH3
heat
H
H
H
EtO2C
H
O
O
CH3
H
CO2Et + CH3CO2H

Fig. 9.11

on heating to form only one alkene. A boat conformation is needed to produce a perfectly planar arrangement of the ester group and the hydrogen atom, but as shown by the Newman projection, a synclinal (*gauche*) relationship in the chair form would bring the two entities close together, sufficiently close, indeed, for reaction to proceed. In keeping with this view that a *gauche* interaction is sufficient for reaction to take place, the methyl xanthate shown in Fig. 9.12 forms equimolar amounts of two products since hydrogen atoms on both adjacent carbon atoms are accessible in this case.

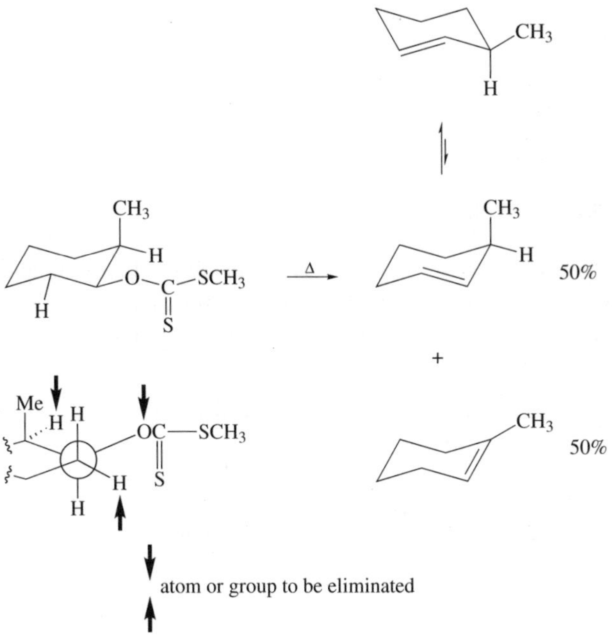

Fig. 9.12

9.2 Wittig and related reactions

The reaction between a phosphorane (or phosphonium ylide) and an aldehyde or ketone to form a phosphine oxide and an alkene is known as the Wittig reaction after the German chemist Georg Wittig. In contrast with the majority of the elimination reactions discussed above, the Wittig reaction gives rise to alkenes in which the position of the double bond is predetermined and unambiguous (Fig. 9.13).

The Wittig reagent is prepared by treating a phosphonium salt with base (Fig. 9.13) to give an ylide. (An ylide is a compound with opposite charges on adjacent, covalently bound atoms, each of which has an octet of electrons.) Reaction then takes place by attack of the carbanionoid carbon atom of the ylide on the

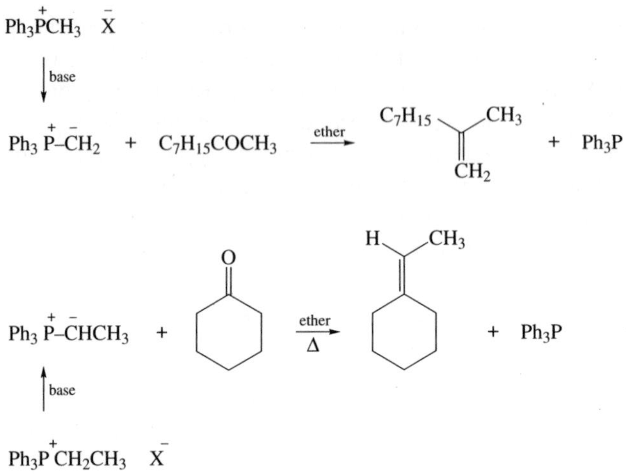

$$Ph_3\overset{+}{P}CH_3 \quad \overset{-}{X}$$

$$\downarrow \text{base}$$

$$Ph_3 \overset{+}{P}-\overset{-}{C}H_2 \quad + \quad C_7H_{15}COCH_3 \quad \xrightarrow{\text{ether}} \quad$$

C_7H_{15}\ \ \ CH_3 with CH_2 + Ph_3P

$$Ph_3 \overset{+}{P}-\overset{-}{C}HCH_3 \quad + \quad$$

(cyclohexanone) $\xrightarrow[\Delta]{\text{ether}}$ (ethylidenecyclohexane) H, CH_3 + Ph_3P

$$\uparrow \text{base}$$

$$Ph_3\overset{+}{P}CH_2CH_3 \quad \overset{-}{X}$$

Fig. 9.13 The Wittig reaction.

electrophilic carbonyl carbon atom to give a betaine which collapses to the products by way of a four-membered cyclic transition state. The driving force for the forward reaction is provided by the formation of the very strong phosphorus–oxygen bonds (Fig. 9.14).

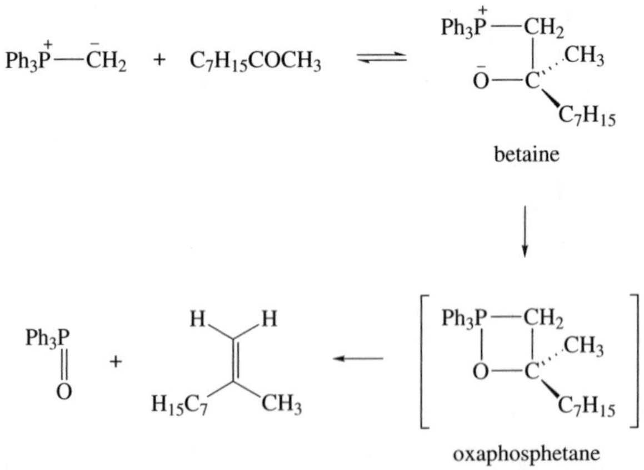

$$Ph_3\overset{+}{P}-\overset{-}{C}H_2 \quad + \quad C_7H_{15}COCH_3 \quad \rightleftharpoons \quad$$

Ph_3P—CH_2 / O—C with CH_3 and C_7H_{15}

betaine

$$\downarrow$$

Ph_3P / O + H_{15}C_7 / CH_3 alkene ← [Ph_3P—CH_2 / O—C with CH_3 and C_7H_{15}]

oxaphosphetane

Fig. 9.14

The ylide $Ph_3\overset{+}{P}-\overset{-}{C}H_2$ is stabilized by back-donation of the negative charge on the carbon atom into an available empty d orbital around the phosphorus atom, so that the formula $Ph_3P=CH_2$ can also be used to describe this Wittig reagent. Further stabilization of the negative charge on the ylide can be provided by other substituents on the carbon atom, that is, substituents which have the capacity to delocalize the negative charge, such as a carbonyl group. These stabilized phosphoranes, for example $Ph_3\overset{+}{P}-\overset{-}{C}HCO_2Et$, react relatively slowly with aldehydes and ketones. This tardiness is overcome by employing a valuable alternative procedure that involves phosphonate esters. Reaction of such phosphonates with a suitable base gives the corresponding carbanions (Fig. 9.15) which are more nucleophilic than the related phosphoranes since the negative charge on the carbon atom is no longer dispersed by delocalization into d orbitals of the adjacent phosphorus atom. This variation on the Wittig reaction (the Wadsworth–Emmons modification) is widely used to prepare αβ-unsaturated esters and ketones.

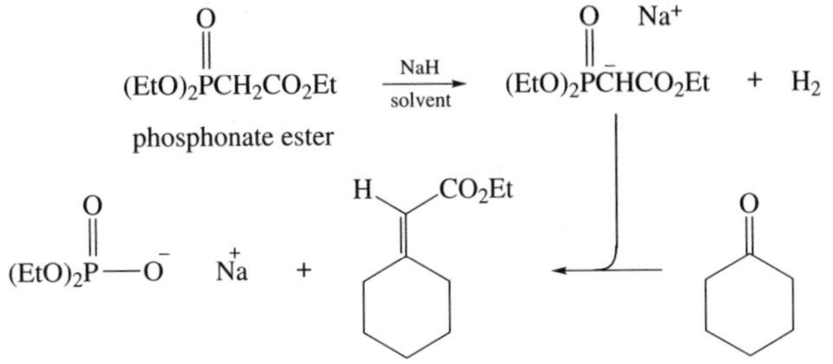

Fig. 9.15

The main disadvantage with the Wittig reaction is that, for the synthesis of non-terminal, disubstituted alkenes, as well as trisubstituted and tetrasubstituted systems, both *Z*- and *E*-alkenes can be formed. This can be a nuisance since alkene isomers are difficult to separate. However, we find that by choosing the reacting partners carefully we can exert some degree of control; thus in the reaction of phosphonate anions and resonance-stabilized ylides with aldehydes the *E*-alkene generally predominates. Non-stabilized ylides, on the other hand, usually give more of the *Z*-alkene (Fig. 9.16) (for specific examples see the prostaglandin syntheses in Chapter 16).

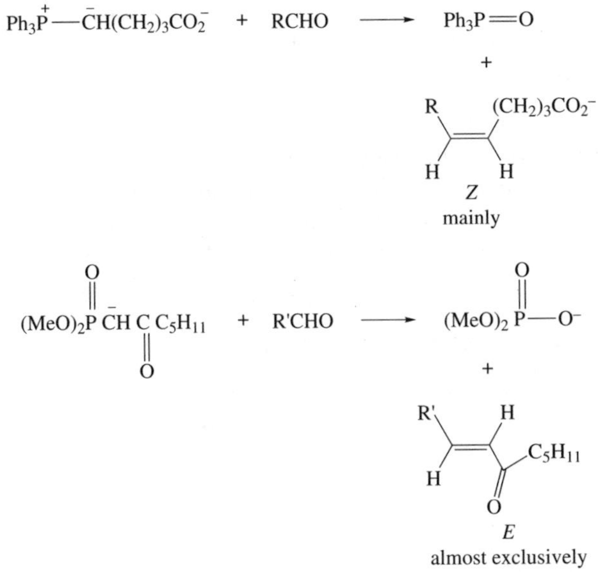

Fig. 9.16

With stabilized ylides the preference for the *E*-isomer is a result of the fact that the formation of the intermediate betaines is reversible. The kinetically formed *erythro*-betaine is less stable than the *threo* form which, after a short time, becomes the predominant diastereoisomer. This betaine collapses to provide the *E*-alkene (Fig. 9.17). With the non-stabilized ylide the formation of the betaine is essentially irreversible and conversion into the alkene proceeds mainly from the *erythro* isomer.

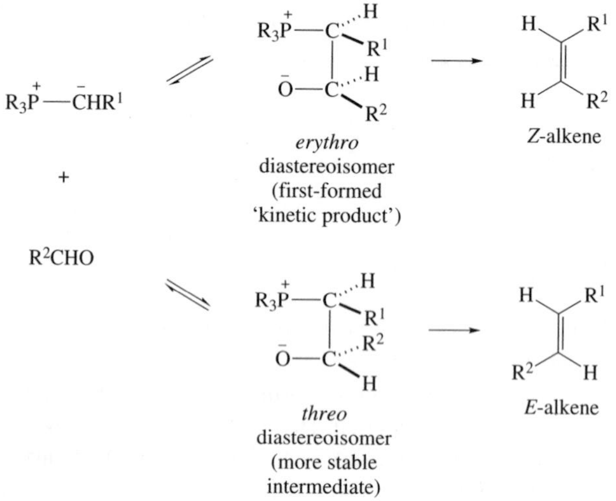

Fig. 9.17

Q. 9.2 Draw the *erythro* and *threo* forms of the betaines featured in Fig. 9.17 in Fischer projections and compare these projections with those for the sugars erythrose and threose (Chapter 3).

The selectivity for the formation of Z-alkenes from non-stabilized systems is potentiated by the use of non-polar solvents and 'salt-free' conditions, to further 'hurry' the reaction through the increasingly unstable intermediate.

The silicon version of the Wittig reaction is known as the Peterson reaction, and entails the elimination of trimethylsilanol (Me_3SiOH) from a β-hydroxy-alkyltrimethylsilane (Fig. 9.18). It is interesting to note that the steric course of

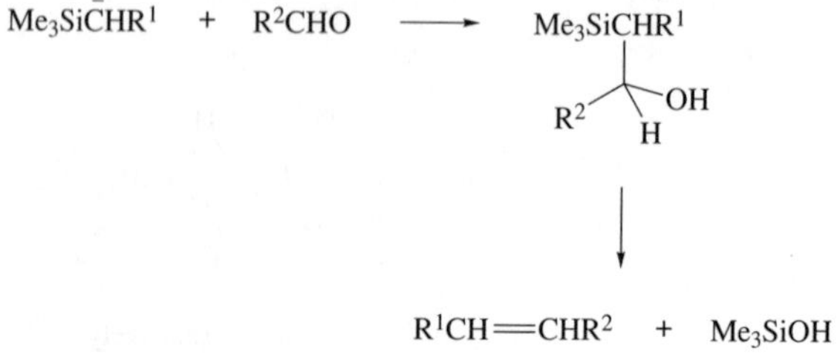

Fig. 9.18 The Peterson reaction.

the elimination reaction can be controlled in such a way as to give either the Z-alkene or the E-alkene from the same alcohol. This is useful because the Peterson reaction almost always gives a mixture of *threo* and *erythro* forms of the β-hydroxyalkylsilane, which can often be separated. Using basic conditions the *threo* isomer eliminates trimethylsilanol to give predominantly (95%) the E-alkene by way of *syn*-elimination. Under Lewis acidic conditions *anti*-elimination takes place and the Z-alkene is formed. Contrariwise the *erythro*-hydroxyalkylsilane gives the Z-alkene with base and the E-alkene with acid (Fig. 9.19).

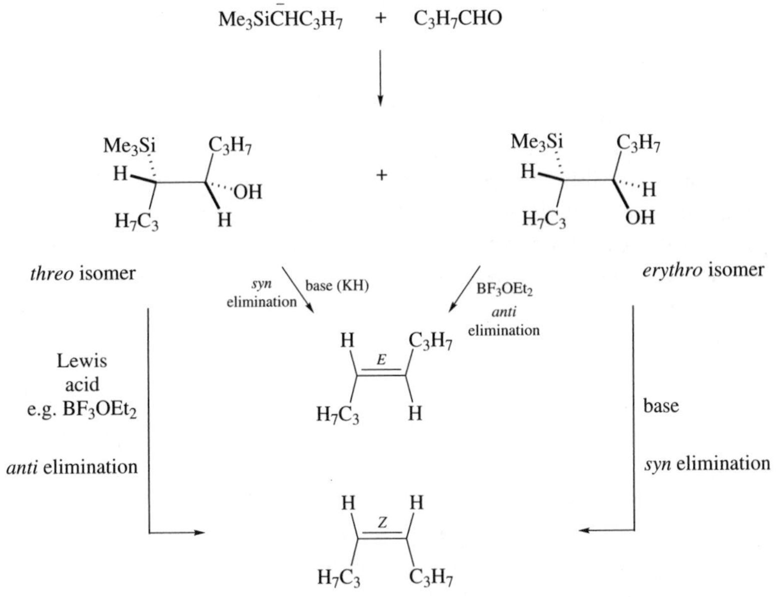

$$Me_3Si\overline{C}HC_3H_7 \quad + \quad C_3H_7CHO$$

Fig. 9.19

> **Q. 9.3** Give plausible mechanisms for the conversion of the *threo*-hydroxysilane described in Fig. 9.19 into the E-alkene and the Z-alkene using base and acid, respectively.

9.3 Reactions of sulfones

Alkyl sulfones are readily deprotonated and the derived carbanions react readily with aldehydes and ketones to afford α-hydroxy sulfones. Formation of the corresponding acetates or toluene-*p*-sulfonates (Fig. 9.20) gives substrates which undergo reductive cleavage with sodium amalgam in methanol to form alkenes. Di-, tri- and tetra-substituted alkenes can be formed in this way.

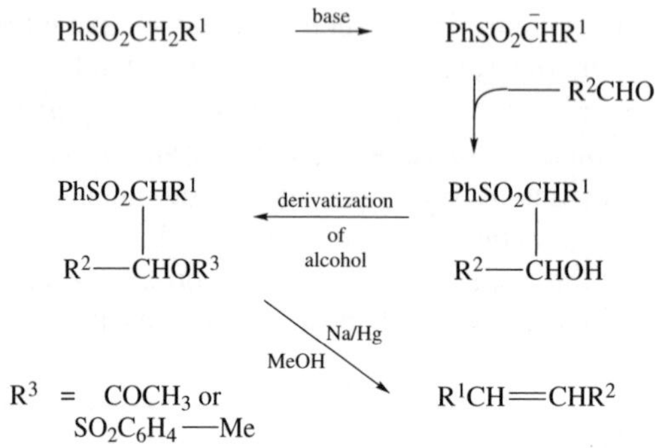

Fig. 9.20

Reactions leading to 1,2-disubstituted alkenes give the *E*-isomers almost exclusively, irrespective of the configuration (*erythro* or *threo*) of the hydroxy-alkyl sulfone. Possibly the reductive cleavage of the phenylsulfonyl group generates an anion which, whatever the original configuration, is sufficiently long-lived to allow bond rotation to provide the low energy conformation (R^1 and R^2 well separated) from which the *E*-alkene is formed by loss of the nucleofuge (Fig. 9.21).

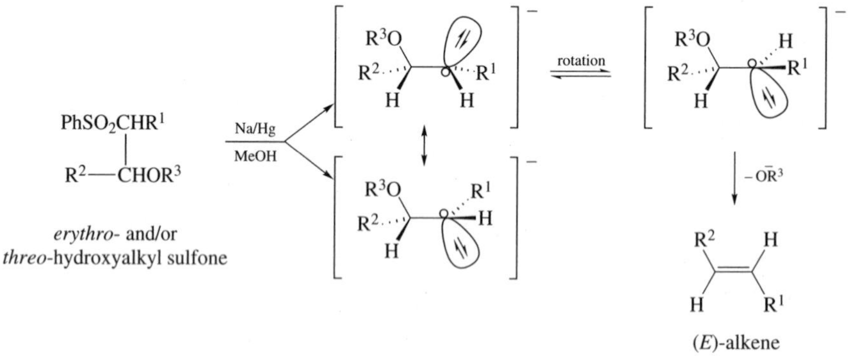

Fig. 9.21

Answers to questions

Q. 9.1   Hofmann eliminations tend to form the less highly substituted (less thermodynamically stable) alkene due to the very large size of the quaternary ammonium system. For trimethyl-(2-butyl)-ammonium hydroxide formed under the reaction conditions in reaction (b) (Fig. 9.3), the two possible arrangements for *anti*-elimination to give but-2-ene give rise to a severe, unfavourable inter-action. In contrast, *anti*-elimination to give but-1-ene can take place without severe steric compression. The relatively small halide atoms (Br, I) or sulfonates (O–SO$_2$R) do not experience the same steric interactions and the more highly substituted butene is formed.

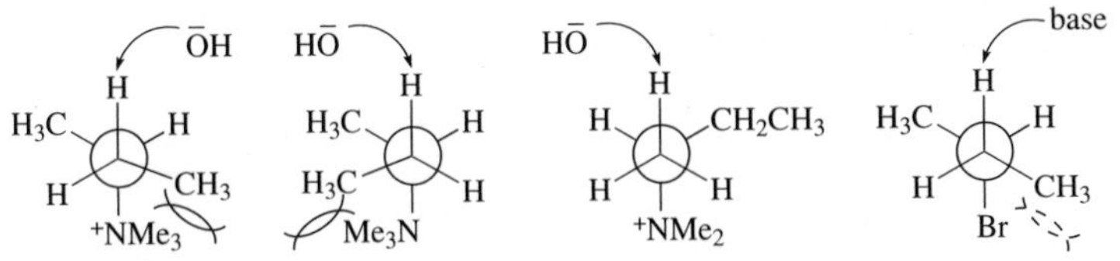

H3C, OH HO, H3C, H HO, H H3C, base

disfavoured   disfavoured   favoured   favoured

Q.9.2

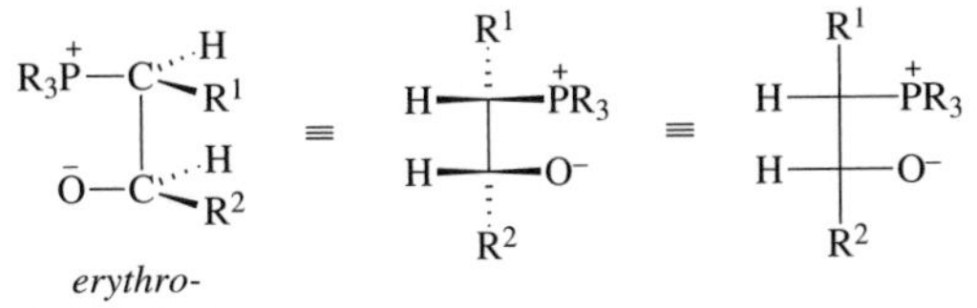

$R_3\overset{+}{P}-C\overset{H}{\underset{R^1}{}}$ ≡ ... ≡ ...

erythro-
diastereoisomer

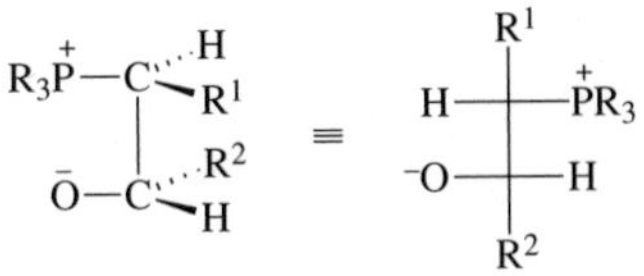

$R_3\overset{+}{P}-C\overset{H}{\underset{R^1}{}}$ ≡ ...

threo-
diastereoisomer

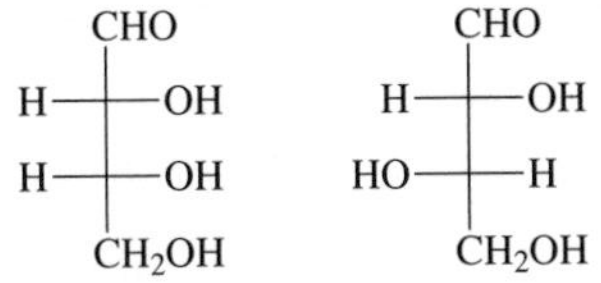

CHO   CHO
H—OH   H—OH
H—OH   HO—H
CH2OH   CH2OH

(−)-erythrose   (+)-threose

Note that, arbitrarily, R^1 was chosen as the main chain in the Wittig intermediates. If R^2 had been the more appropriate group the *erythro*-intermediate
would have corresponded to (+)-erythrose and the *threo*-intermediate would
have corresponded to (−)-threose.

Q. 9.3 The base-catalysed *syn*-elimination involves attack by the nucleophilic
oxygen atom on the adjacent silane moiety:

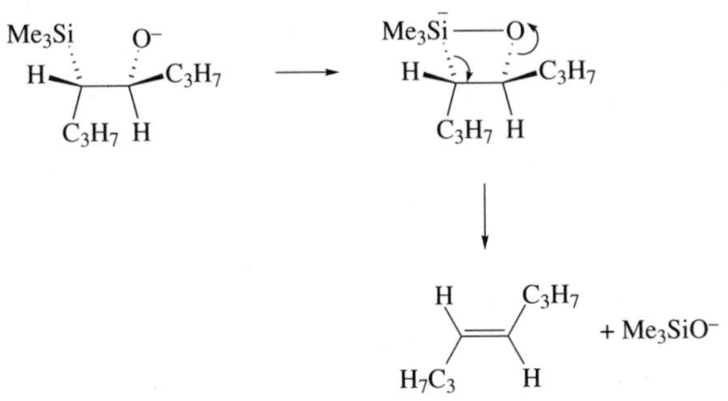

Me_3Si O^- → Me_3Si—O

H C_3H_7

C_3H_7 H

H C_3H_7
        + Me_3SiO^-
H_7C_3 H

Acid catalyses an E2-type elimination:

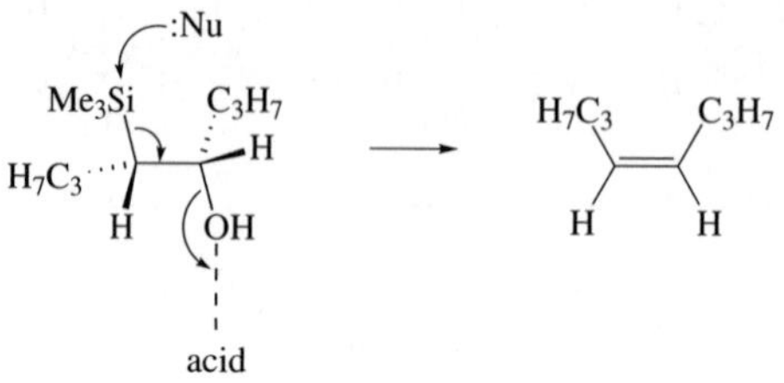

10 Some reactions of simple alkenes

In the last chapter we saw that there are a number of reliable methods for the construction of alkenes. Having made a selected alkene it is possible further to derivatize the molecule by taking advantage of the weakness of the π-bond (Chapter 1). Often the reaction commences with attack of an electrophile on the electron-rich double bond; the stereochemical consequences of the reaction depend on the geometry of the alkene and the mechanism of the reaction under study, as illustrated in the following sections.

10.1 Bromination reactions

A monosubstituted alkene, for example but-1-ene (Fig. 10.1) has two enantiotopic faces; one is designated the *Re* face and the other the *Si* face. Such

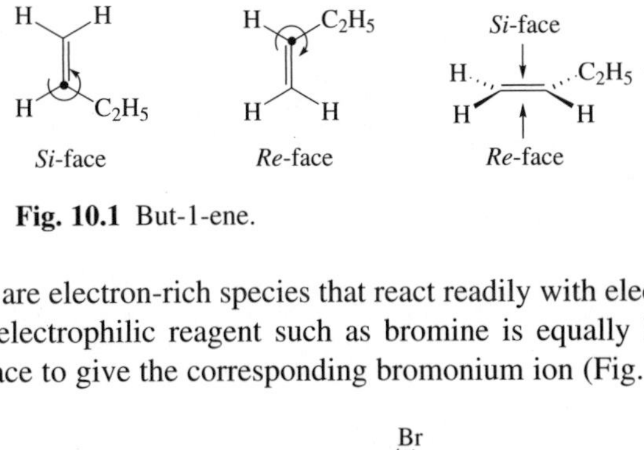

Fig. 10.1 But-1-ene.

alkenes are electron-rich species that react readily with electrophilic reagents. A simple electrophilic reagent such as bromine is equally likely to attack from either face to give the corresponding bromonium ion (Fig. 10.2).

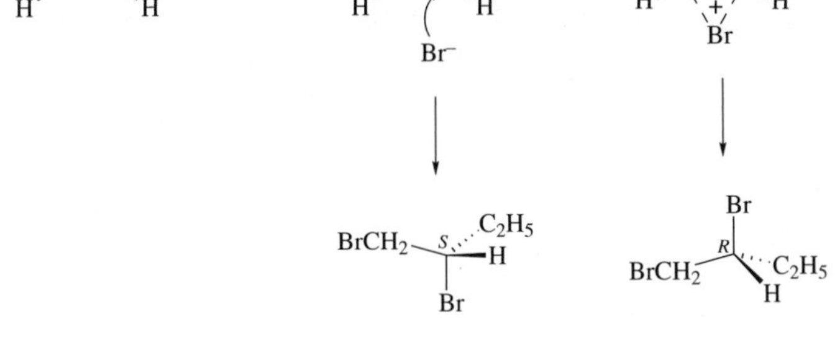

Fig. 10.2

Subsequent attack by the counter-ion gives an equimolar mixture of (*R*)- and (*S*)-1,2-dibromobutane, that is, a racemic mixture. Note that the anion attacks at the more electrophilic C2 atom. Any leakage *via* attack at C1 will be exactly the same for both bromonium ions, still resulting in equal amounts of *R* and *S* product.

Disubstituted alkenes can, in most cases, be classified as *cis* or *trans* alkenes depending on whether the substituents are on the same side or the opposite side of the double bond (Fig. 10.3). However, for tri- and tetra-substituted alkenes the designation of *cis* or *trans* stereochemistry becomes less clear-cut. For all these alkenes (di-, tri- and tetra-substituted) the better system to adopt is that based on the Cahn–Ingold–Prelog sequence rule as described in Chapter 5. To recap, the substituents at the two ends of the double bond are ranked according to this rule. If the substituents of higher priority at each end of the double bond are close together the alkene is designated *Z* [Zusammen ≡ together (German)]; if the two higher priority substituents are on the opposite sides of the double bond then the alkene is designated *E* [Entgegen ≡ opposite (German)] (Fig. 10.3).

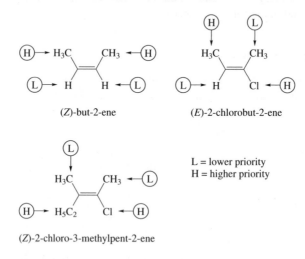

(*Z*)-but-2-ene

(*E*)-2-chlorobut-2-ene

(*Z*)-2-chloro-3-methylpent-2-ene

L = lower priority
H = higher priority

Fig. 10.3

The two faces of a polysubstituted alkene can be described in terms of the *Re/Si* nomenclature, but the carbon atom from which the designation is derived must be noted. Thus for (*E*)-2-chlorobut-2-ene (Fig. 10.4), the 'top face' is the *Re* face for C2 and the *Si* face for C3.

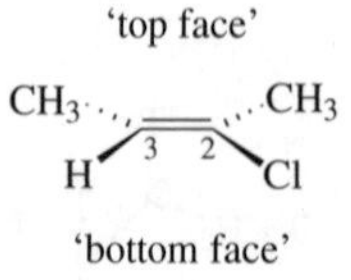

'top face'

CH₃·····═····CH₃
H   3   2   Cl

'bottom face'

Fig. 10.4 (*E*)-2-chlorobut-2-ene.

The reactions of disubstituted alkenes are often highly stereocontrolled, the geometry of the alkene dictating the stereochemistry of the product(s). For example, (Z)-but-2-ene reacts with bromine *via* the formation of a bromonium ion. As for but-1-ene, the bromine atom can sit above or below the plane of the pre-existing double bond, as viewed in Fig. 10.5.

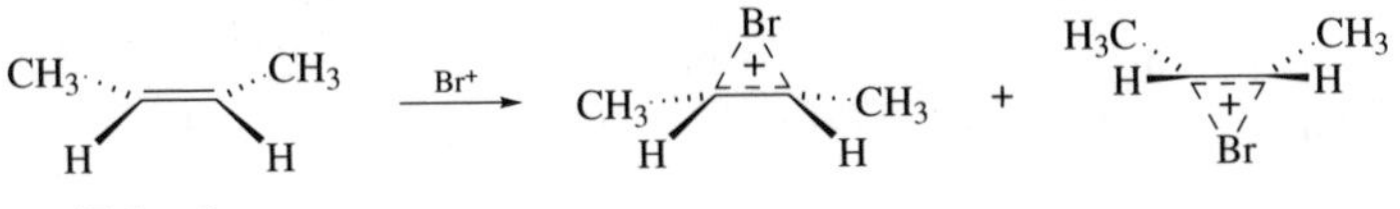

Fig. 10.5

'Backside' attack by the attendant bromide anion on the bromonium ion can take place at one of two positions (Fig. 10.6). Path (a) furnishes the *RR*-isomer of 2,3-dibromobutane: path (b) affords (2*S*,3*S*)-dibromobutane.

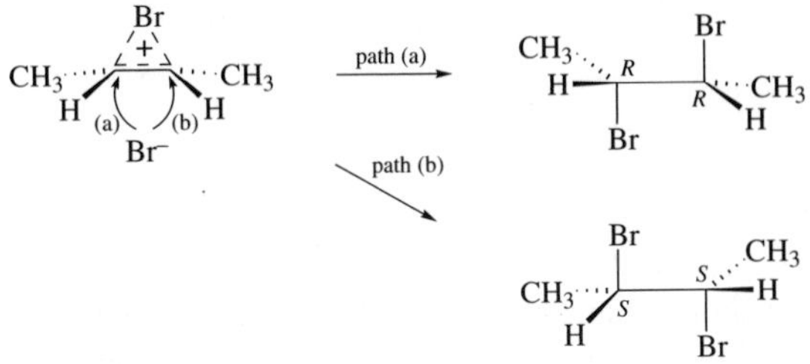

Fig. 10.6

The same analysis can be made of the fate of the alternative bromonium ion (Fig. 10.7). Path (c) produces the *SS* isomer while path (d) produces the *RR* isomer. That is, the same two products are formed irrespective of whether the bromonium ion is formed on the 'top' or the 'bottom' face of the alkene. The products are related as object to mirror image and are formed in equal amounts since there is absolutely no favouritism for paths (b) (c) over (a) (d) and *vice versa*. In other words, bromination of (Z)-but-2-ene gives (±)-2,3-dibromo-butane.

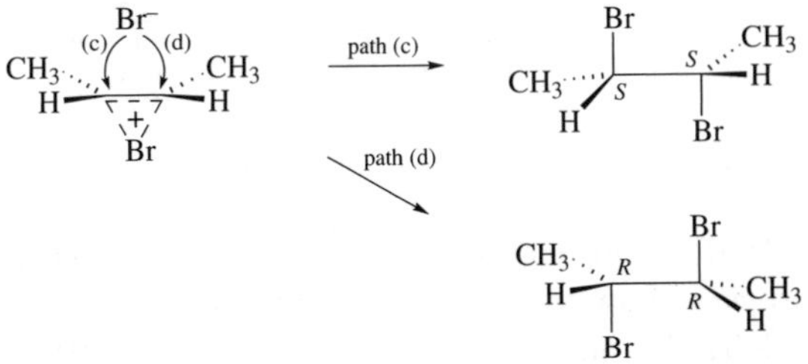

Fig. 10.7

Dibromobutane has two chiral centres. Each chiral centre accommodates the same four substituents: thus there is one other isomer possible for 2,3-dibro-mobutane and that is the *R,S* or *meso* form (cf. butane-2,3-diol or tartaric acid, Chapter 2). This diastereoisomer of 2,3-dibromobutane is produced on bromination of (*E*)-but-2-ene (Fig. 10.8). The bromonium ions formed from (*E*)-but-2-ene can be envisaged to react with bromide ion through attack by Br⁻ along pathways (a)–(d). In fact these routes lead to the same product.

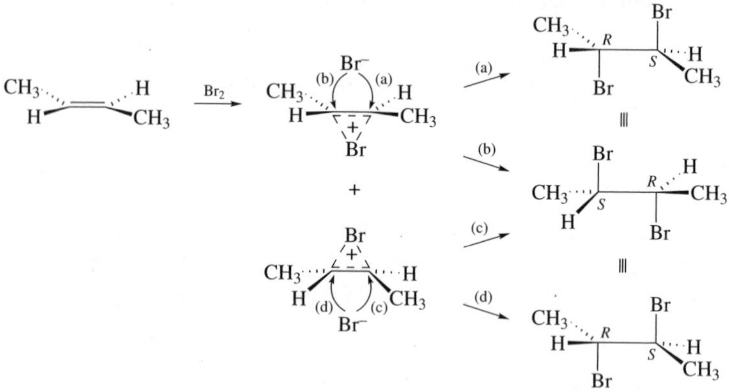

Fig. 10.8

The situation changes when (*Z*)- or (*E*)-pent-2-ene is brominated. Two diastereoisomers are produced in each case. (*Z*)-Pent-2-ene gives a racemic mixture of (2*S*,3*S*)-dibromopentane and (2*R*,3*R*)-dibromopentane (Fig. 10.9). The newly formed chiral centres are dissimilar. In a complementary fashion, (*E*)-pent-2-ene reacts with bromine to give (2*S*,3*R*)- and (2*R*,3*S*)-dibromopentane (Fig. 10.9). Note that the bromination reactions of pent-2-enes are stereospecific.

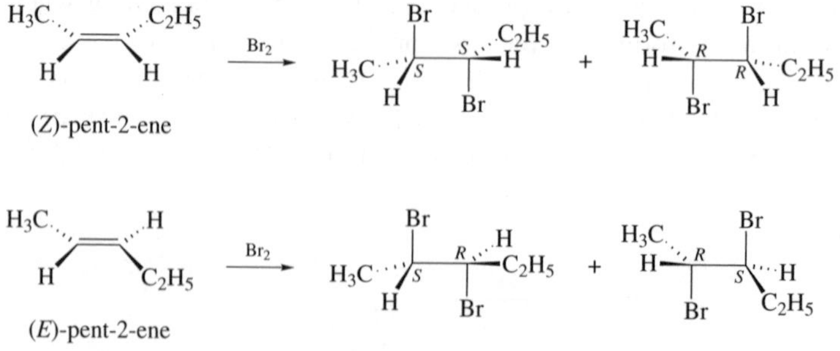

Fig. 10.9

(2*S*,3*S*)- and (2*S*,3*R*)-dibromopentane can be written out in slightly different ways (Fig. 10.10) to show how the two substituents (the bromine atoms) are mutually disposed along the main carbon backbone. The (2*S*,3*S*)-compound is known as the *syn*-isomer and also as the *threo*-isomer on the basis of the similarity to the four-carbon sugar threose (Fig. 10.10). The (2*S*,3*R*)-compound is known as the *anti*-diastereoisomer or the *erythro* form due to the similarity to erythrose. The *anti/syn* nomenclature applied to diastereoisomers parallels the *anti/syn* nomenclature in mechanistic organic chemistry. However, the connections between the two descriptions are not straightforward [*anti*-addition of bromine to (Z)-pent-2-ene gives a racemic mixture of the *syn*-diastereo-isomers.]

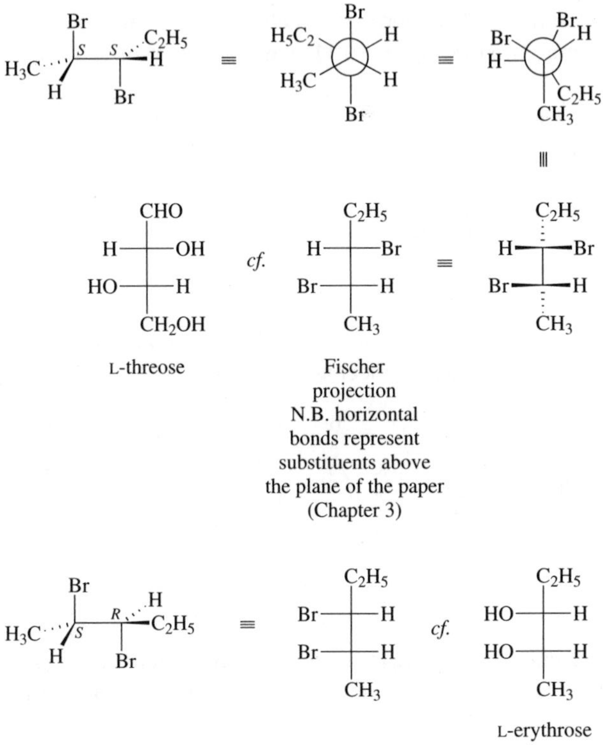

L-threose   *cf.*   Fischer projection
N.B. horizontal bonds represent substituents above the plane of the paper
(Chapter 3)

L-erythrose

Fig. 10.10

Q. 10.1 The prefixes *syn* and *anti* can be applied to many types of acyclic compound having substituents along the backbone of the molecule. It is sometimes easier to write the main carbon chain as a zig-zag (see below) before labelling substituents *syn* or *anti*.

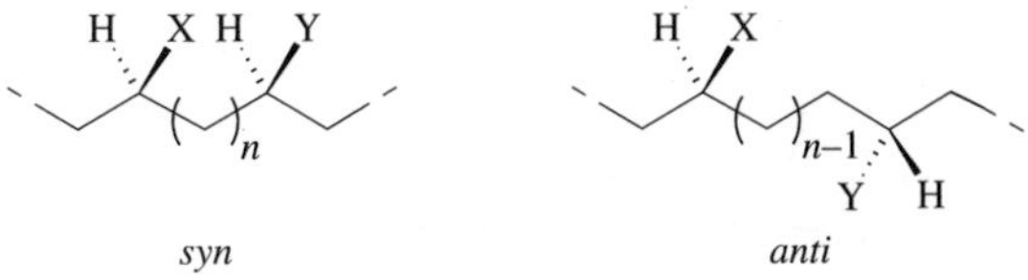

syn   *anti*

Given this information, write out 2,3-dibromopentane in zig-zag style to see that the *anti*-diastereoisomer on this system equates to the *erythro*-diastereoisomer on the Fischer system. Then draw 2,4-*anti*-2,5-*syn*-octane-2,4,5-triol.

Two dissimilar chiral centres are also produced when (Z)- or (E)-but-2-ene is brominated by slow addition of bromine to a solution of the alkene in water or methanol. In the latter case, methanol provides the nucleophilic component instead of bromide ion and (Z)-but-2-ene furnishes (2S,3S)-2-bromo-3-methoxybutane and (2R,3R)-2-bromo-3-methoxybutane in equal proportions (that is, the racemic mixture) (Fig. 10.11) while reaction of (E)-but-2-ene with

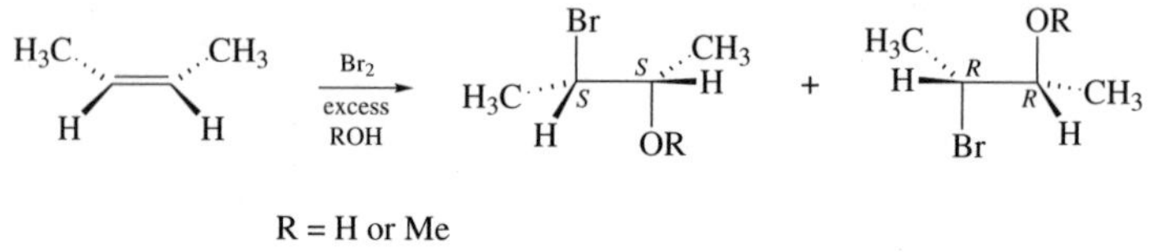

R = H or Me

Fig. 10.11

bromine in methanol produces (2S,3R)-2-bromo-3-methoxybutane and (2R,3S)-2-bromo-3-methoxybutane as the racemate (Fig. 10.12).

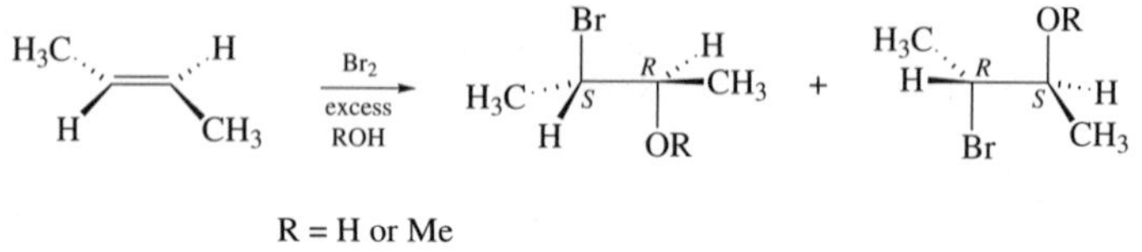

R = H or Me

Fig. 10.12

Like (Z)-but-2-ene, cyclopentene gives a racemic mixture of (1R,2R)- and (1S,2S)-1-bromo-2-methoxycyclopentane on bromination in methanol (Fig. 10.13).

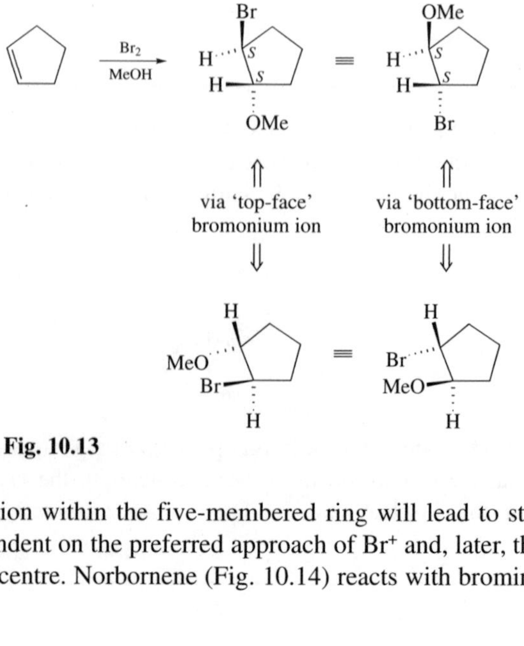

Fig. 10.13

Substitution within the five-membered ring will lead to structure patterns that are dependent on the preferred approach of Br+ and, later, the nucleophile to the reaction centre. Norbornene (Fig. 10.14) reacts with bromine in the presence of

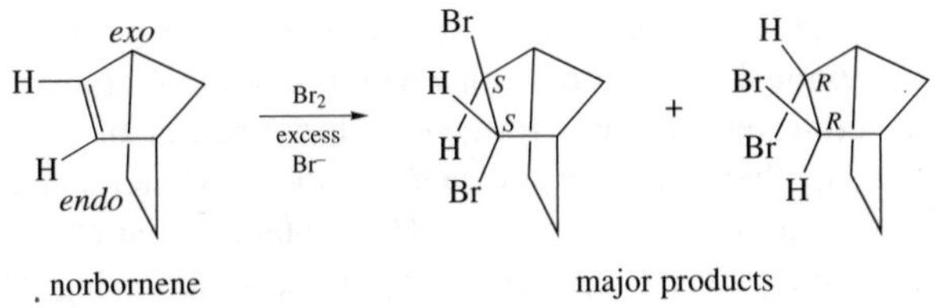

Fig. 10.14

excess bromide ion in such a way that the bromonium ion is preferentially formed on the more exposed *exo*-face of the molecule. To complete the process of *anti*-addition, the nucleophile is forced to attack from the more hindered *endo*-face to give, as a major product, a racemic mixture of 2-*exo*, 3-*endo*-dibromonorbornane.

The bicyclic system described in Fig. 10.15 is isomeric with norbornene. However, unlike norbornene bicyclo[3.2.0.]hept-2-ene is chiral. The enantiomer

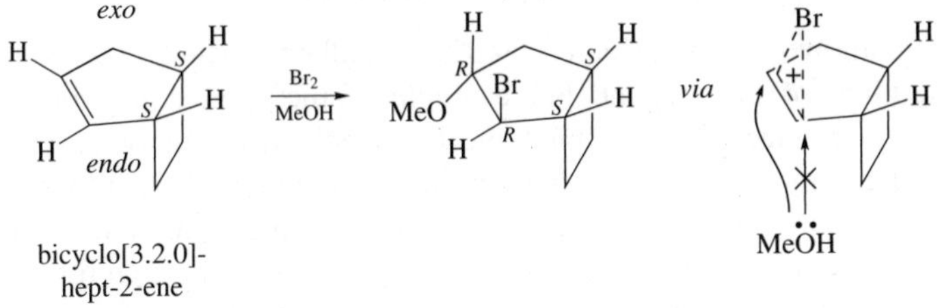

Fig. 10.15

shown reacts with methanolic bromine to give one product. This is because the bromonium ion is formed, once again, on the more exposed *exo*-face of the molecule. The attendant nucleophile, in this case methanol, has difficulty in gaining access to the carbon atom adjacent to the four-membered ring, due to steric hindrance. As a result nucleophilic attack takes place almost exclusively at the carbon centre furthest from the four-membered ring. In this case the

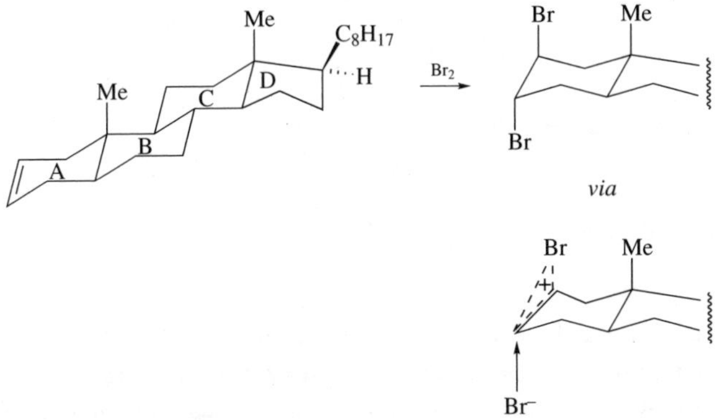

Fig. 10.16

four-membered ring induces a pattern of substitution and a particular stereo-chemistry at two previously achiral positions within the five-membered ring.

Analysis of the preferred products formed on bromination of cyclohexenyl units in steroid systems led to further information about the stereochemistry of such reactions. The steroid nucleus (Fig. 10.16) contains three chair-type six-membered rings fused together (rings ABC). A double bond in ring A reacts to give the diaxial dibromo compound, since the preferred reaction pathway involves the dispositions illustrated in Fig. 10.16. This diaxial arrangement of substituents is also formed in other closely related cases.

Q. 10.2 Predict the major product formed on bromination of 4-*tert*-butyl-cyclohex-1-ene.

10.2   Reactions involving osmium tetraoxide

Up to this point we have considered reactions of alkenes which proceed by way of *anti*-addition of the reagent. For example, if the reagent is bromine, Br$^+$ attacks from one face of the double bond Br$^-$ attacks from the other. However, there are important reactions that lead to the *syn*-addition of substituents to the alkene unit, in particular the reaction of an alkene with osmium tetraoxide.

For example, maleic acid reacts with osmium tetraoxide through formation of an osmium(VI) ester (Fig. 10.17). At first sight two osmate esters seem possible, but the two intermediate species pictured in Fig. 10.17 are identical. Decomposition of the ester gives *meso*-(R,S)-tartaric acid.

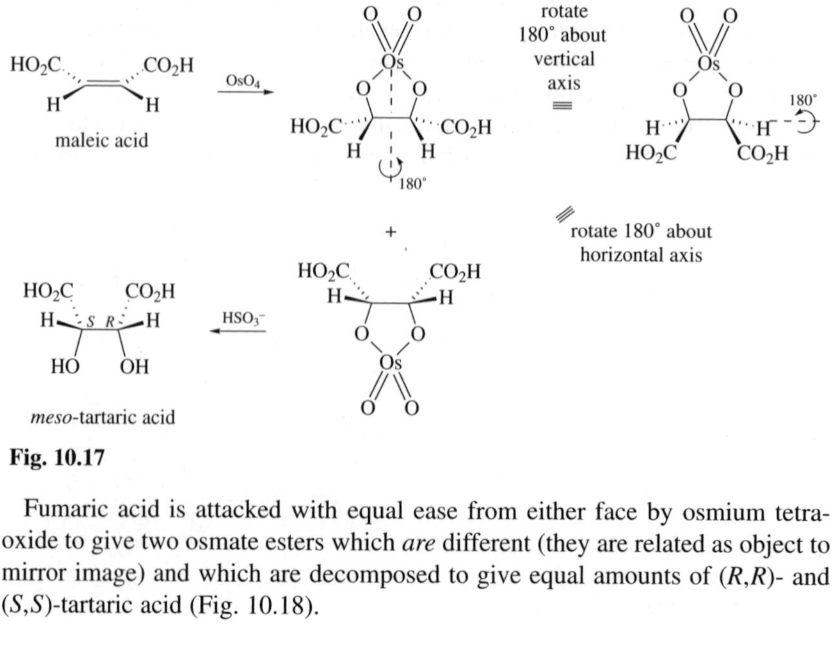

Fig. 10.17

Fumaric acid is attacked with equal ease from either face by osmium tetra-oxide to give two osmate esters which *are* different (they are related as object to mirror image) and which are decomposed to give equal amounts of (R,R)- and (S,S)-tartaric acid (Fig. 10.18).

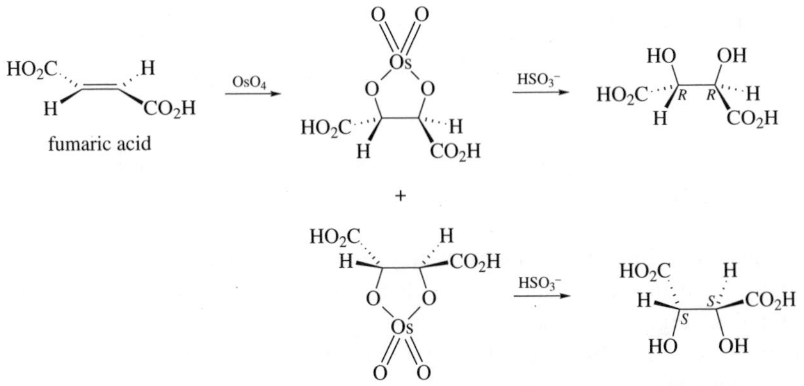

Fig. 10.18

In the presence of amines the osmate ester is modified by the attachment of one or more ligands and this forms the basis of the Sharpless asymmetric dihydroxylation procedure (Chapter 15).

10.3 Epoxidation of alkenes

Addition of an oxygen atom to a carbon–carbon double bond can be effected using a peracid (RCO_3H) (Fig. 10.19). The concerted formation of the two new

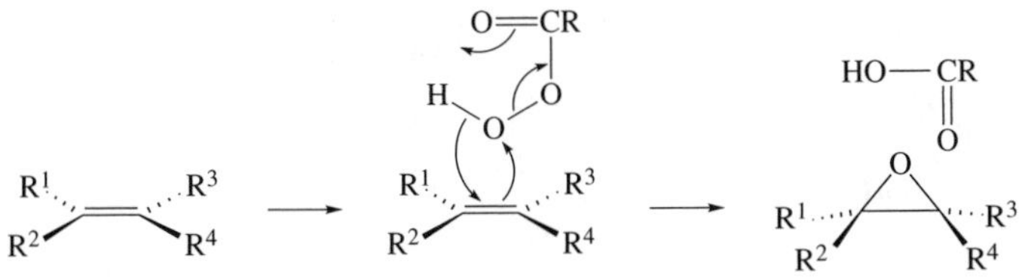

Fig. 10.19

carbon–oxygen bonds ensures that the arrangement of substituents around the alkene is reflected in the stereochemistry of the product, as illustrated in Fig. 10.20.

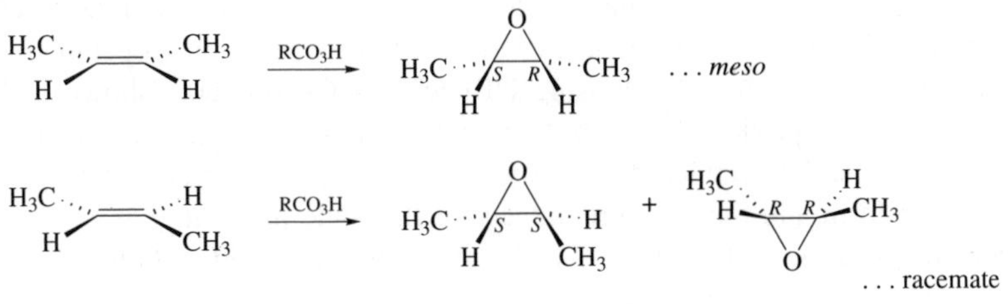

Fig. 10.20

The approach of the peracid to the alkene unit is influenced by steric factors (Fig. 10.21) and electronic factors (Chapter 15). Thus in Fig. 10.21 oxidation of

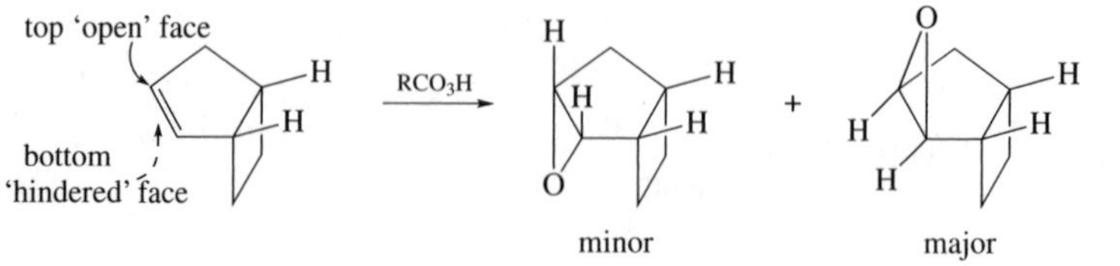

top 'open' face

bottom 'hindered' face

RCO₃H

minor   major

Fig. 10.21

the alkene unit in an unsymmetrical bicycloalkene is seen to take place preferentially from the less hindered face. It is noteworthy that the minor product featured in Fig. 10.21, namely the *endo*-epoxide, may be obtained from the bicycloheptene in a better yield using a two-step process with the bromohydrin as an intermediate (Fig. 10.22).

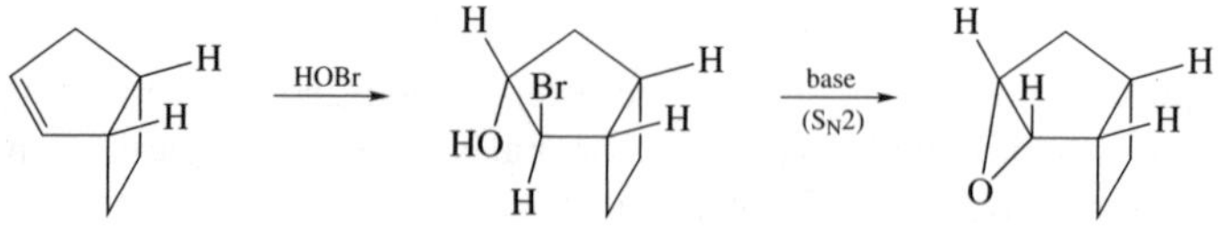

HOBr   base (S$_N$2)

Fig. 10.22

Epoxides (oxiranes) are valuable compounds in organic synthesis because they undergo ring opening on attack by a wide variety of nucleophiles (Fig. 10.23). The regioselectivity of the ring-opening reaction is controlled by a

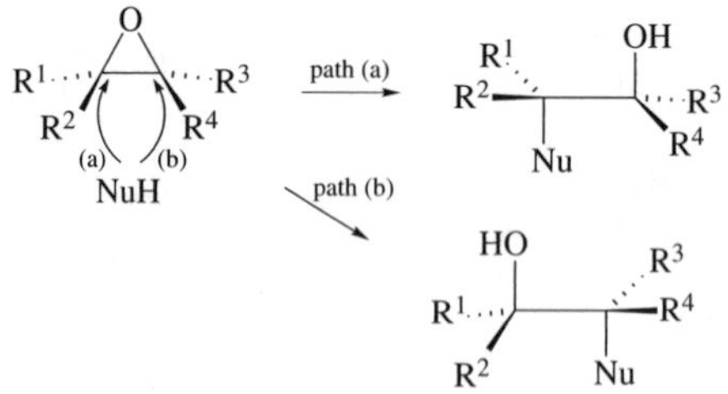

path (a)

path (b)

NuH

Fig. 10.23

number of factors, including the nucleophile being employed, the size and electronic characteristics of the adjacent atoms or groups $R^1 \rightarrow R^4$, and the reaction conditions (for example, use of acid catalysis). For example, ring opening of the epoxides shown in Fig. 10.21 using HBr affords the products shown in Fig. 10.24. The *exo*-epoxide reacts through protonation of oxygen and attack by bromide at the less-hindered position; the *endo*-epoxide reacts through the protonated form and displays preference for the route involving an axial–axial arrangement of the embryonic C—Br bond and the residual C—O bond.

Q. 10.3 Predict the product formed on reaction of *cis*-4-*tert*-butylcyclohexene oxide with lithium aluminium deuteride.

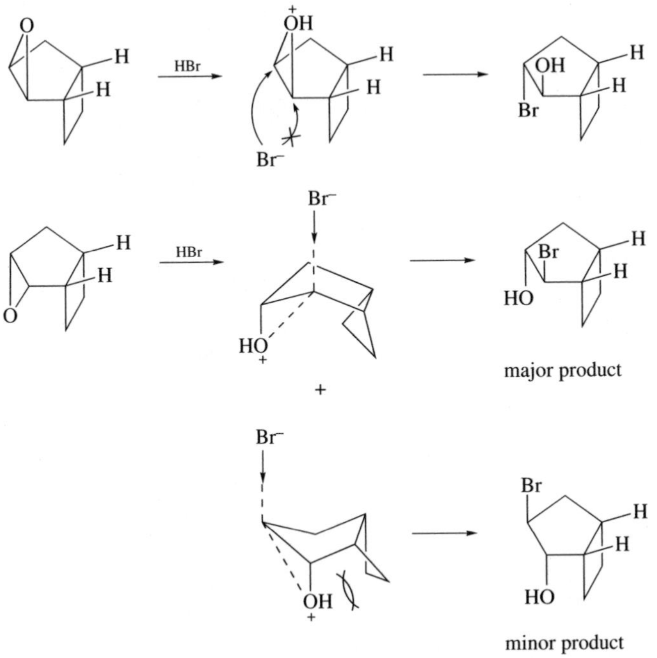

Fig. 10.24

10.4 Reaction of alkenes with carbenes

Cyclopropanes can be prepared by addition of a carbene R^1R^2C: to an alkene. Carbenes can be of the spin-paired (singlet) or parallel-spin type (Fig. 10.25).

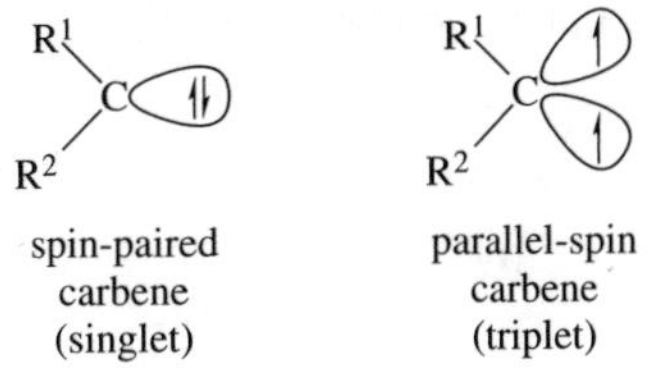

spin-paired
carbene
(singlet)

parallel-spin
carbene
(triplet)

Fig. 10.25

Simultaneous formation of two new carbon–carbon bonds is feasible from the spin-paired (singlet) carbene and thus stereochemical features of the alkene are reliably translated to the cyclopropane unit. Triplet (parallel-spin) carbenes react to give a diradical in which rotation can take place before the second bond is formed (Fig. 10.26).

One of the simplest cyclopropanation reactions is the Simmons–Smith reaction involving the generation of a H_2C:-type species (a zinc carbenoid) through the action of zinc on diiodomethane. Addition of $:CH_2$ often occurs with retention of stereochemistry of the alkene moiety.

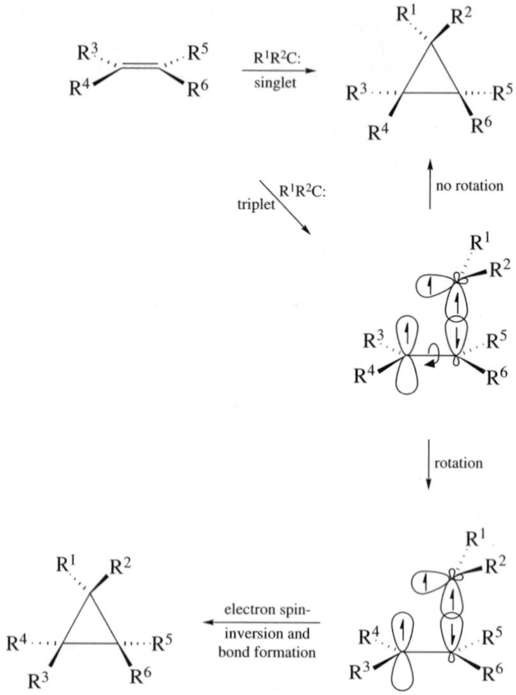

Fig. 10.26

Answers to questions

Q. 10.1   The *threo*-isomer of 2,3-dibromopentane can be manipulated as follows:

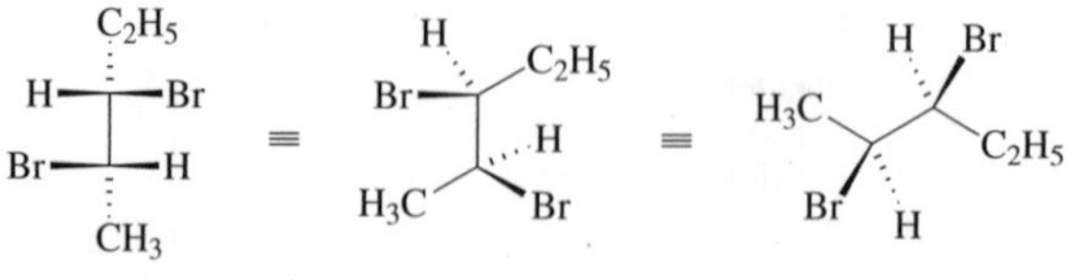

threo   *syn*

Similarly, *erythro*-2,3-dibromopentane is seen to be the *anti*-diastereoisomer utilizing the zig-zag system.

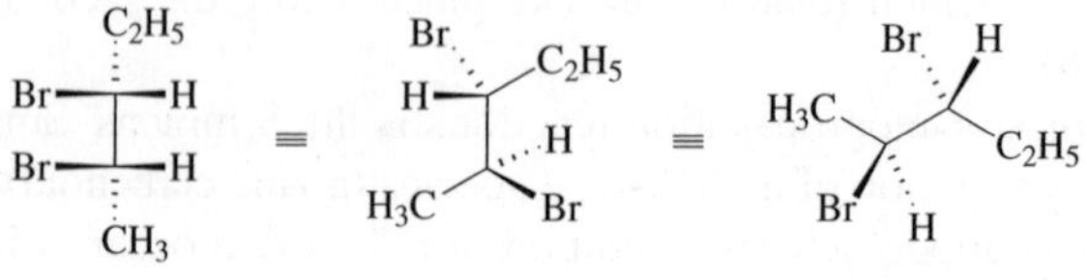

erythro   *anti*

The arrangement for 2,4-*anti*-2,5-*syn*-octane-2,4,5-triol is as follows:

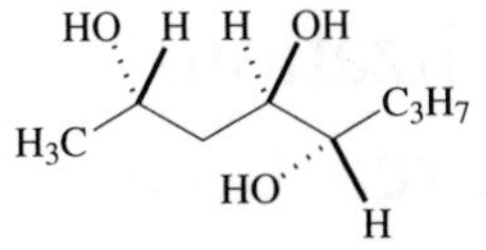

Q. 10.2   4-*tert*-Butylcyclohex-1-ene exists preferentially with the bulky *tert*-butyl group occupying an equatorial situation (Chapter 2). Diaxial bromination then takes place as described below:

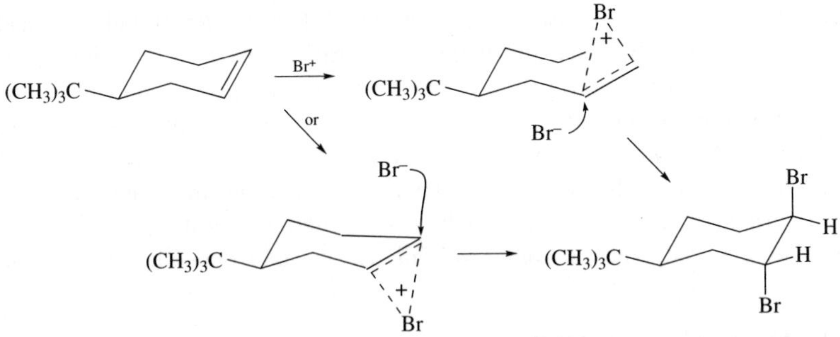

Q. 10.3   The product formed is *c*-4-*tert*-butyl *c*-2-deuteriocyclohexan-*r*-ol: the bulky *tert*-butyl group occupies the equatorial position. The opening of the three-membered ring takes place to give the *trans*-diaxial product (Fürst–Plattner rule).

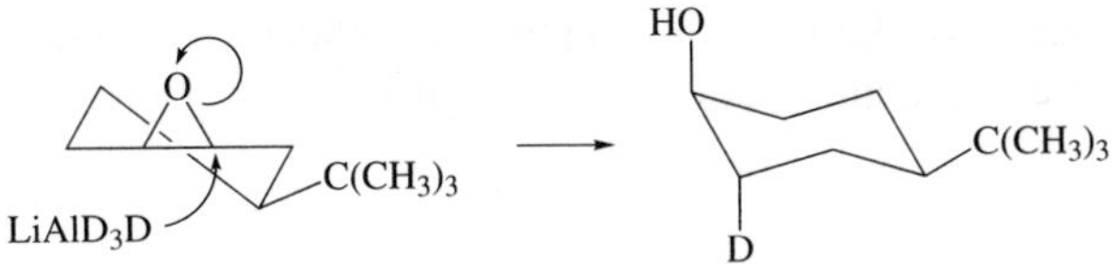

11   Some important cyclizations involving pericyclic reactions

Pericyclic reactions are those reactions in which two or more bonds are made and/or broken simultaneously through a cyclic shift of electrons. A wide variety of pericyclic reactions give rise to products under strict stereochemical control. Because two or more bonds are made and/or broken in a synchronous or quasi-synchronous fashion, stereochemical features in the reactants are transposed into the products. Four of the most common pericyclic reactions in organic chemistry are described in this chapter, namely Diels–Alder reactions, [2 + 2] cyclo-additions, electrocyclic reactions involving trienes and Cope–Claisen reactions.

11.1   Diels–Alder reactions

This is the reaction between a diene and a dienophile ('diene seeker'). The simplest conceivable Diels–Alder reaction is described in Fig. 11.1; the curly arrows denote the movement of pairs of electrons in valence bond terms. However, it is easier to grasp the various stereochemical points if you consider the reaction in terms of the overlap of molecular orbitals. To do this we must elaborate on some of the information given in Chapter 1.

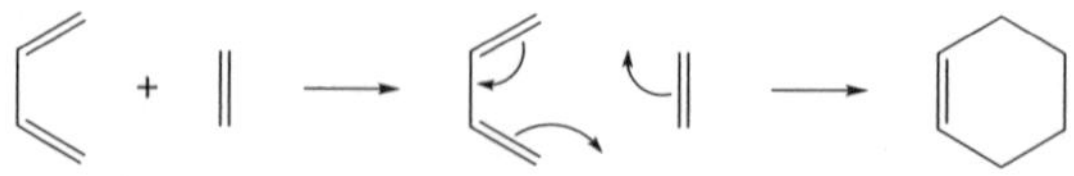

Fig. 11.1

The p orbitals of ethene are identical in shape above and below the plane containing the four hydrogen atoms and the two carbon atoms. However, the two lobes of the p orbitals differ in phase and to distinguish the two phases of a particular orbital in diagrams, the two lobes are usually coloured differently. In this book one lobe of the p orbital is shaded while the other is left blank.

In Chapter 1 we discussed how the two 'p' orbitals of ethene overlap to form a bonding molecular orbital. In fact the *two* atomic orbitals containing the 'p' electrons give rise to *two* molecular orbitals, a low-energy bonding molecular orbital (MO) which contains the two electrons (this MO has been referred to in our previous discussions) and a high-energy antibonding molecular orbital (Fig. 11.2). In the bonding molecular orbital (ψ_1) the phases of the atomic orbitals are matched while in the antibonding system (ψ_2) these phases are mismatched; the phases of the orbitals are distinguished by shading as mentioned above. Similarly, butadiene should not be considered as two isolated double bonds but as a smear of negative charge surrounding the four carbon atoms. The *four* atomic orbitals form *four* molecular orbitals (ψ_1–ψ_4) and the four electrons occupy the two orbitals of lowest energy (Fig. 11.2).

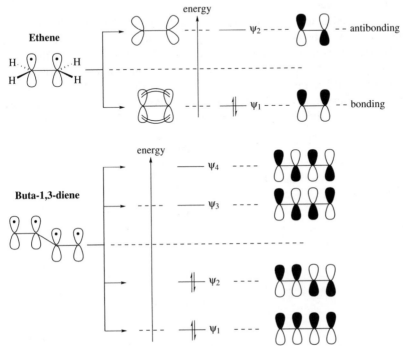

Fig. 11.2

The most common Diels–Alder reactions take place by interaction of the lowest unoccupied molecular orbital (LUMO) of the dienophile (usually an electron-poor alkene) and the highest occupied molecular orbital (HOMO) of the diene. The phases of these so-called 'frontier' orbitals are correctly matched as shown by the dotted lines in Fig. 11.3. The reaction is designated a [4 + 2] cyclo-addition, since four centres in one component and two centres in the other

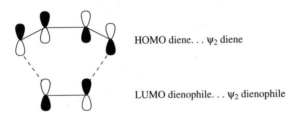

HOMO diene. . . ψ_2 diene

LUMO dienophile. . . ψ_2 dienophile

Fig. 11.3

component are involved in the transformation. The concerted formation of bonds in the Diels–Alder reaction allows the preparation of a wide variety of six-membered ring compounds with remarkable stereoselectivity. It should be noted, however, that the high stereoselectivity applies only to the initial kinetically controlled reaction, and may be lost in cases where there is easy dissociation of the adduct. Repeated addition/dissociation will then lead to thermodynamic control of the reaction.

In 1937 Alder and Stein put forward the principle that the relative stereo-chemistries of substituents in both the dienophile and the diene are retained in the adduct. Thus the relative configuration of the substituents in the 1- and 4-positions of the diene is retained in the Diels–Alder product. An (E,E)-1,4-disubstituted diene gives rise to adducts in which the 1- and 4-substituents are *cis* to each other and a (Z,E)-diene gives adducts with *trans* substituents (Fig. 11.4). Stereochemical features of the dienophile are also preserved in the course

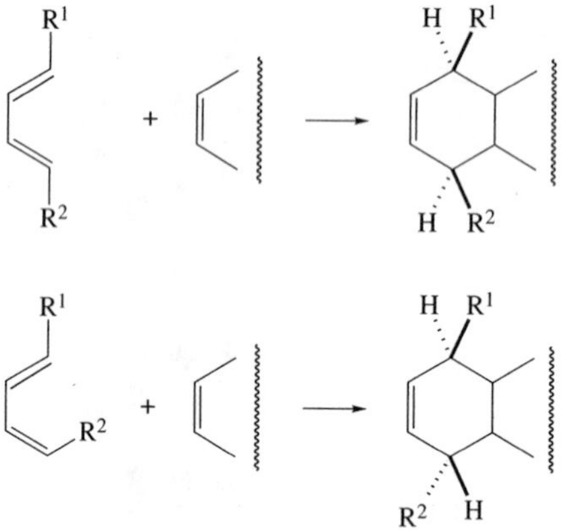

Fig. 11.4

of the reaction. Thus cyclopentadiene reacts with dimethyl maleate to give the two *cis* adducts (Fig. 11.5) while dimethyl fumarate reacts with the same diene to give the *trans* adduct (Fig. 11.6).

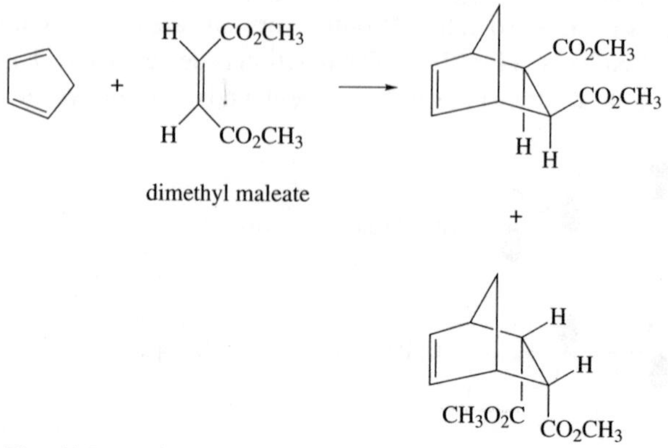

dimethyl maleate

Fig. 11.5

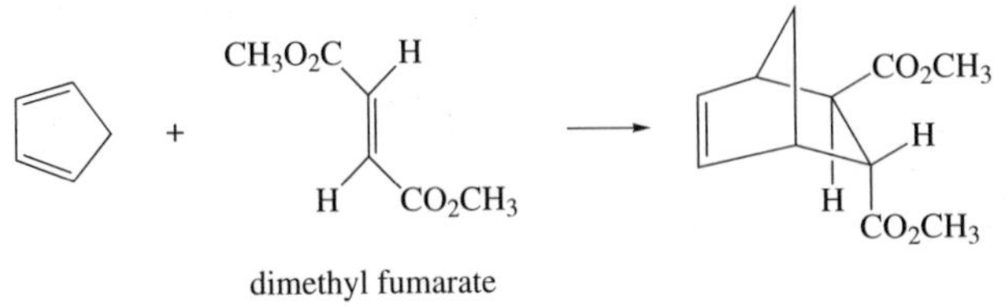

Fig. 11.6

> **Q. 11.1** Explain why (*E*)-penta-1,3-diene reacts with tetracyanoethene much faster than (*Z*)-penta-1,3-diene which, in turn, reacts with the same dienophile much faster than (*Z*,*Z*)-hexa-2,4-diene.

The intramolecular reaction featured in Fig. 11.7 nicely illustrates the Alder–Stein principle for both the diene and the dienophile. Note also that the dienophile orients under (rather than over) the diene moiety to minimize unfavourable interactions with the ethyl group.

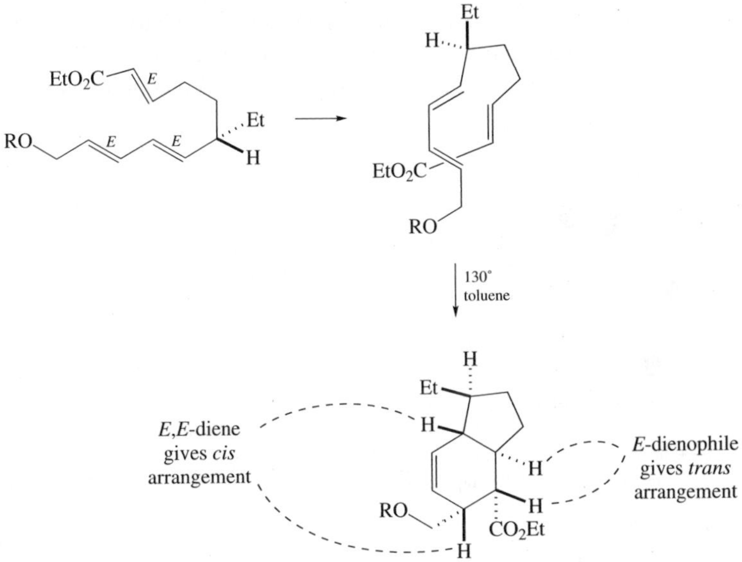

Fig. 11.7

Left to stand at room temperature cyclopentadiene reacts with itself (one molecule reacting as a diene, the second as a dienophile) to give *endo*-dicyclopentadiene (Fig. 11.8). The description '*endo*' indicates that the bulk of the dienophile lies in the more hindered region of the molecule, adjacent to the ethene bridge (Chapter 5). The *exo*-isomer (with the substituent adjacent to the smaller bridge) has fewer unfavourable steric interactions. The reason for the formation of *endo*-dicyclopentadiene as the kinetically favoured product was

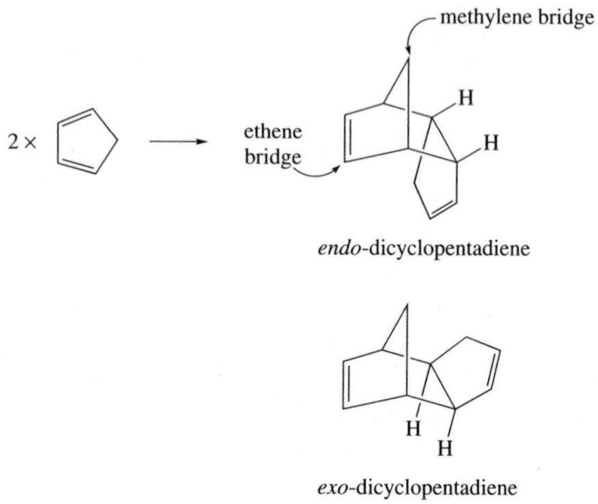

methylene bridge

2 ×

ethene
bridge

H

H

endo-dicyclopentadiene

H

H

exo-dicyclopentadiene

Fig. 11.8

explained by Woodward and Hofmann: in the dimerization of cyclopentadiene, favourable *secondary* orbital interactions, represented by dashed lines in Fig. 11.9, lower the energy of the *endo* transition state (shown) relative to that of the *exo*-transition state where these secondary interactions are absent. Hence the *endo*-adduct is the one obtained under kinetically controlled conditions.

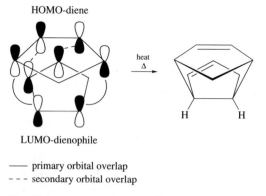

HOMO-diene

heat
Δ

H   H

LUMO-dienophile

——— primary orbital overlap
– – – secondary orbital overlap

Fig. 11.9

Similarly, cyclopentadiene and maleic anhydride form the *endo*-adduct almost exclusively. In contrast, the cycloaddition involving furan and maleic anhydride does not obey the '*endo* rule'. The reason is that the initially formed *endo*-adduct easily dissociates at moderate temperatures, facilitating conversion of the kinetic *endo*-adduct into the thermodynamically more stable *exo*-isomer (Fig. 11.10).

The adducts obtained from *acyclic* dienes and *cyclic* dienophiles are frequently formed in accordance with the '*endo* rule'. However, in the cycloaddition of *acyclic* dienophiles to *cyclic* dienes the *endo* rule is not always

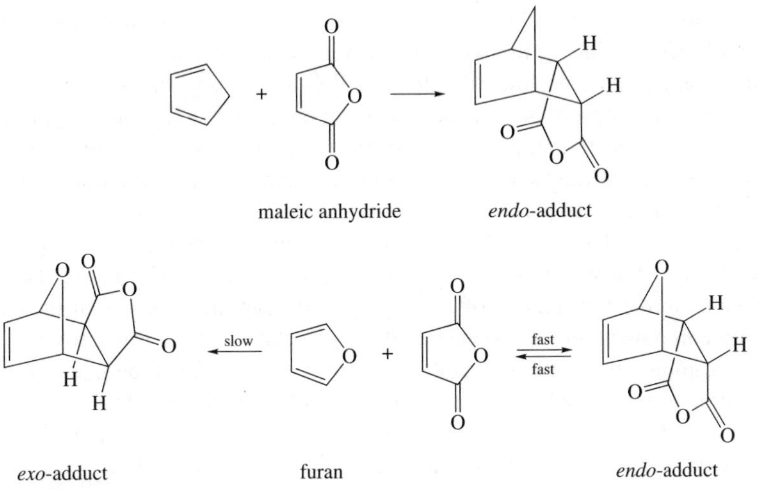

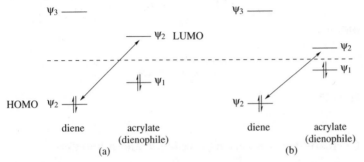

Fig. 11.10

obeyed and the *exo:endo* composition of the mixture obtained depends on the precise structure of the dienophile and on the reaction conditions. Thus addition of acrylic acid to cyclopentadiene gives a 75:25 ratio of *endo* and *exo* products. The proportion of *endo*-adduct is noticeably increased by the presence of Lewis acid catalysts; this is a fairly general phenomenon and gives an indication as to why it has become popular to catalyse Diels–Alder reactions in this way. Indeed, Lewis acid complexation of the dienophile can enhance the rate of a Diels–Alder reaction by as much as a factor of 10^6, allowing the reaction to proceed at low temperatures.

The mode of action of the Lewis acid appears to involve coordination of the carbonyl group of the unsaturated compound to the electrophilic metal, thus causing the carbonyl group to become more electron withdrawing. This has the effect of lowering the energy level of the LUMO of the dienophile, enabling more efficient overlap with the HOMO of the diene (Fig. 11.11). Reaction therefore proceeds with greater ease. Such catalysis serves to increase the proportion of the kinetically favoured *endo*-isomer to >99%.

Fig. 11.11 Effect of Lewis acid catalysts on the interaction of the frontier orbitals of a diene and an acrylate. (a) Non-catalysed reaction; (b) catalysed reaction.

As nicely illustrated in Fig. 11.7, the dienophile will approach the diene from the less hindered side. Similarly, if the dienophile has a different pattern of substitution on its two faces then the diene will be directed to the more open face. Indeed if the dienophile is chiral and the reaction gives rise to a new asymmetric centre, then the preferential approach of the diene from one direction will result in the formation of the two diastereoisomeric forms of the new chiral molecule in an unequal ratio. In practice, the normal thermal reaction gives only modest optical yields and best results have been obtained in reactions catalysed by Lewis acids at low temperatures. The absolute stereochemistry of the predominant isomer formed can often be predicted from the structure of the chiral starting component and a knowledge of the orientation of the transition state.

For example, the chiral acrylate described in Fig. 11.12 can be manufactured from a readily available natural product called camphor. The dienophilic carbon–carbon double bond and the neopentyl unit [$CH_2C(CH_3)_3$] are oriented as shown. Approach of a diene, such as cyclopentadiene to the *Re* face of the acrylate double bond is hindered by the *tert*-butyl group. There is little hindrance to the back (*Si*) face so that under Lewis acid catalysis at low temperature one diastereoisomer is formed almost exclusively. Reduction of this ester with lithium aluminium hydride gives the norbornenylmethanol in an optically pure state and regenerates the camphor-derived chiral auxiliary. Another example of an asymmetric Diels–Alder reaction is described in Chapter 15.

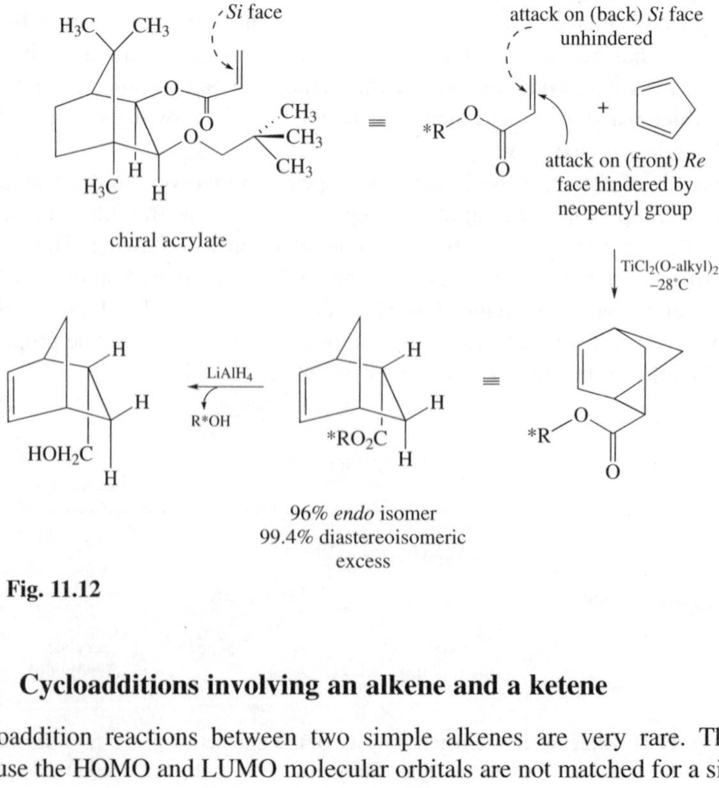

Fig. 11.12

11.2 Cycloadditions involving an alkene and a ketene

Cycloaddition reactions between two simple alkenes are very rare. This is because the HOMO and LUMO molecular orbitals are not matched for a simple (suprafacial) coupling (Fig. 11.13) and the alternative crossed (antarafacial) approach is prohibited by unfavourable steric interactions.

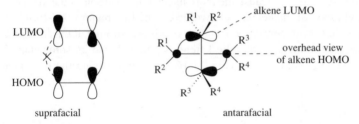

Fig. 11.13

These problematic steric interactions are minimized when a ketene interacts with an alkene (Fig. 11.14) and thermal cycloaddition reactions take place readily, particularly when an electron-withdrawing substituent is present on the ketene. With two participating atoms in each component, the reaction is called a [2 + 2]-cycloaddition.

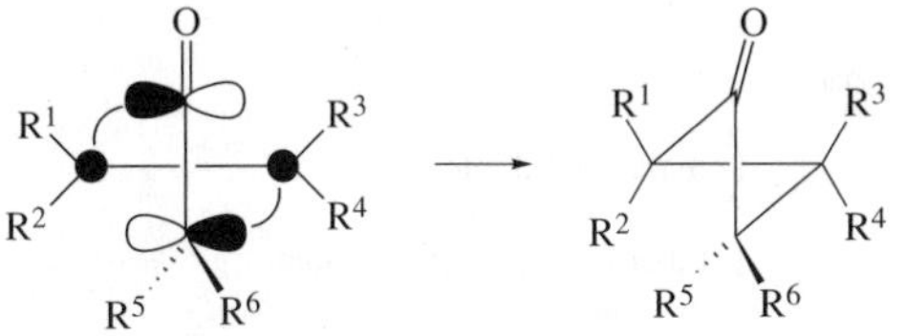

Fig. 11.14

This is a very simple method for the synthesis of highly substituted cyclo-butanones. The antarafacial addition of the ketene and the alkene has profound stereochemical consequences when the cycloaddition involves a substituted alkene and an unsymmetrically substituted ketene. Consider the cycloaddition of chloroketene and cyclopentadiene (Fig. 11.15). The ketene orients its carbonyl

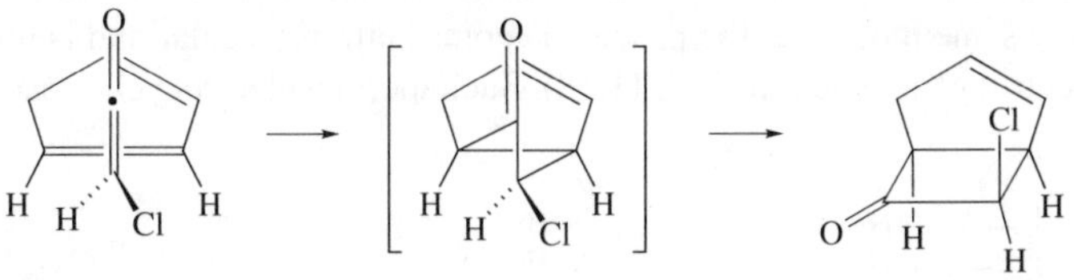

Fig. 11.15

group over the face of the diene and has the less bulky substituent pointing in the direction of the coupling partner. After formation of the two new carbon–carbon bonds the twisted cyclobutanone ring flattens out and this twisting motion places the bulky substituent (in this case the chlorine atom) adjacent to the five-membered ring. This 'masochistic' effect almost invariably produces the thermodynamically less favoured adduct. The orientation of the addition of chloroketene to cyclopentadiene is also interesting. The carbonyl carbon atom becomes bonded to the carbon atom that was the terminal carbon atom in the diene moiety. This is because the more nucleophilic carbon of the alkene and the carbonyl group of the ketene get involved in the first major bond-forming inter-action. (While both new carbon–carbon bonds are forming at the same time, one is ahead of the other.) Thus the ketene is better drawn as somewhat skewed across the alkene framework (Fig. 11.16) and, emphasizing the prominent role

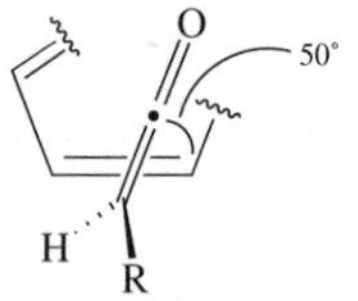

Fig. 11.16

played by the carbonyl carbon atom, the preferred interaction of orbitals is now believed to be that shown in Fig. 11.17. Though more difficult to perceive, the stereochemical implications regarding the ketene–alkene cycloadditions remain the same.

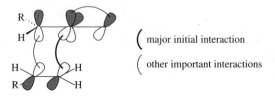

(major initial interaction

(other important interactions

Fig. 11.17

11.3   Electrocyclic reactions

The Diels–Alder reaction is a pericyclic reaction that involves the redistribution of six π-electrons via a cyclic transition state. If the six electrons are contained within the same molecule, the process becomes intramolecular and is referred to as an electrocyclic reaction (Fig. 11.18). Such thermal electrocyclic reactions are

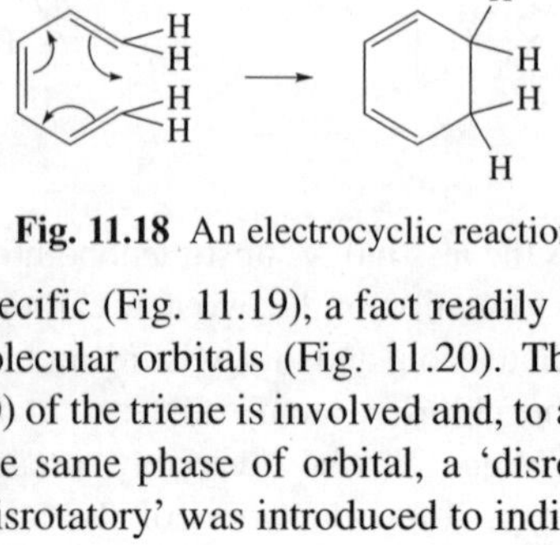

Fig. 11.18 An electrocyclic reaction.

stereospecific (Fig. 11.19), a fact readily explained by consideration of the rele-vant molecular orbitals (Fig. 11.20). The highest occupied molecular orbital (HOMO) of the triene is involved and, to allow the new sigma bond to be formed using the same phase of orbital, a 'disrotatory' movement is mandatory. The word 'disrotatory' was introduced to indicate that the substituents at the ends of

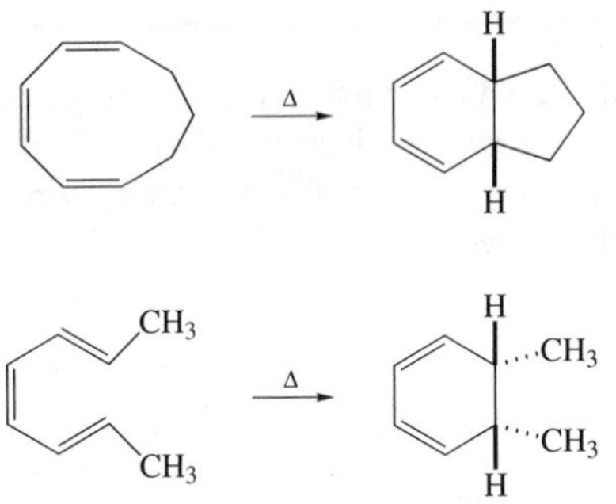

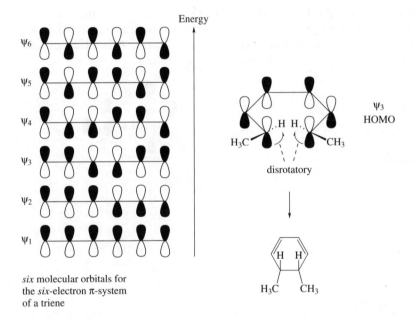

Δ
H
H
CH3
CH3
CH3
CH3

Fig. 11.19

Energy
ψ6
ψ5
ψ4
ψ3
ψ2
ψ1
ψ3
HOMO
H H
H3C
CH3
disrotatory
H H
H3C
CH3

six molecular orbitals for the *six*-electron π-system of a triene

Fig. 11.20

the triene unit move in different directions, one clockwise, one anticlockwise (Fig. 11.21).

Electrocyclic ring closure may also be brought about by photochemical means. In such cases the stereochemistry of the products is the opposite of that obtained by thermal cyclization (Fig. 11.22). Irradiation of the triene promotes an electron from the molecular orbital ψ_3 to the molecular orbital ψ_4. The latter

H3C
CH3
H3C
CH3

Fig. 11.21

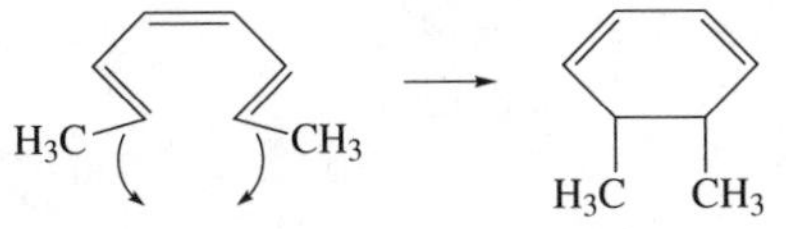

Q. 11.2 The dimethylcyclobutene (**A**) is unstable, the ring opening at 200°C to give a hexa-2,4-diene. Given that electrocyclic reactions can be considered as reversible (acyclic ↔ cyclic) predict the stereochemistry of the hexa-2,4-diene formed in the above reaction.

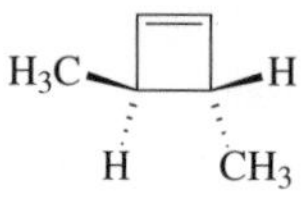

A

becomes the highest occupied molecular orbital and is the 'frontier' orbital in the reaction leading to a 'conrotatory' motion of the termini of the triene and the formation of *trans*-5,6-dimethylcyclohexa-1,3-diene. Note that both methyl groups rotate in an anticlockwise sense (conrotation).

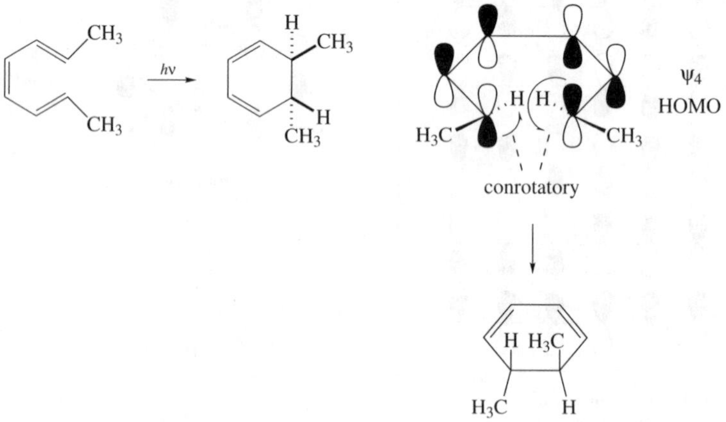

Fig. 11.22

The final type of pericyclic process to be considered here is the [3,3]-Cope rearrangement. There are three participating centres in each component of this six-electron reaction (Fig. 11.23). The rearrangement is generally stereospecific,

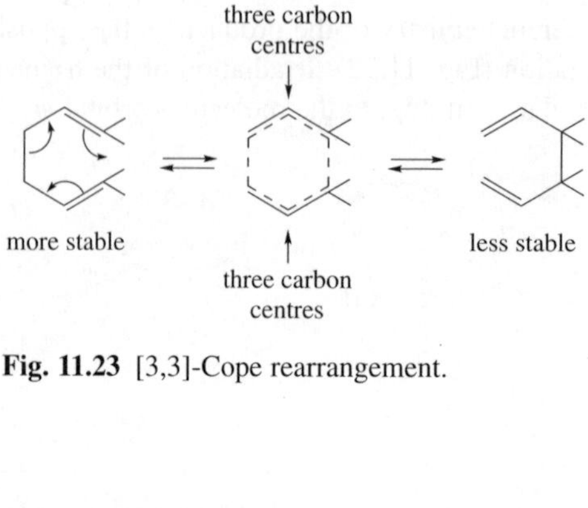

Fig. 11.23 [3,3]-Cope rearrangement.

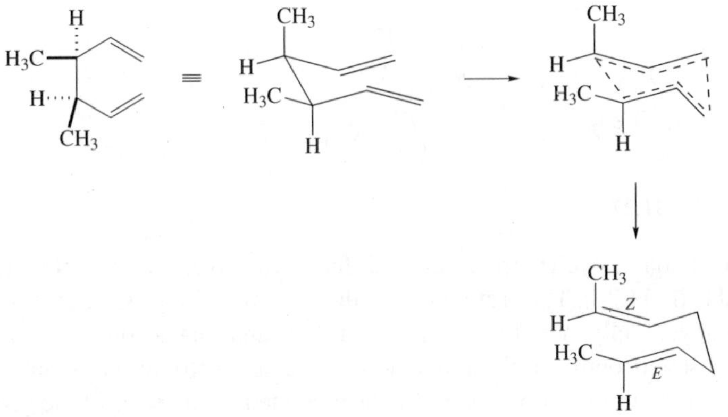

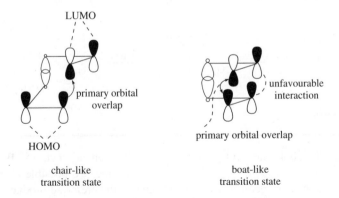

Fig. 11.24

proceeding through a chair-type transition state whenever possible (Fig. 11.24). The boat-like transition state is less favoured, a fact explicable in terms of molecular orbitals. Thus if the formation of the new single bond is regarded as a HOMO–LUMO interaction, the formation of the boat-like transition state requires an unfavourable orbital interaction between C2 and C5; such an unfavourable interaction is absent in the chair-like transition state (Fig. 11.25).

LUMO

primary orbital
overlap

HOMO

chair-like
transition state

unfavourable
interaction

primary orbital overlap

boat-like
transition state

Fig. 11.25

Involvement of an oxygen atom in the chain linking the two alkene units gives rise to a similar rearrangement, this time called the Claisen rearrangement (Fig. 11.26). The energy requirements and stereochemical features of the Cope and Claisen reactions are virtually identical and, in both reactions, examples of the

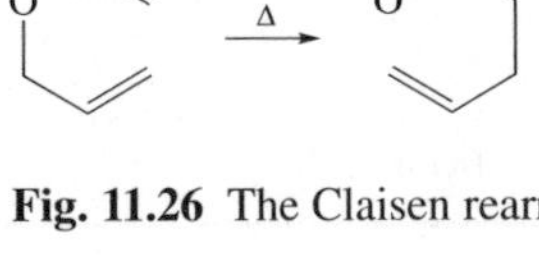

Fig. 11.26 The Claisen rearrangement.

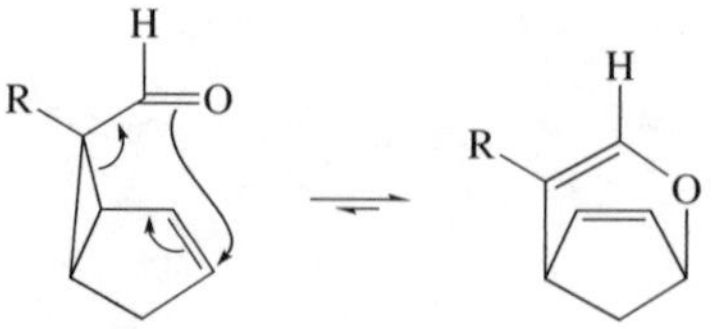

Fig. 11.27

interconversion of cyclic structures of different ring-sizes and functionality can be found (Fig. 11.27). The preference for the chair form in the Claisen rearrangement of acyclic allyl vinyl ethers is about 95% and this accounts for the high degree of stereocontrol in these reactions. Thus the distribution of substituents about the new carbon–carbon bond can be predicted on the basis of the geometry of the alkene bonds in the allyl vinyl ether. This principle is illustrated in Fig. 11.28 where the *EE* and *ZZ* isomers give predominantly (>95%) the *threo-*diastereoisomer whereas the *EZ* isomers give predominantly (again >95%) the *erythro*-diastereoisomer.

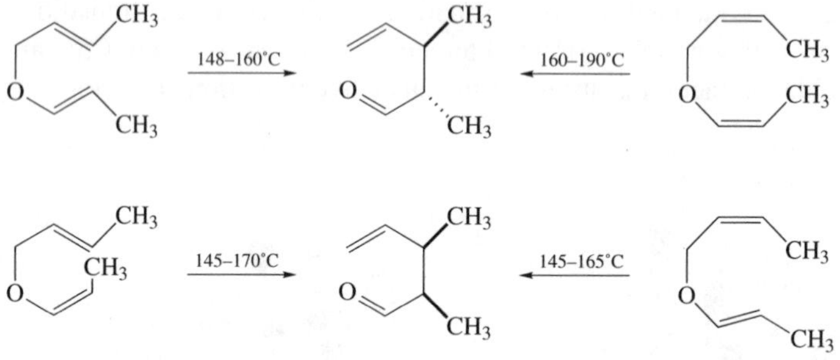

Fig. 11.28

Q. 11.3 The enol ether (**A**) gives one of the two enantiomers (*S*)-**B** or (*R*)-**B** as the major (>96%) product. By drawing the two possible chair-like transition states deduce which enantiomer is the preferred product.

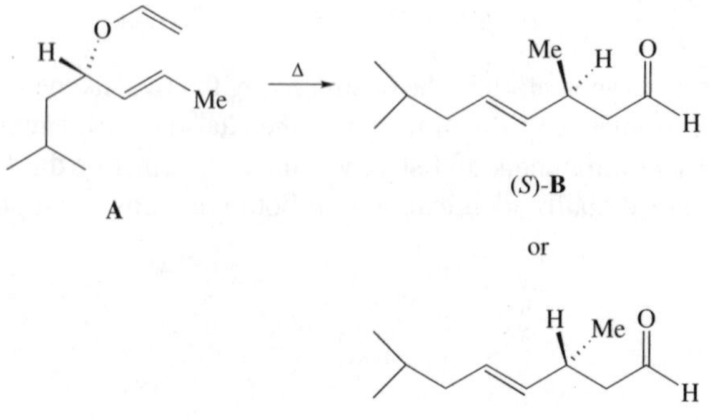

Answers to questions

Q. 11.1   (*E*)-penta-1,3-diene can form the s-*cis* conformer much more readily than (*Z*)-penta-1,3-diene. The s-*cis*-arrangement needs to be accessible for effective overlap of the molecular orbitals in the pericyclic reaction. The s-*cis*-arrangement is very difficult to form for (*Z*,*Z*)-hexa-2,4-diene due to the severe interactions of the adjacent methyl groups.

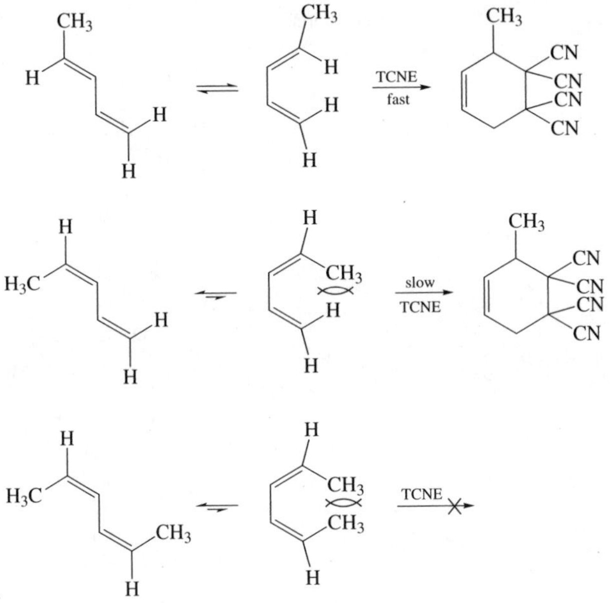

Q. 11.2

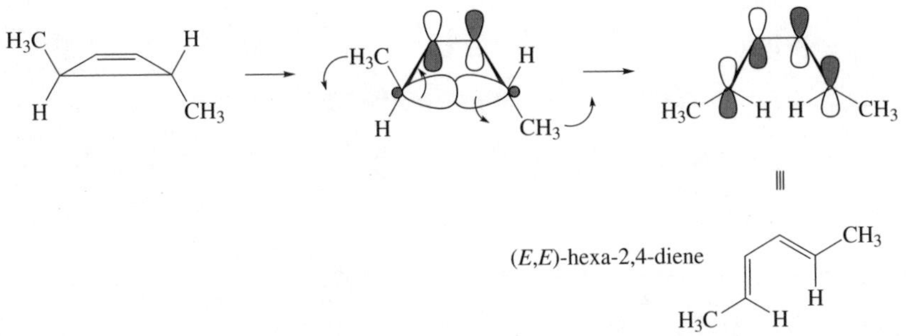

(*E*,*E*)-hexa-2,4-diene

The carbon–carbon sigma bond with two electrons comprises two sp³ hybrid orbitals of matching phase. To form a molecular orbital (with two electrons) spanning the four ring-carbons, the hybrid orbitals must break their bond and must move to match the unoccupied antibonding orbital of the alkene unit, thus forming the HOMO containing two electrons. A conrotatory motion is required, giving the *E*,*E*-diene as the major product.

Q.11.3   (*S*)-**B** is formed as the major product since in this pericyclic reaction an unfavourable axial orientation of the *sec*-butyl group is avoided.

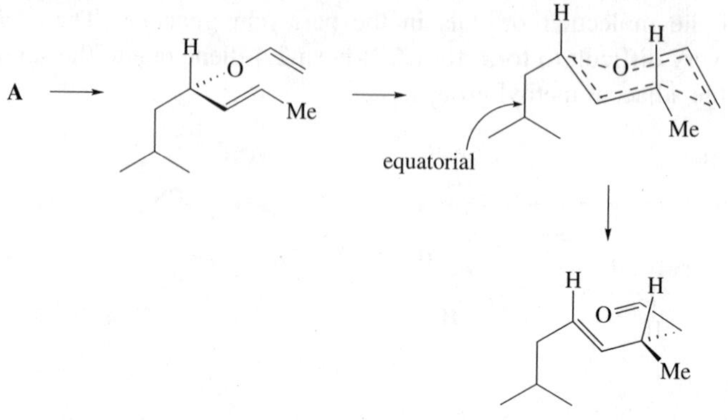

equatorial

(*S*)-**B**

axial

slow

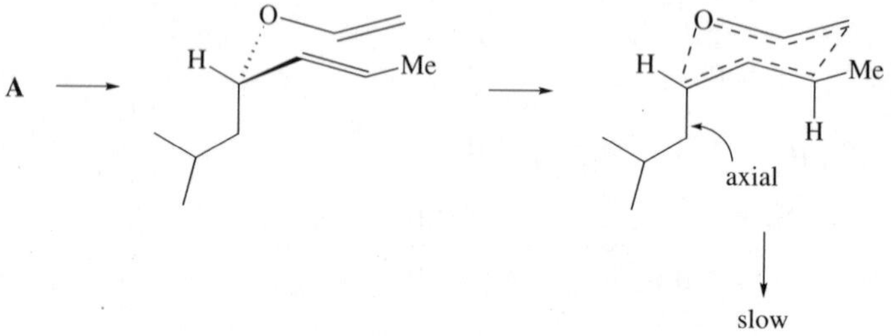

12 Cyclization reactions that proceed through high energy intermediates

This chapter will introduce reactions of the type indicated in Fig. 12.1, that is, cyclization reactions that proceed through unstable, high energy intermediates, be they carbocations, carbanions or radicals. Many reactions of this type have been studied, and a rationalization of the patterns of reaction observed and the stereocontrol offered by many of these processes has been put forward by Jack Baldwin (Oxford, UK) and Athelstan Beckwith (Australian National University, Canberra), *inter alia*.

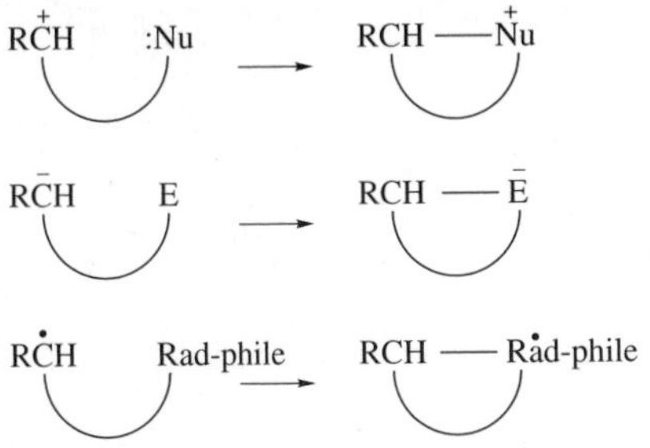

$$R\overset{+}{C}H \quad :Nu \quad \longrightarrow \quad RCH\!-\!\overset{+}{Nu}$$

$$R\overset{-}{C}H \quad E \quad \longrightarrow \quad RCH\!-\!\overset{-}{E}$$

$$R\overset{\cdot}{C}H \quad Rad\text{-phile} \quad \longrightarrow \quad RCH\!-\!\overset{\cdot}{Rad}\text{-phile}$$

Nu = nucleophile; E = electrophile; Rad-phile = radicalophile

Fig. 12.1

12.1 Intramolecular attack by a nucleophile

The rationalization put forward by Baldwin is best illustrated by using examples involving the attack by nucleophiles at an electrophilic centre. S_N2 reactions proceed through rear-side attack by the nucleophile at the sp^3 hybridized tetrahedral (*tet*) carbon centre (Fig. 12.2). Attack by the nucleophile at an sp^2 hybridized trigonal (*trig*) centre (such as the carbonyl carbon) takes place at an angle of 109° to the electron acceptor (the Bürgi–Dunitz trajectory). For an attack by a nucleophile at an sp hybridized digonal (*dig*) carbon centre (for example, a nitrile) it has been proposed that the corresponding angle is 60°.

When the nucleophile is tethered to the electrophilic component by a series of n atoms then the length of this chain will be a factor in determining whether the reaction is favoured or disfavoured. An additional factor is whether the electron flow of the reaction is external (*exo*) to the ring being formed or whether it is endocyclic (*endo*) (Fig. 12.3). It is therefore possible to describe reactions of this

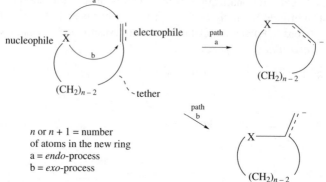

Fig. 12.2

Fig. 12.3

type in terms of the following:

(1) the number of atoms in the new ring (n)
(2) whether the electrophilic centre under attack is *tet*, *trig* or *dig*
(3) whether the electron flow is *exo* or *endo*.

and it is found that the various reaction types are favoured or disfavoured as shown in Table 12.1.

Table 12.1

n	exo/endo	tet, trig or dig	favoured/disfavoured
3–7	exo	tet	favoured
3–6	endo	tet	disfavoured
3–7	exo	trig	favoured
6–7	endo	trig	favoured
3–5	endo	trig	disfavoured
5–7	exo	dig	favoured
3-7	endo	dig	favoured
3,4	exo	dig	disfavoured

In practice the disfavoured reactions are not impossible but they are more difficult to achieve. Baldwin's rules apply strictly to first-row elements (C, N, O);
second-row elements, being larger and more easily polarized, can achieve angles
of attack not available to the first-row elements. Some examples of favoured
reactions involving a nucleophilic attack on an electrophilic centre are shown in
Fig. 12.4.

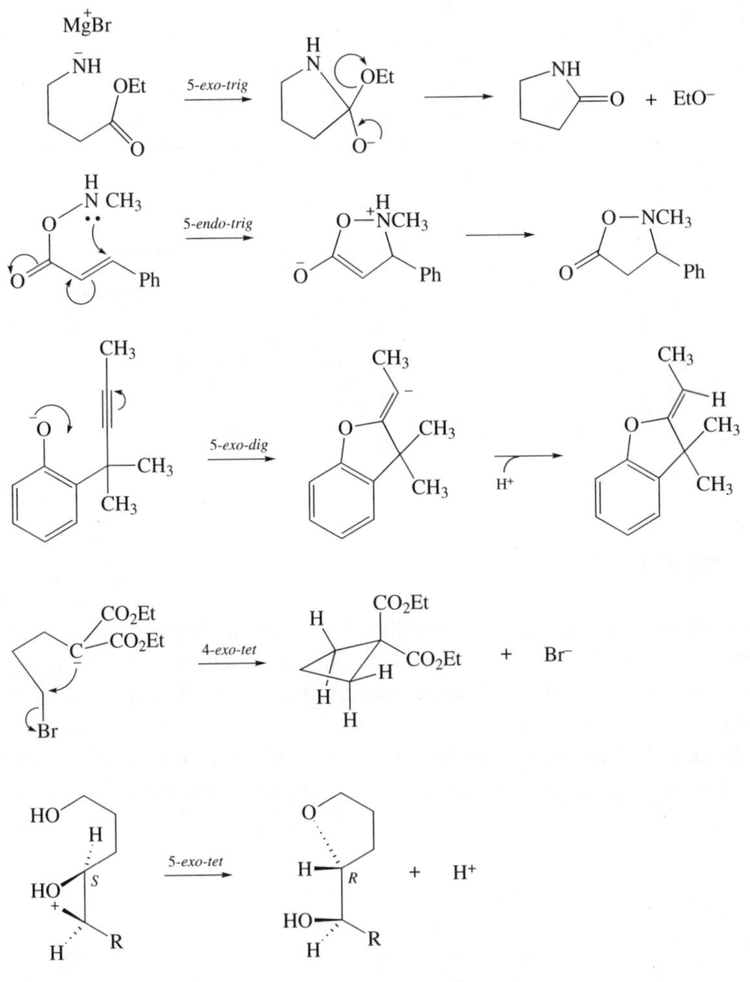

Fig. 12.4

The acid-catalysed ring opening of an epoxide unit by an adjacent nucleophile
(final example in Fig. 12.4) has a strong analogy to iodolactonization reactions
(for example, see Fig. 12.5) which have become very popular in synthetic
chemistry because they construct oxygen-containing ring systems with good
stereocontrol.

Such halolactonization reactions are very often interpreted in terms of
Baldwin's rules but, since they proceed *via* cationic intermediates, the rules are

Q. 12.1 The iodide (**A**) can be induced to undergo a 6-*endo-dig* cyclization (with initial loss of the iodine atom). Give the structure of the product.

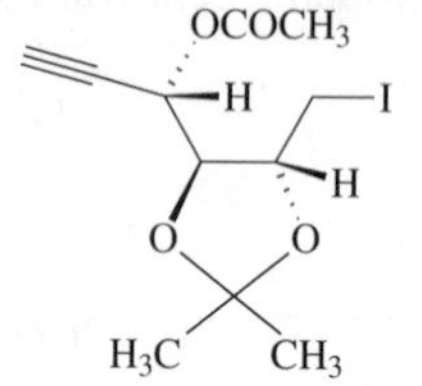

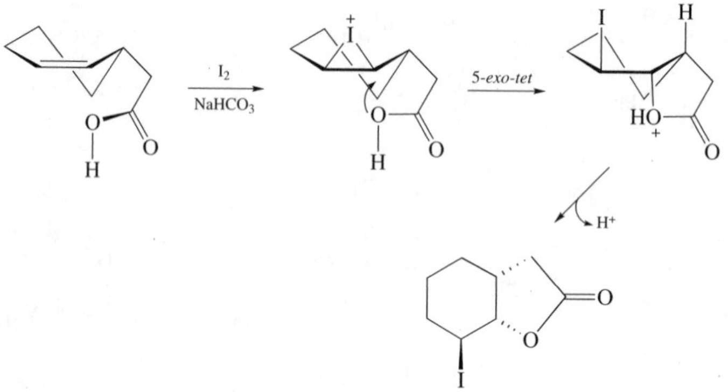

Fig. 12.5

not always strictly applicable. Nevertheless, some analogy is often evident and reactions can be devised to produce highly functionalized heterocyclic compounds from alkenoic acids and alkenols (Fig. 12.6). The transformations featured in Fig. 12.6 demonstrate that *exo*-reactions are almost always preferred when the reactions are performed under kinetic conditions and overwhelmingly so when steric or electronic factors are absent. The preferred transition states for

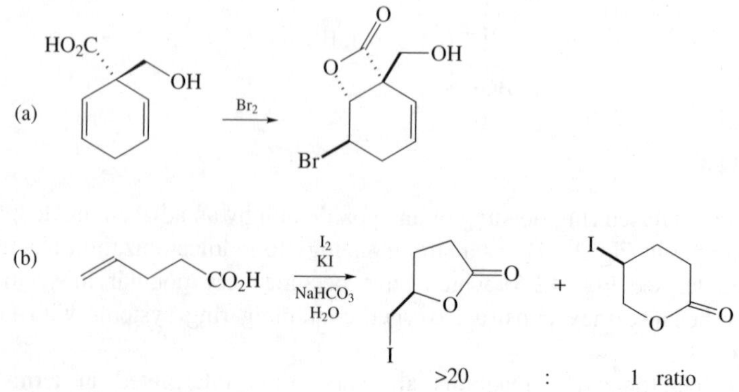

Fig. 12.6

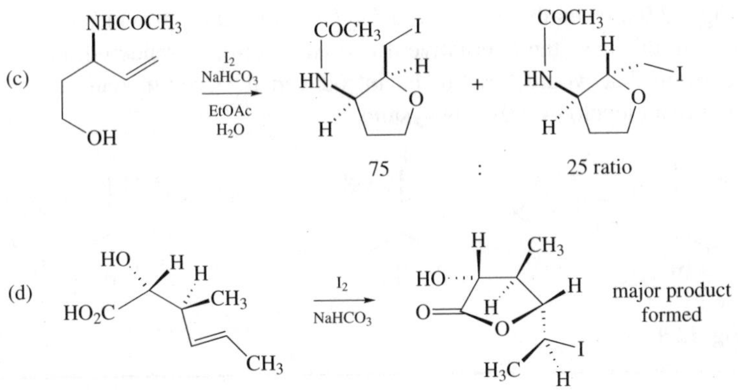

Fig. 12.6 *cont'd*

the last two reactions in Fig. 12.6 are shown again in Fig. 12.7 where the reacting alkene unit can be seen to be synclinal to the bulkier (than hydrogen) amide (a) or methyl group (b).

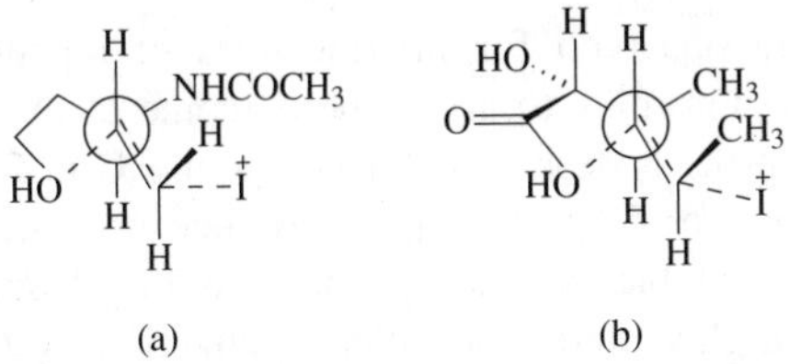

(a)   (b)

Fig. 12.7

Reactions run under kinetic conditions and thermodynamic conditions (Fig. 12.8) can lead to different ratios of products. Iodolactonization of (S)-4-methyl-hex-5-enoic acid gives the *cis*-lactone through a transition state similar to the one featured in Fig. 12.7(a). Under thermodynamic control, the adjacent substituents in the newly formed ring system will be *trans*-oriented, since this is the more favourable arrangement with both substituents in equatorial positions.

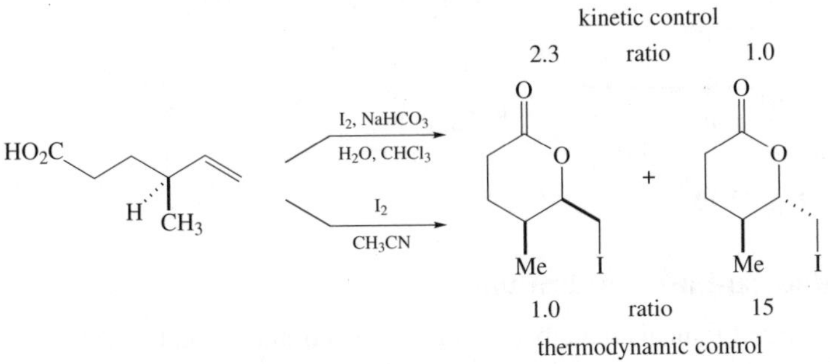

Fig. 12.8

In Fig. 12.9 a formally disfavoured 5-*endo-tet* reaction provides the observed product. In this case the alternative 4-*exo-tet* reaction is rendered unlikely due to the strain that would have to be introduced to accommodate a *trans*-ring junction in a bicyclo [4.2.0]-ring system.

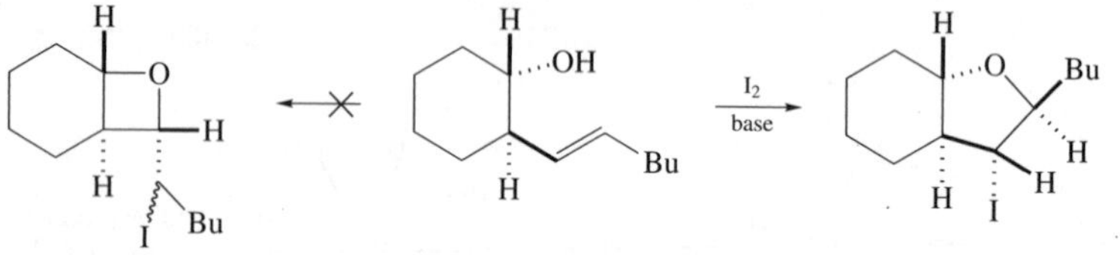

Fig. 12.9

Q. 12.2 Draw out the transition state (involving an iodonium ion) which explains the stereochemistry of the product in the transformation shown in Fig. 12.9.

12.2 Cyclizations involving carbocations

Similar reactivity patterns are observed for intermediates of higher energy, though other factors can come into play to affect the outcome of the reaction. For instance, the 6-*endo-trig* process described for compound **C** in Fig. 12.10 is preferred over the alternative 5-*exo-trig* pathway because the tertiary carbocation so formed is more stable than the primary cation. A spectacular example of a cation-based, highly stereocontrolled cyclization is Johnson's reaction of a polyene to give a steroidal structure (Fig. 12.11).

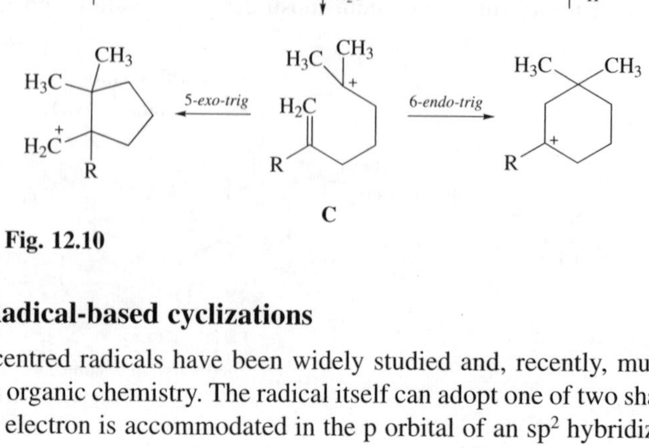

Fig. 12.10

12.3 Radical-based cyclizations

Carbon-centred radicals have been widely studied and, recently, much used in synthetic organic chemistry. The radical itself can adopt one of two shapes. If the unpaired electron is accommodated in the p orbital of an sp² hybridized system

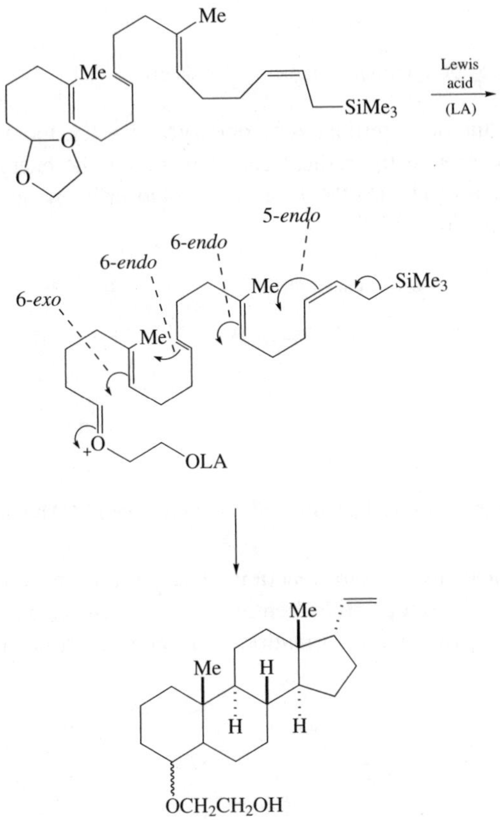

Fig. 12.11

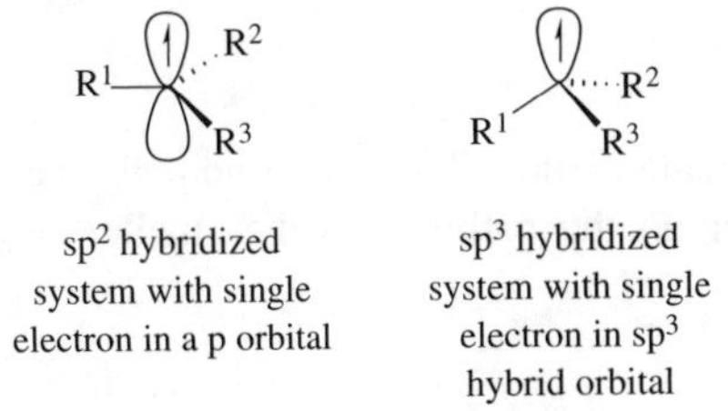

sp² hybridized
system with single
electron in a p orbital

sp³ hybridized
system with single
electron in sp³
hybrid orbital

Fig. 12.12

then the radical would have a planar arrangement of substituents (Fig. 12.12). If the unpaired electron occupies one of the four sp³ hybridized orbitals a pyramidal structure is formed. Spectroscopic evidence points to simple alkyl radicals having a planar geometry. However, if the carbon-centred radical contains a highly electronegative substituent, the configuration becomes more pyramidal and this deviation increases with the number of electronegative substituents (Fig. 12.13). For the pyramidal radicals the barrier to inversion is low.

Alkyl radicals are nucleophilic and will attack simple compounds such as dimethyl fumarate $[(E)\text{-MeO}_2\text{CCH=CHCO}_2\text{Me}]$. The relative rates of reaction

$$Me^{\bullet} \; < \; FH_2C^{\bullet} \; < \; F_2HC^{\bullet} \; < \; F_3C^{\bullet}$$

Fig. 12.13 Degree of pyramidal character of radicals.

of alkyl radicals follow the order tertiary > secondary > primary. Increasing the number of alkyl groups around the radical centre increases the energy of the singly occupied molecular orbital (SOMO) bringing it closer in energy terms to the LUMO of the alkene (Fig. 12.14).

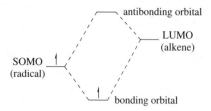

Fig. 12.14 Interaction of nucleophilic radical SOMO with LUMO of alkene.

In the context of radical ring-closure reactions, nucleophilic (alkyl) radicals react preferentially with electropositive centres in a 5-*exo-trig* mode (Fig. 12.15). Given a choice between the formation of a larger or a smaller ring,

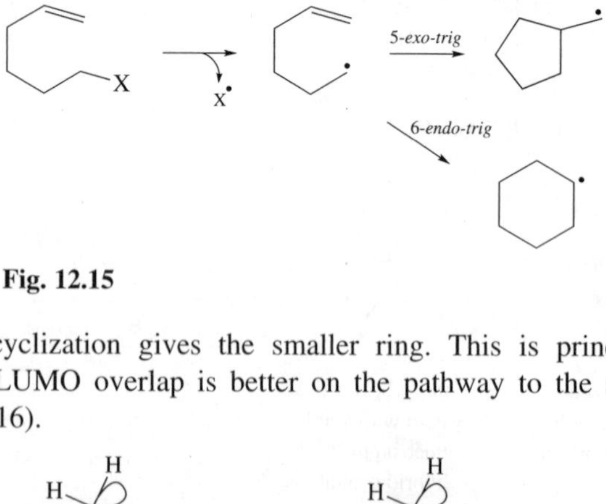

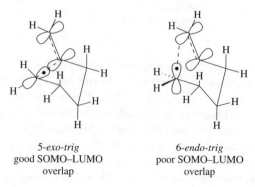

Fig. 12.15

radical cyclization gives the smaller ring. This is principally because the SOMO–LUMO overlap is better on the pathway to the smaller ring system (Fig. 12.16).

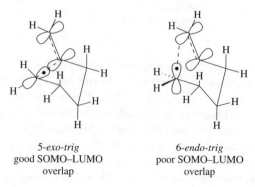

5-*exo-trig*
good SOMO–LUMO
overlap

6-*endo-trig*
poor SOMO–LUMO
overlap

Fig. 12.16

The degree of stereocontrol in such cyclization reactions does depend on the precise substitution pattern in the acyclic precursor. Some simple cyclopentane derivatives are produced with poor stereocontrol (Fig. 12.17) but with

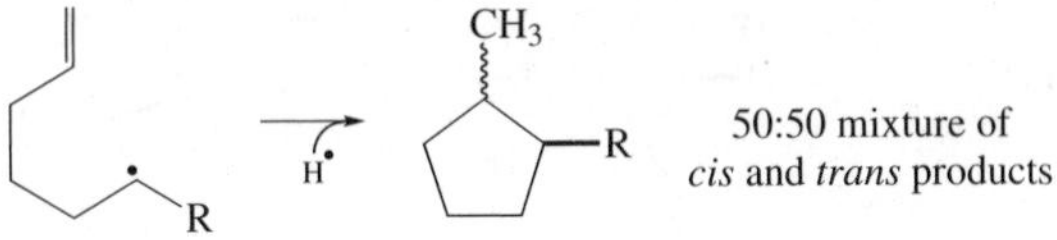

50:50 mixture of
cis and *trans* products

Fig. 12.17

more heavily substituted systems excellent control can be achieved. The example featured in Fig. 12.18 shows that an acetal unit can be used to good effect to direct and control the stereochemistry of a radical-based cyclization.

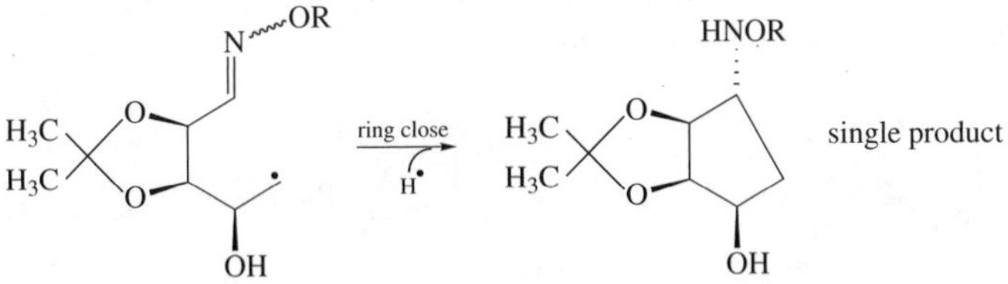

single product

Fig. 12.18

As with the carbocation-promoted cyclizations (Fig. 12.11), some notable radical-based polyene cyclizations have been designed that proceed with excellent stereocontrol (Fig. 12.19).

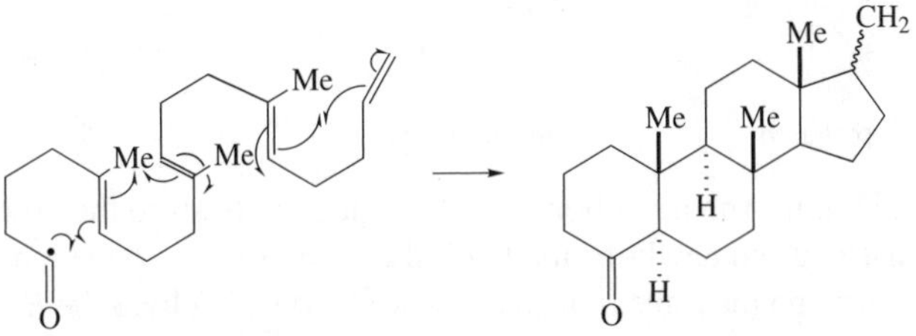

Fig. 12.19

Q. 12.3 It has been mooted that the addition of a hydrogen atom (H·) to the carbon-centred radical **A** passes through a transition state that is very similar to the Felkin–Anh transition state for the delivery of hydride ion to the ketone PhCH(Me)COMe (see section 8.2.2). If this is so and if the configuration at the stereogenic centre in **A** is (*R*), predict the stereochemistry of the product.

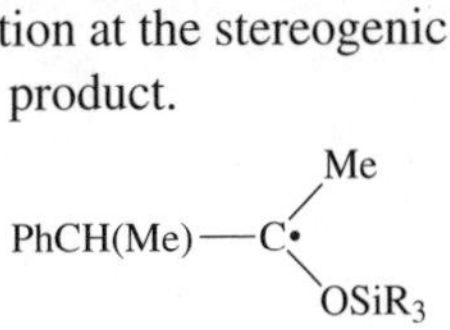

A

Answers to questions

Q. 12.1

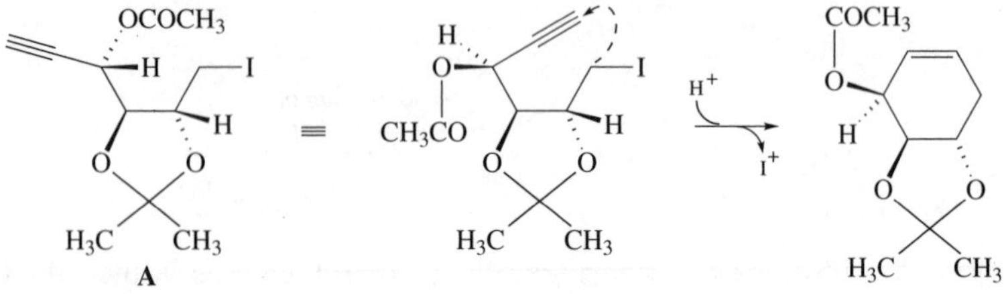

OCOCH$_3$
—H
—I
H
O   O
H$_3$C   CH$_3$
A

≡

H.
O
CH$_3$CO
—I
H
O   O
H$_3$C   CH$_3$

H$^+$
I$^+$

COCH$_3$
O
H
O   O
H$_3$C   CH$_3$

Q. 12.2
The iodoetherification reaction proceeds through a transition state in which the six-membered ring is chair-like and the developing five-membered ring has an envelope conformation.

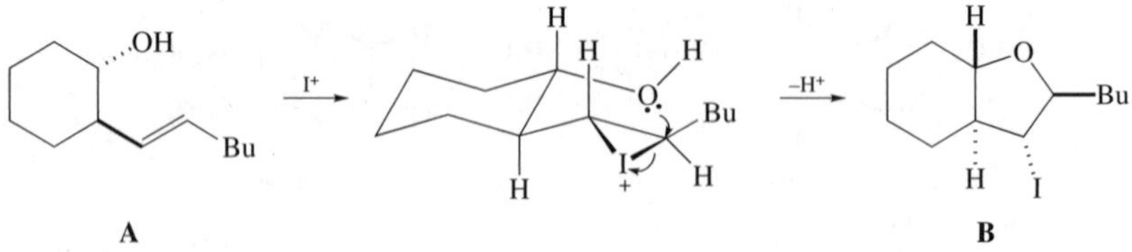

...OH
Bu
A

I$^+$

H
H   H
O.
Bu
I
H   H

−H$^+$

H
O
Bu
H   I
B

Q. 12.3

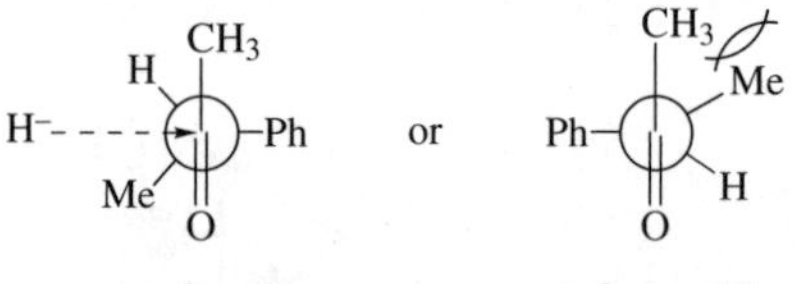

H CH$_3$
H– – – – ⊕ –Ph
Me   O

preferred

or

CH$_3$
Me
Ph– ⊕
H
O

unfavourable

The preferred Felkin–Anh transition state for reduction leads to the product with *S* stereochemistry at the newly formed chiral centre. Similarly, the radical reaction proceeds through the transition state shown below to deliver the *RS* product.

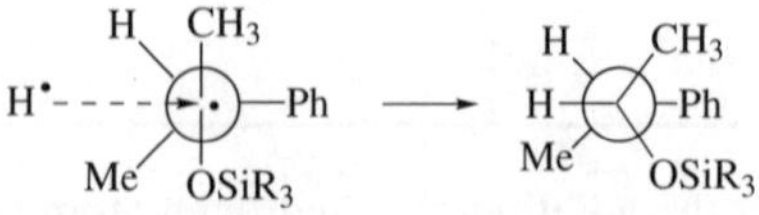

H CH$_3$
H• – – – – ⊕ –Ph
Me   OSiR$_3$

H   CH$_3$
H–⊕–Ph
Me   OSiR$_3$

13 Stereochemistry of selected polymers

Polymers can be natural (for example, proteins, carbohydrates) or can be man-made (for example, polyesters). We need to consider these materials in this book because the properties of the polymer are often influenced by the stereo-chemistry of substituents that are present along the backbone.

13.1 Synthetic polymers

Polypropylene is the simplest polymer for which bulk properties of the material are influenced by stereochemistry. When propylene (prop-2-ene) is polymerized, the methyl groups on the polymer backbone can be oriented syndiotactically, isotactically or in a random (atactic) way (Fig. 13.1).

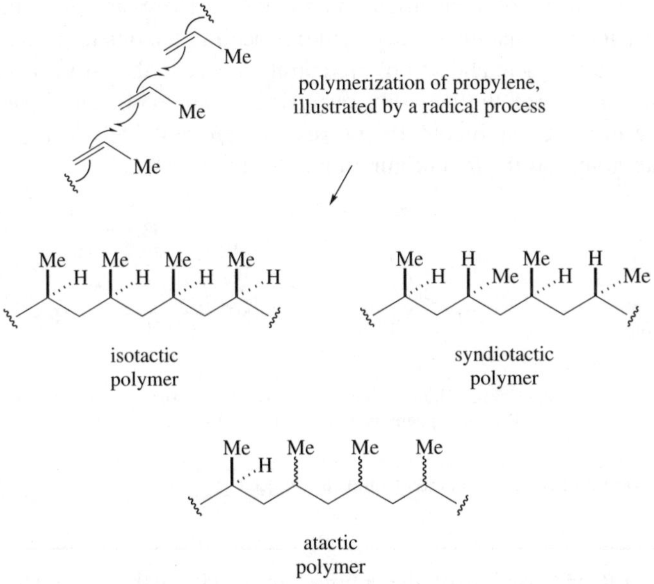

polymerization of propylene, illustrated by a radical process

isotactic polymer

syndiotactic polymer

atactic polymer

Fig. 13.1 Isotactic, syndiotactic and atactic forms of polypropylene.

An isotactic polymer is often produced by polymerizing the monomer using a Ziegler–Natta catalyst (a complex metal-based initiator prepared from trialkyl-aluminium and titanium trichloride). The identical groups (methyl groups in polypropylene) point in the same direction along a stereoregular, conformation-ally fixed backbone. Van der Waals interactions between chains with this regu-lar arrangement are stronger than those between chains with randomly oriented groups (atactic polymer). An increase in tacticity of a polymer is accompanied by a marked increase in crystallinity. Thus isotactic polymers are often brittle, high-melting solids whereas the corresponding atactic polymers are often rub-bery, low-melting solids or syrupy liquids.

When the monomer is asymmetric and optically active, the derived polymer possesses the same two properties. For example, the polymerization of (*R*)-propylene oxide leads to isotactic, optically active poly-(*R*)-propylene oxide (Fig. 13.2).

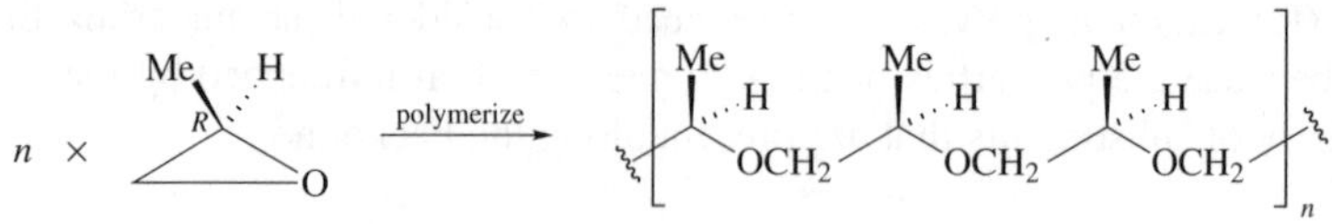

Fig. 13.2 Isotactic poly-(*R*)-propylene oxide.

13.2 Proteins

Proteins are made of condensed amino acids. There are 20 amino acids commonly found in nature. One (glycine) is achiral, the others (alanine, serine, cysteine, glutamic acid and so on) are chiral and are in the L-series (Fig. 13.3). The vast majority of the naturally occurring amino acids are in the *S* series according to the Cahn–Ingold–Prelog rules *except* cysteine which (because of the sulfur atom) has the *R* configuration (Chapter 3).

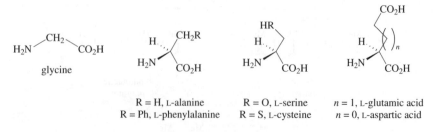

R = H, L-alanine   R = O, L-serine   n = 1, L-glutamic acid
R = Ph, L-phenylalanine   R = S, L-cysteine   n = 0, L-aspartic acid

Fig. 13.3 Some of the amino acids found in proteins.

Q. 13.1 Condensation of the amino group within the methyl ester of phenylalanine with the α-carboxylic acid group of aspartic acid gives a dipeptide. The *RS*, *SR*, and *RR* diastereoisomers are bitter to the taste but the *SS*-diastereoisomer is intensely sweet and is sold under the trademark Aspartame. Draw the structure of aspartame.

The condensation of a series of amino acids to form a peptide chain, or polypeptide, gives the primary structure of the protein (Fig 13.4). One of the simplest proteins is silk, which is a polymer of the amino acids glycine, alanine

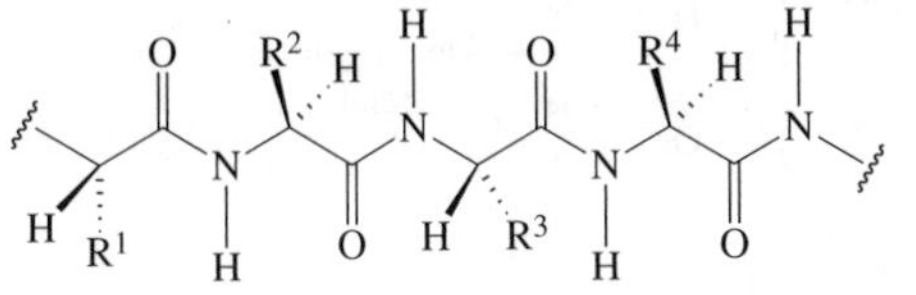

For silk
R^1, R^2, R^3...
= H or CH_3 or CH_2OH

Fig. 13.4 Primary structure of a protein.

and serine. Other structural proteins are more complex; wool, for example, contains cysteine residues, cross-linked by the formation of disulfide (S—S) bonds (Fig. 13.5).

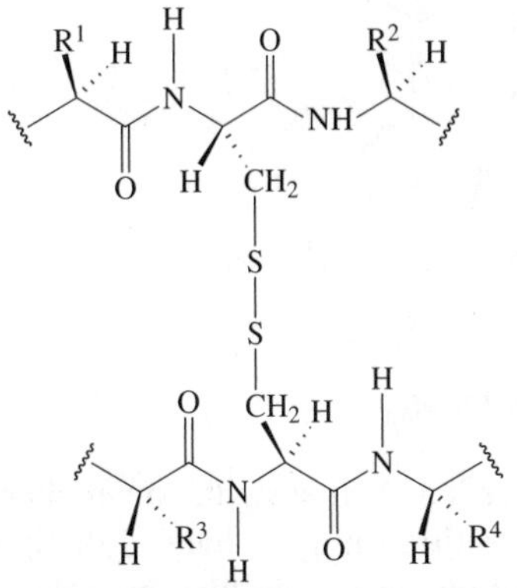

Fig. 13.5 Typical cross-linking of polypeptide chains, e.g. in wool.

Considering that at each position in Fig. 13.4 R^1, R^2, ... can be H or one of 19 groups and that a small protein can often contain more than 200 amino-acid residues, it is easy to see that the variation possible for these natural polymers is immense. When methyl groups and hydrogen atoms are the only substituents on the backbone (R^1, R^2 ... = Me or H), there is little interaction between them. However, for all the larger groups, there are *repulsive* interactions that result in the twisting of the backbone chain so that the alkyl groups rotate away from each other, reducing unfavourable steric effects. This deviation from a totally planar arrangement results in what is known as pleating.

When adjacent peptide chains are oriented in opposite directions (Fig. 13.6) they are properly positioned for *inter*molecular hydrogen bonding with other peptide backbones. This orientation of strands is called the antiparallel arrangement. Each strand is called a β-strand and is 'pleated', having the groups R^1, R^2, etc. successively a little below and above the plane of the β-sheet. In Fig. 13.7 the alternative sheet arrangement is shown, with the amino acids running in a parallel arrangement.

To effect *intra*molecular hydrogen bonding between adjacent peptide units a

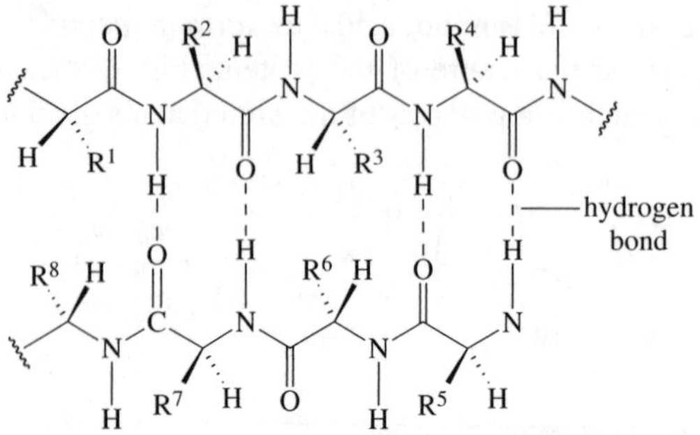

Fig. 13.6 Antiparallel pleated sheets of protein.

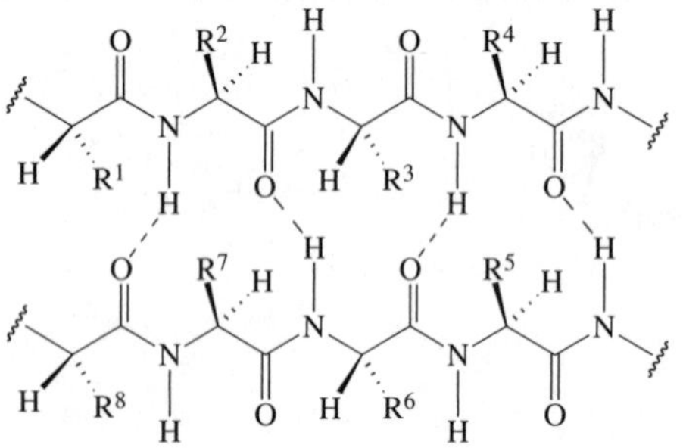

Fig. 13.7 Parallel sheet structure for protein.

three-dimensional helical structure is needed. At first sight, you might expect the chain to coil equally well in either direction (in a left-handed or right-handed fashion). However, the twisting to allow intramolecular hydrogen bonding places the R side chains of the amino acids (R^1, R^2, ...) in distinctly different positions. In the right-handed helix (Fig. 13.8) the substituents of the α-amino acids are oriented, more or less, away from the helical structure, with the hydrogen atoms pointed to the interior. The left-handed helix would result in unfavourable interactions between the substituents on the main chain.

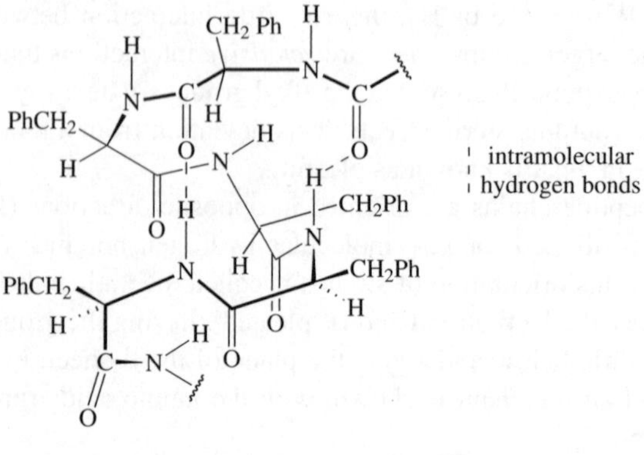

Fig. 13.8 The α-helix of a protein made up of phenylalanine residues.

These arrangements of a protein chain in sheets and helices are called the secondary structure of the protein. The complete protein structure is made up of sheets, helices, loops and turns that give it a complete tertiary structure, usually a quite beautiful arrangement as shown by the X-ray crystallographic picture of a protein in Fig. 13.9.

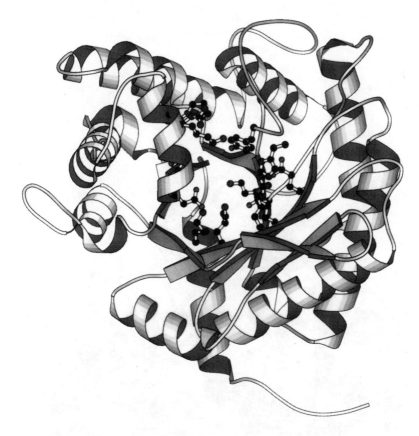

Fig. 13.9 Structure of protein (human aldolase A) showing helical regions, sheet structure (arrows) and non-defined structure (tubes).

As well as being structural features in microorganisms, plants, mammals, etc., proteins also have a functional role as enzymes, catalysing the transformation of a substrate (such as an ester) in aqueous solution into products (such as an alcohol and a carboxylic acid). The enzyme, being a true catalyst, is regenerated unchanged after the reaction (Fig. 13.10).

$$R^1CO_2R^2 \xrightarrow[H_2O]{enzyme} R^1CO_2H + R^2OH$$

Fig. 13.10 Hydrolysis of an ester catalysed by a hydrolase enzyme.

In the particular case shown in Fig. 13.11, the hydrolase enzyme has a serine residue at the active site. The protein catalyses the reaction through the formation of an acyl–enzyme complex which is rapidly hydrolysed.

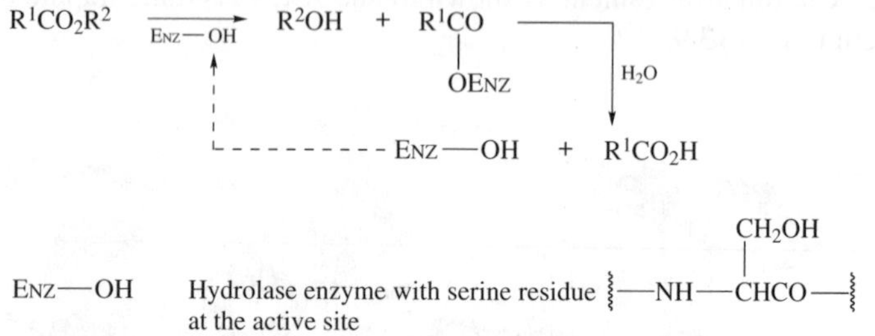

$$R^1CO_2R^2 \xrightarrow{\text{Enz—OH}} R^2OH \ + \ R^1CO$$

$$\text{OEnz} \quad \Big| \, H_2O$$

$$\text{Enz—OH} \ + \ R^1CO_2H$$

Enz—OH   Hydrolase enzyme with serine residue —NH—CHCO— at the active site, with CH$_2$OH

Fig. 13.11 Mechanism of hydrolysis of an ester by an enzyme having a serine residue at the active site.

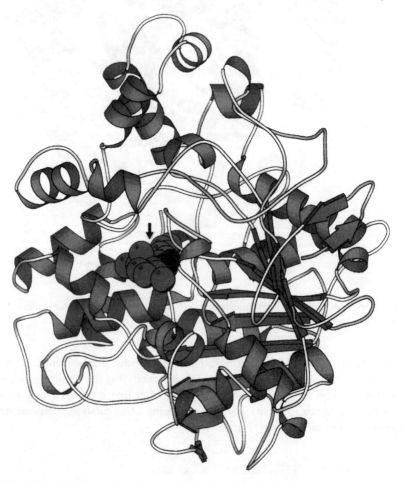

Fig. 13.12 *Candida rugosa* lipase complexed with (1*R*)-menthyl hexyl phosphonate (arrowed).

In such enzyme-catalysed reactions the substrate (for example, an ester) fits snugly into the active site of the protein and often a variety of favourable interactions (electrostatic, hydrogen bonding, van der Waals, etc.) orient the substrate so that acylation of the serine residue takes place readily. Figure 13.12 shows the complementarity of the shape of the hydrolase enzyme and its partner, like a lock and key, as first pointed out by Emil Fischer. Very often conformational changes within the protein and in the partner take place to optimize the intermolecular binding. For further discussion on enzyme action and, in particular, the use of hydrolytic and other enzymes in synthetic organic chemistry, see Chapter 15.

The association between an enzyme and its substrate leads to a transformation of the latter species (for example, an ester is hydrolysed to the corresponding carboxylic acid and alcohol). Another association of a protein and a small molecule is evident in receptor–(ant)agonist binding. In this case the large molecule/small molecule association involves non-covalent binding (electrostatic, hydrophobic, hydrogen-bonding phenomena, etc.) and the small molecule is not changed as a result of the association. The action of receptors is so intimately associated with conformational (that is, stereochemical) changes (as opposed to reactions), it is important to digress a little at this stage to discuss the function of these key macromolecules.

There are membrane-bound receptors in the mammalian system for a wide variety of substances, ranging from small molecules such as acetylcholine, through histamine and adrenaline to higher molecular weight polypeptides such as substance P (Fig. 13.13). The natural compounds shown in Fig. 13.13 act as agonists to their receptor. Agonist–receptor binding results in a change in conformation of the receptor both outside and inside the cell, causing either the intracellular activation of an enzyme or the opening of an ion channel (Fig. 13.14). The transformation of substrate A into product B or the movement of ions (such as Na^+, K^+, Ca^{2+}) along the channel can lead to a physiological response. For example, activation of adrenaline receptors on the heart causes an increase in the rate and force of contraction of this muscle.

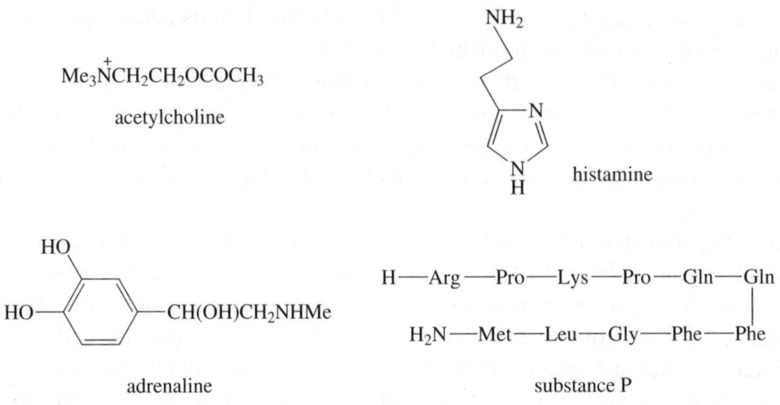

Fig. 13.13 Structures of some neurotransmitters.

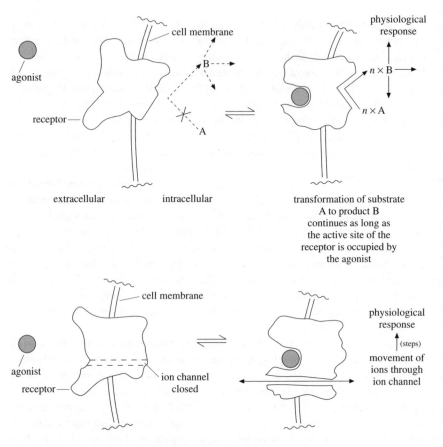

Fig. 13.14 Diagrammatic illustration of the change in conformation of a protein (receptor) on non-covalent binding of an agonist.

Activation of adrenaline receptors on smooth muscle of the bronchial system causes a relaxation and widening of the airways. The mode of action of the anti-asthmatic salbutamol (Ventalin) (Fig. 13.15) is based on its adrenergic (adrenaline releasing) activity on bronchial smooth muscle.

An antagonist is a molecule that binds strongly to a receptor but does not elicit a response (any change in conformation of the receptor is not sufficient to activate the intracellular enzyme or open the ion channel). The tight binding of the receptor–antagonist system prevents the natural agonist from activating the system (Fig. 13.16).

Antihypertensive drugs such as propranolol (Fig. 13.15) block the action of adrenaline on the heart muscle, effectively lowering blood pressure by reducing the rate and force of contraction. As a second example, ranitidine (Zantac) (Fig. 13.15) antagonizes the action of histamine on the parietal cells in the stomach. Histamine usually stimulates the release of HCl into the stomach so, by blocking this action, Zantac reduces the level of acid, allowing ulcers in the gastrointestinal tract to heal more readily.

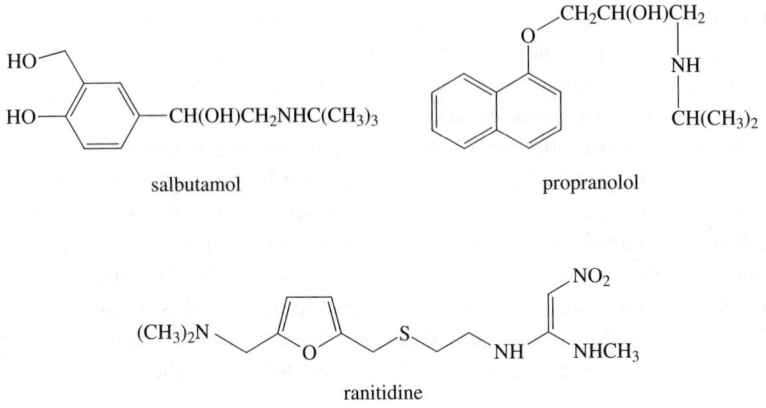

HO
HO
CH(OH)CH2NHC(CH3)3
salbutamol
O
CH2CH(OH)CH2
NH
CH(CH3)2
propranolol
NO2
(CH3)2N
O
S
NH
NHCH3
ranitidine

Fig. 13.15 Structures of some important drugs.

antagonist
A
agonist
receptor
cell membrane
A
B
A
agonist

Fig. 13.16 The action of an antagonist.

13.3 Carbohydrates

Carbohydrates are also found in polymeric form and are particularly important as structural units in plants. For example, α-amylose (Fig. 13.17) is a polymeric form of glucose. Like proteins, the polymer is shaped in the form of a helix with

C1,C4-linkage
α-configuration at C1
OH
O
HO
OH
6
4
1
1

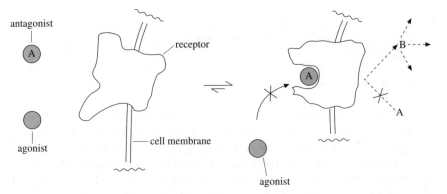

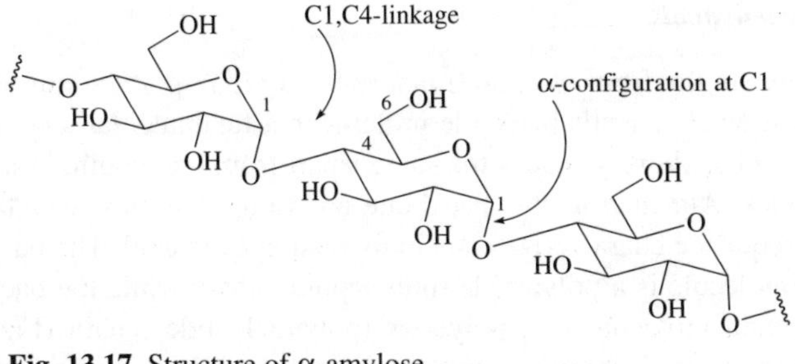

Fig. 13.17 Structure of α-amylose.

the formation of hydrogen bonds between adjacent rings on the same polymer chain. Starch is a polymer closely related to α-amylose except that, instead of having all α(1–4) linkages it has α(1–4) and α(1–6) junctions. This water-soluble polymer contains up to 4000 monosaccharide units.

Cellulose is also a polymer made up of condensed glucose residues. In this case 3000–5000 glucose residues couple through hydroxy groups in the β-configuration at the anomeric centre (Fig. 13.18) to give water-insoluble material. The water insolubility is partly a result of the fact that the repeating β-linkage means that different polymer chains can fit together much better and intermolecular hydrogen bonding occurs between the individual chains. Peracetylation of cellulose gives cellulose acetate which is much more soluble in organic solvents; such solutions can be processed into thin sheets for use in photographic film.

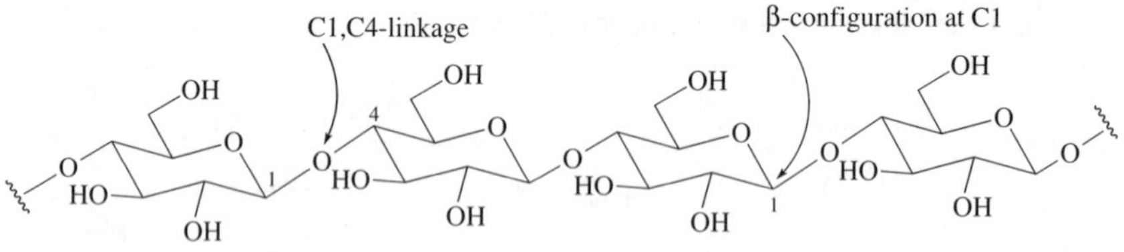

Fig. 13.18 Structure of cellulose.

Q. 13.2 Chitin is a polysaccharide that occurs widely in nature (for example, in the shells of crabs). It is a polymer of 2-acetamido-2-deoxy-D-glucose, a compound which has the NHCOMe unit replacing the OH group in the 2-position of glucose. The carbohydrate residues are connected by β-1,4-glucosidic linkages. Draw the structure of chitin.

Cyclic carbohydrates such as cyclodextrin (Fig. 13.19) are interesting structures with a hydrophilic exterior and a hydrophobic cavity in the interior. These hydrophobic cavities make good 'hiding places' for organic molecules in aqueous solution. As noted previously, cyclodextrins have been bound to solid supports to provide chiral stationary phases for chromatography (Chapter 6).

13.3.1 Nucleic acids

Carbohydrates in the form of pentafuranose units are important components of nucleic acids. Nucleic acids resemble proteins in a fundamental way: there is a long chain – a backbone – that is the same, apart from the length, in all nucleic acid molecules. Attached to this backbone are various groups, which by their nature and sequence characterize each individual nucleic acid. The backbone of the protein molecule is a polyamide (polypeptide) chain while the backbone of the nucleic acid molecule is a polyester (polynucleotide) chain (Fig. 13.20), comprising sugars and phosphate with bases attached. In the case of RNA the

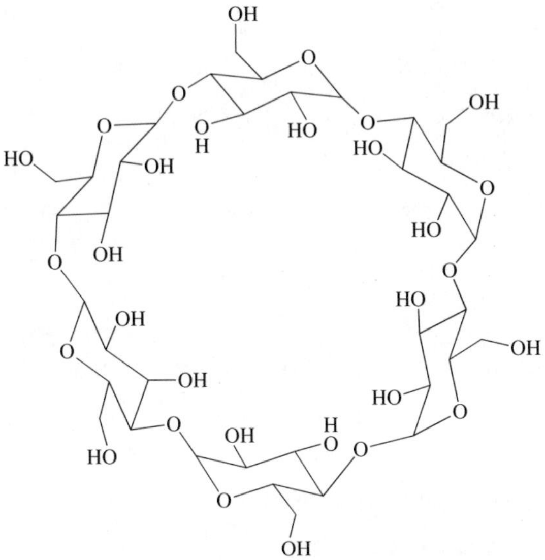

Fig. 13.19 Structure of α-cyclodextrin.

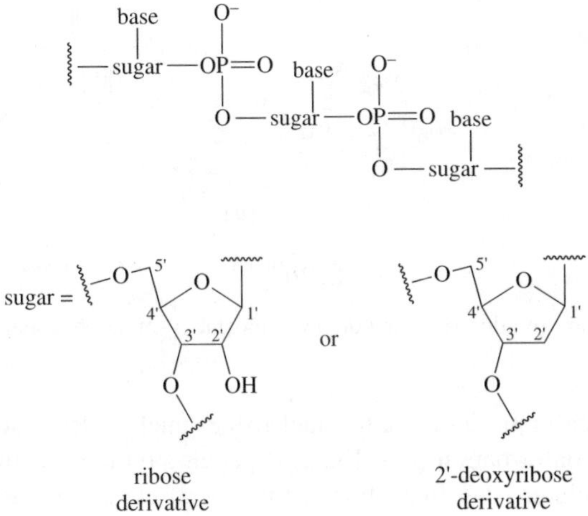

Fig. 13.20 Basic structure of a nucleic acid.

polymer has D-ribose as the sugar component; DNA has deoxyribose as the sugar. Attached to C1′ of each sugar, through a C—N linkage, is one of five bases: adenine, cytosine, thymine, uracil and guanine (Fig. 13.21). Adenine, cytosine and guanine are common to DNA and RNA. Uracil is peculiar to RNA, thymine is peculiar to DNA.

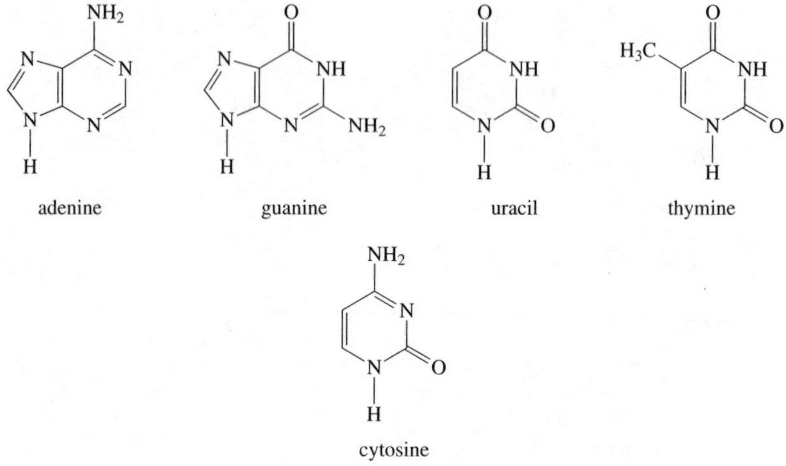

Fig. 13.21 Bases present in nucleic acids.

The base–sugar combination is called a nucleoside while a base–sugar–phosphate unit is called a nucleotide. Adenosine is a typical nucleoside, deoxyguanosine 5′-monophosphate is a typical nucleotide (Fig. 13.22).

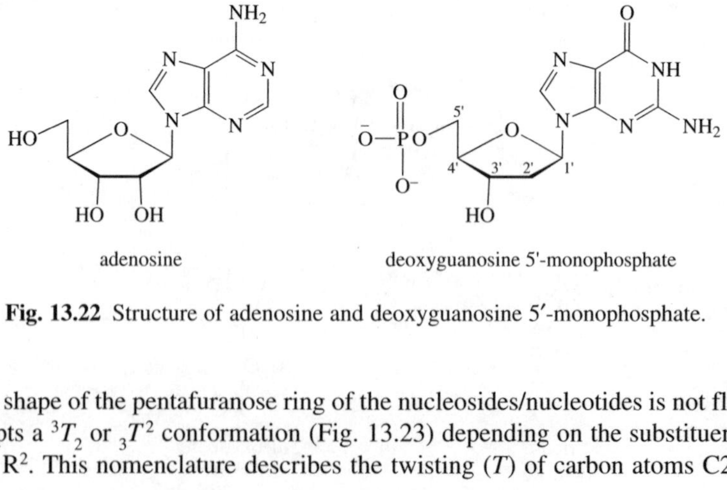

Fig. 13.22 Structure of adenosine and deoxyguanosine 5′-monophosphate.

The shape of the pentafuranose ring of the nucleosides/nucleotides is not flat but adopts a $^{3}T_{2}$ or $_{3}T^{2}$ conformation (Fig. 13.23) depending on the substituents R^{1} and R^{2}. This nomenclature describes the twisting (T) of carbon atoms C2′ and

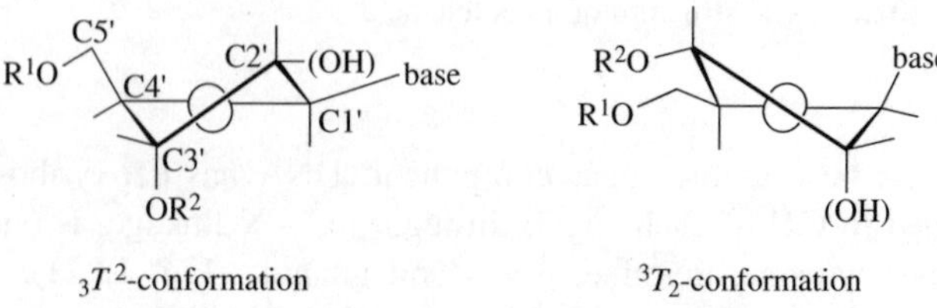

$_{3}T^{2}$-conformation $^{3}T_{2}$-conformation

Fig. 13.23 Conformations of deoxyribonucleosides.

C3′ above (superscript) or below (subscript) the plane created by the three atoms C1′, O and C4′. If, in the $_3T^2$ arrangement, the upward displacement of C2′ is greater than the downward displacement of C3′ the stereochemistry is termed C2′ *endo*, the term *endo* indicating that C2′ is on the same side of the C1′, O, C4′ plane as the base unit (Fig. 13.24).

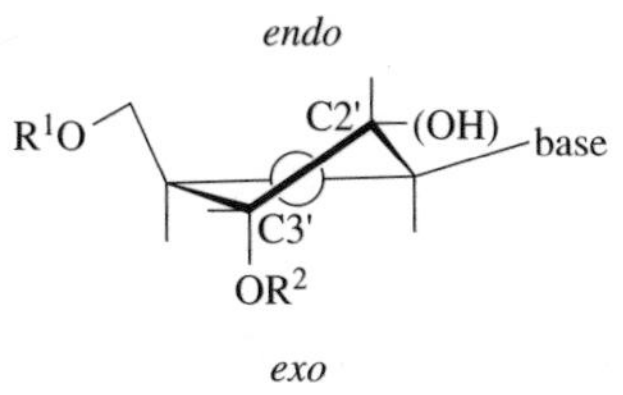

Fig. 13.24 C2′ *endo* conformation of a deoxyribonucleoside or nucleotide.

The proportions of the four bases and the sequence in which they follow each other along the polynucleotide chain differ from one nucleic acid to another. The 'primary sequence' determines the genetic information contained within the DNA/RNA molecule. As with proteins, the secondary structure of DNA is not linear. It is made up of two polynucleotide chains wound around each other to form a double helix 20 Å in diameter (Fig. 13.25). The adenine molecules in

Fig. 13.25 Schematic representation of the double helix structure of DNA.

one chain are hydrogen bonded to the thymine residues in the other, and the guanine base units are linked to the cytosine residues. Both helices are right-handed and both have 10 or 11 nucleotide units for each complete turn which occurs every 33 Å or 36 Å* along the axis. The two chains head in opposite directions, that is, the deoxyribose units are oriented in opposite ways so that the sequence is C3′,C5′ in one chain and C5′,C3′ in the other (Fig. 13.26). The double helical structure was first proposed by Watson and Crick in 1953.

In the secondary structure of RNA double-stranded helices are again involved, but this time they are made up from the same chain doubling back on itself (Fig. 13.27). In the most commonly found form of RNA there are 11 bases per complete turn and the pitch is 34 Å.

The nucleic acids are often bound to proteins and these nucleoproteins are coiled and folded to make up the chromosomes.

---

*There are two common forms of DNA: A-DNA has 11 residues per turn and a pitch of 36 Å, B-DNA has 10 residues per turn and a pitch of 33 Å.

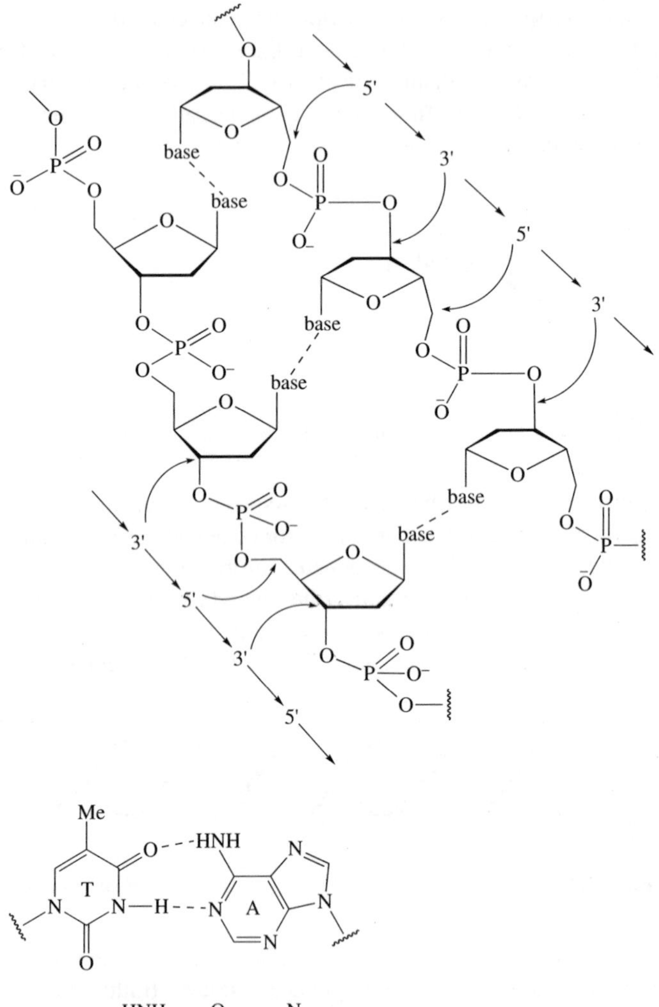

Fig. 13.26 Complementarity of base pairing [thymine(T)–adenine(A); cytosine(C)–guanine(G)] in the DNA double helix. The bases lie inside the helix, the hydrophilic phosphate units are on the exterior.

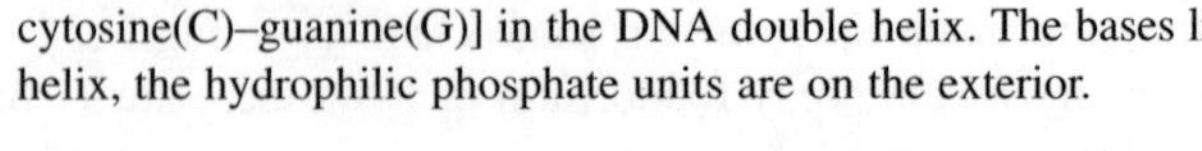

Fig. 13.27 Simple representation of a double helix made up of a single strand of deoxyribonucleotides.

Answers to questions

Q. 13.1   The structure of aspartame is:

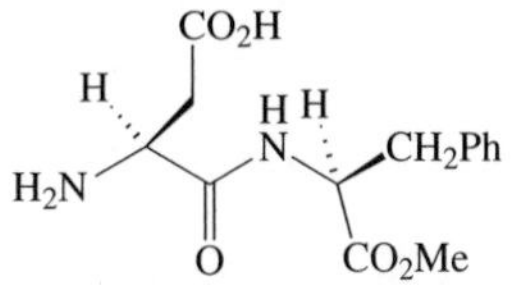

Q. 13.2   The structure of chitin is:

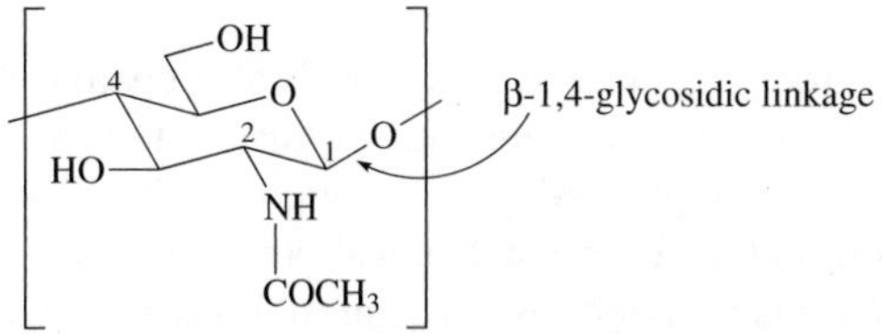

14 Stereochemistry and organic synthesis

14.1 Introduction

One major objective of synthesis in organic chemistry is to prepare useful, high value compounds from cheap, readily available starting materials. The end-products are frequently quite complicated and are often decorated with functional groups which are present at set stereochemical positions relative to each other. These products of organic synthesis are almost invariably useful in a secondary sense; that is in terms of their *biological* activity, perhaps acting as a medicine, as an aid for agriculture (for example, crop enhancement or protection) or as an attractive fragrance or flavour. Other products may be useful in different ways, such as novel polymers or materials with useful electrical properties. The biological and/or physical properties of the molecules are dependent on the presence and relative disposition of the carbon-based framework and the array of heteroatoms in this framework and in the functional groups. Synthetic organic chemists endeavour to set up the correct stereochemical relationships of the functional groups on the carbon backbones in the minimum number of steps.

Chemical purity is always important for the product; there is also an increasing awareness that chiral compounds should be supplied in an optically pure form. In biological terms the criterion of optical purity is an important one. We are chiral, and we are made up of chiral materials such as D-sugars and L-amino acids, so it is not surprising that the enantiomers of a chiral compound interact with biological macromolecules in different ways and can produce different effects, such as physiological responses.

The case of thalidomide is often quoted as a prime example of the need to address the issue of optical purity for compounds that are to be sold commercially. In the case of thalidomide the (+)-isomer (Fig. 14.1) was an effective

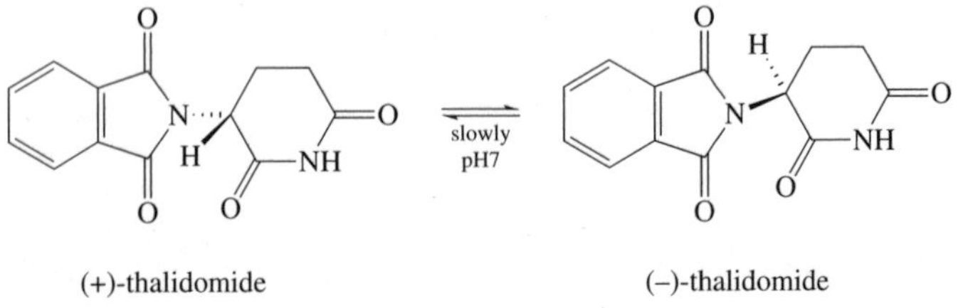

(+)-thalidomide   (−)-thalidomide

Fig. 14.1

antiemetic while the (–)-isomer was responsible for the terrible teratogenic properties. The compound was sold as the racemate with disastrous consequences. Note that even if the (+)-isomer had been used as an optically pure compound it may not have been entirely safe since racemization has been shown to take place (through the enol form) under physiological conditions.

In the case of certain flavours, two optical isomers can have totally different tastes. For example, (R)-limonene (Fig. 14.2) has an orange flavour while (S)-limonene tastes of lemon. Similarly, (R)-carvone tastes of spearmint while the (S)-enantiomer tastes of caraway. The receptors on the tongue are chiral and the different sensations are due to the formation of diastereoisomerically related complexes which cause different messages to be sent to the brain. This extends to unnatural flavours: Aspartame® (Fig. 14.2) is a low-calorie sweetener (see Q. 13.1).

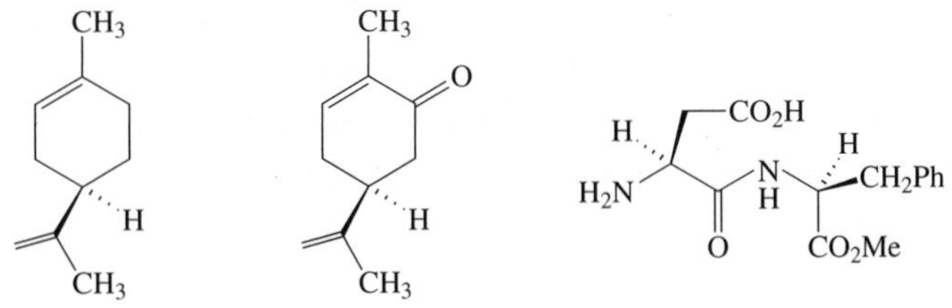

(R)-(+)-limonene   (S)-(+)-carvone   Aspartame®

Fig. 14.2

In compounds for agricultural use optical purity is also of importance, for instance in pheromones used as pest-control agents. Pheromones are often low-molecular weight chiral molecules that are released by insects to attract partners. Sometimes one enantiomer is the active moiety while the mirror image compound actually decreases the effect. For example, the (7R,8S)-(+)-enantiomer of disparlure (Fig. 14.3) acts as the pheromone for the gypsy moth *Lymantria dispar* while the (7S,8R)-(–)-enantiomer is an inhibitor of this biological activity.

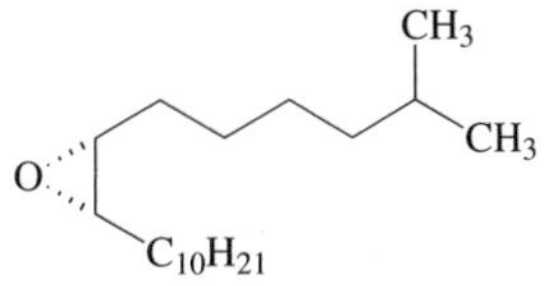

(7R,8S)-disparlure

Fig. 14.3

It is because of the differences in behaviour illustrated by the above examples that organic chemists spend much time in seeking methods for the production of diastereoisomerically pure and optically pure materials. There are several different ways of achieving this goal:

1. using materials from 'the chiral pool'
2. employing classical resolution techniques
3. asymmetric synthesis utilizing
    (a) chiral auxiliaries
    (b) chiral reagents
    (c) chiral catalysts

The first two approaches are discussed in this chapter; the third section embraces a lot of different reactions, techniques and strategies and justifies a separate discussion (see Chapter 15).

14.2   Using materials from the chiral pool

'Chiral pool' is a term used to embrace all natural asymmetric compounds that are readily available to the chemist. Nature provides a wide variety of such materials in optically pure form as constituents of plants (sugars, steroids, alkaloids, terpenes, etc.) or as secondary metabolites produced by microorganisms (for example, penicillins from *Penicillium* fungi) (Fig. 14.4).

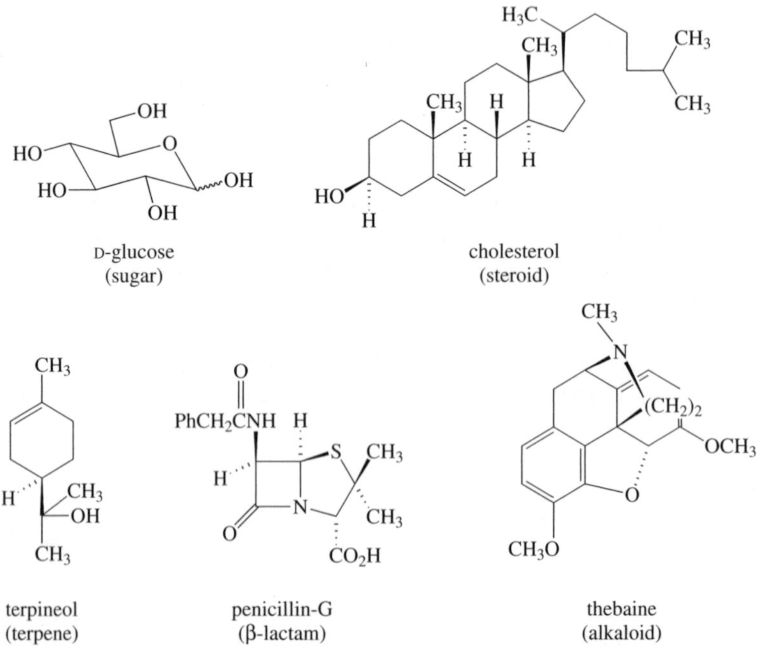

D-glucose
(sugar)

cholesterol
(steroid)

terpineol
(terpene)

penicillin-G
(β-lactam)

thebaine
(alkaloid)

Fig. 14.4 Some naturally occurring chiral compounds.

Functional groups in compounds derived from the chiral pool can be modified and/or the carbon and heteroatom backbones rearranged to provide the desired end products. In some instances only one or two chemical steps are needed for the transformation. For example, penicillin-G is readily converted into 6-aminopenicillanic acid (6-APA) and this compound is acylated using D-phenylglycine to give ampicillin, an antibacterial used to treat urinary tract and respiratory tract infections (Fig. 14.5). The method of production of D-phenylglycine is discussed later in this chapter (see Fig. 14.17).

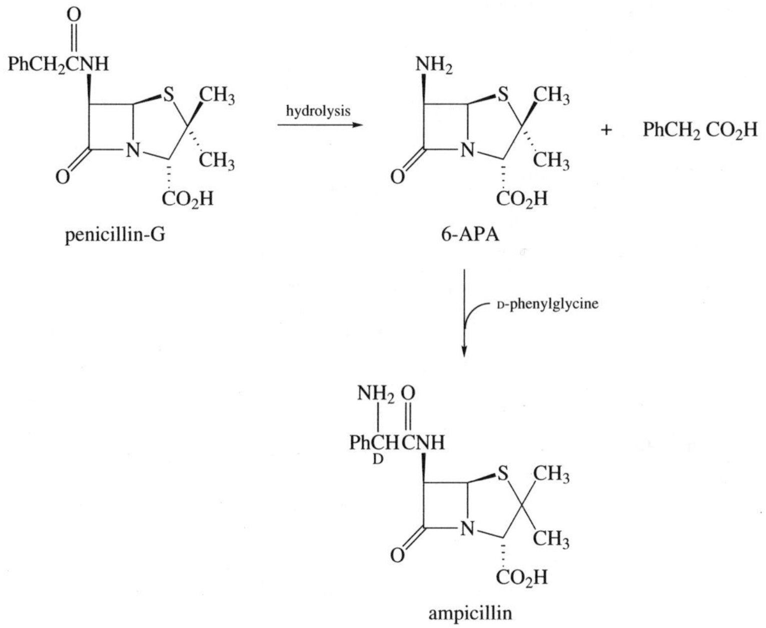

Fig. 14.5

Taxol has become a compound of considerable interest because of its potential as a cancer chemotherapeutic drug (Fig. 14.6). It is only available directly from the bark of the slow-growing Pacific yew (*Taxus brevifolia*): fortunately, the more prolific European yew (*Taxus baccata*) contains substantial quantities of 10-acetylbaccatin III in its leaves, from which taxol and its analogues can be made more readily.

In other cases a considerable number of chemical steps are needed to convert a cheap starting material into the expensive end-product. For instance, hecogenin, a constituent of the widely cultivated sisal plant, has been converted into betamethasone, a powerful steroidal anti-inflammatory agent (Fig. 14.7). Similarly, thebaine, a relative of morphine and isolated from the same source (the poppy *Papaver somniferum*) is used to make the powerful, non-addictive analgesic buprenorphine.

Thus, complex natural products that are readily available can be converted into useful compounds of the same structural type. On the other hand, simpler chiral compounds such as amino acids and sugars have been used to make a diverse range of structural types. For example, adrenaline receptor-blockers

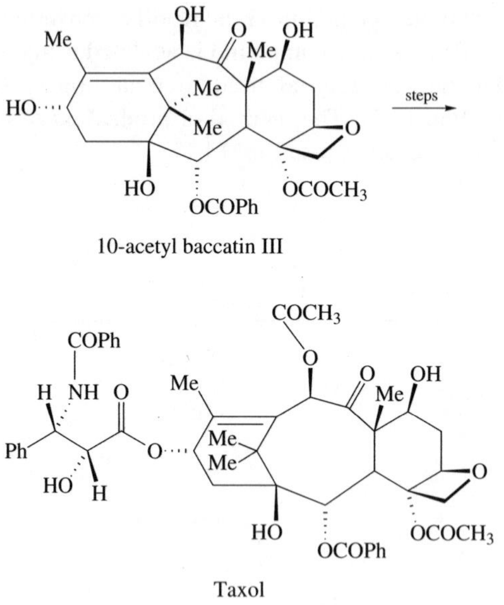
steps
10-acetyl baccatin III
Taxol

Fig. 14.6

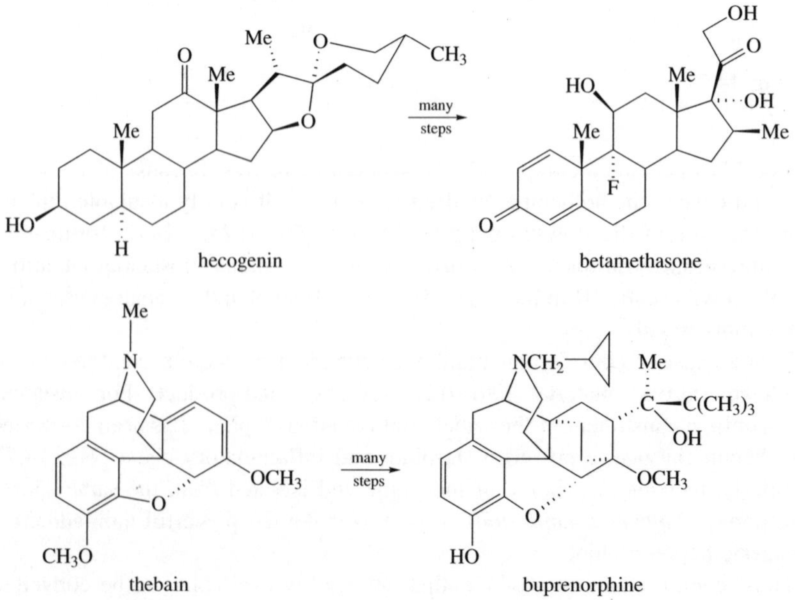
many
steps
hecogenin
betamethasone
thebain
buprenorphine

Fig. 14.7

(Chapter 13) such as propranolol or practolol (used clinically to treat patients suffering from angina and hypertension) are easily made from glycerol (Fig. 14.8).

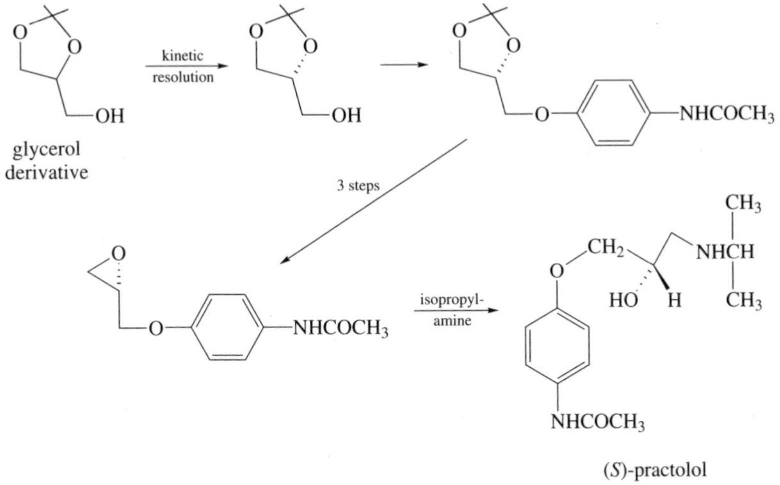

Fig. 14.8

L-(–)-Threonine, a naturally occurring amino acid, has been converted into the antibiotic thienamycin, a bacteriocide with powerful activity against Gram-positive and Gram-negative bacteria. It is noteworthy that the key elements of stereochemistry in and around the four-membered β-lactam ring are set up in the first steps of the sequence. Since this sequence illustrates nicely some of the stereochemical principles discussed in this book, it is worth spending some time on looking at the synthetic route in detail. Thus diazotization of the amino acid (**1**) in the presence of HBr leads to the formation of the bromohydrin **2** *via* neighbouring group participation of the carboxy group (Fig. 14.9). Base treatment of the bromohydrin **2** gives the oxirane **3** *via* an intramolecular S_N2 reaction. Conversion of the carboxylic acid moiety in compound **3** into an amide unit gives compound **4**. Abstraction of a proton from the active methylene group leads to an S_N2 reaction and formation of the β-lactam **5** *via* a 4-*exo-tet*-process. The lactam **5** is converted into the final product without affecting the stereochemical features.

Glucose is readily available, of course, and has been used as a starting material in a number of synthetic routes. Figure 14.10 describes the use of this common carbohydrate in the preparation of deoxymannojirimycin. Deoxymannojirimycin is a piperidine derivative and is an inhibitor of an enzyme that trims sugar chains attached to the surface of cells. The compound may have important implications for both anticancer and antiviral chemotherapy. The synthetic sequence may look a bit daunting at first but most of it does involve quite simple chemistry as follows.

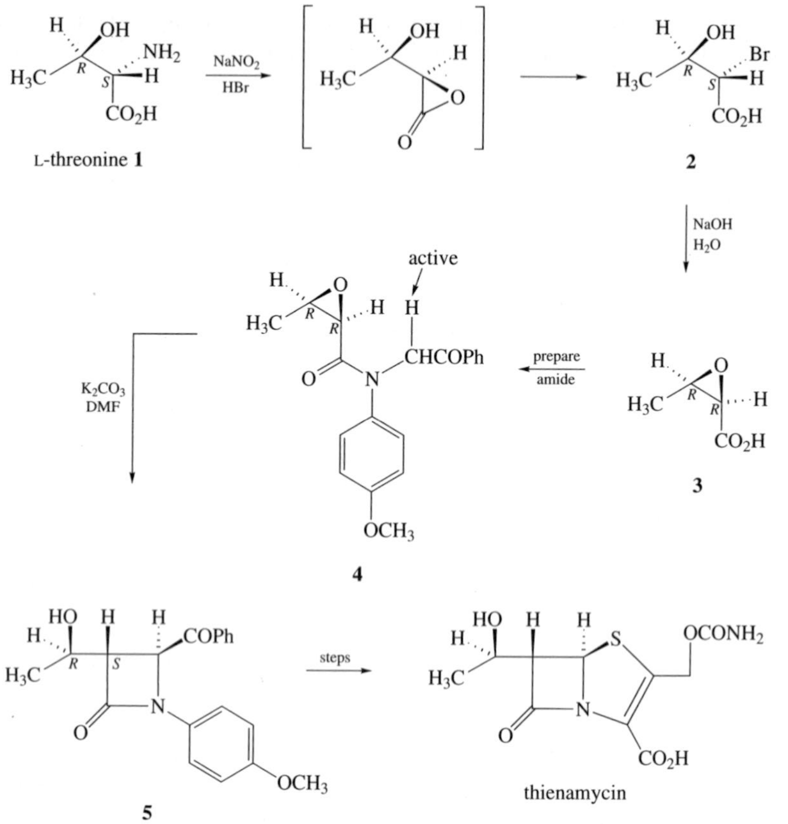

Fig. 14.9 Synthesis of thienamycin from L-threonine.

Under acid catalysis glucose reacts with acetone *via* the furanose form to give the glucose diacetonide **6** (Chapter 3). Under mild conditions the acetal unit at C5, C6 can be hydrolysed selectively, leaving the remaining protecting group spanning the hydroxy groups attached to C1 and C2. Reaction with methane-sulfonyl chloride involves only the less hindered primary hydroxy group at C6; the hydroxy groups at C3 and C5 are then protected with two benzyl groups. Displacement of the methane sulfonate unit using azide ion is followed by methanolysis of the residual acetal group. The hydroxy group at C2 is activated towards S_N2 substitution by derivatization as the trifluoromethanesulfonate; reduction of the azide group to the corresponding amine using triphenyl-phosphine is followed by spontaneous nucleophilic attack at C2 to afford the bicyclic compound **7**. Protection of the secondary amine by benzyloxycarbonyl-ation precedes acid-catalysed cleavage of the acetal group. Reduction of the aldehyde unit and deprotection leads to the target molecule.

Three of the four stereogenic centres in deoxymannojirimycin (*) are present in the correct configuration within glucose: the fourth centre (†) is set up by an S_N2 reaction.

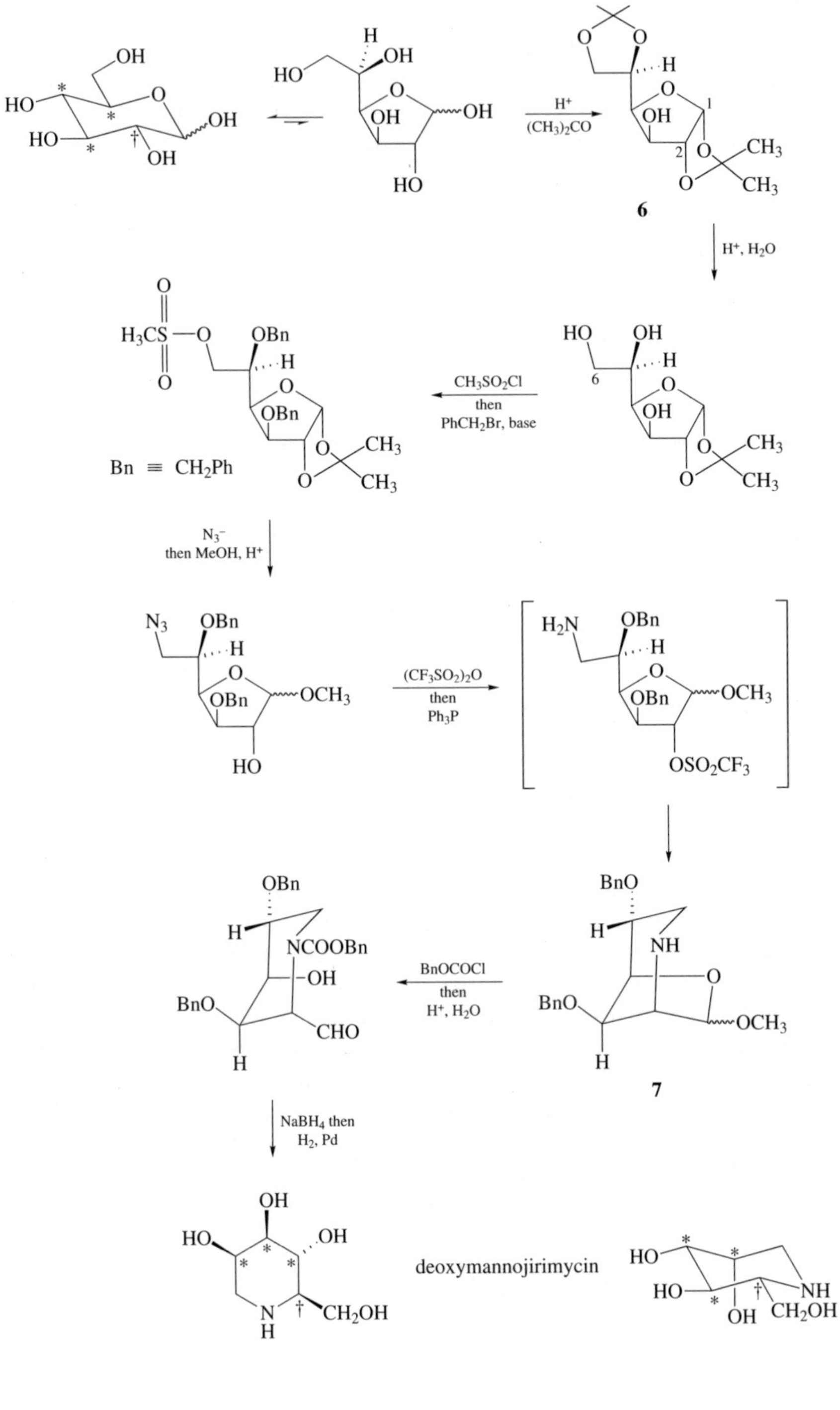

Fig. 14.10 Synthesis of deoxymannojirimycin from glucose.

Q. 14.1 The carboxylic acid unit participates, as a neighbouring group, in the conversion of compound **1** into compound **2** (Fig. 14.9). Neighbouring group participation is well established for other types of functional group. Consider how a neighbouring group effect could explain the high-yield conversion of compound **A** into compound **B**.

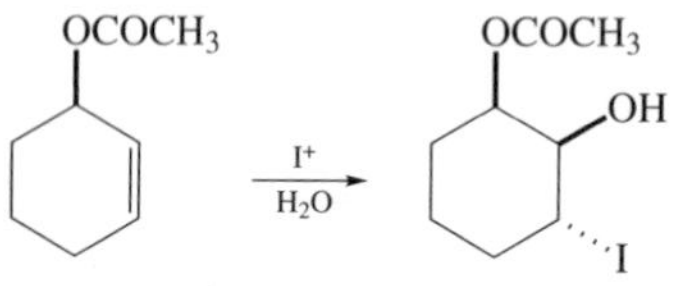

A   **B**

Q. 14.2 Label the stereocentres in glucose marked * in Fig. 14.10 and the corresponding centres (*) in deoxymannojirimycin as (R) or (S) using the Cahn–Ingold–Prelog sequence rules.

14.3 Using classical resolution techniques

The most common methods for the production of optically pure compounds, particularly on a large scale, involve the resolution of racemates. Direct crystallization of an optically active compound from a racemate can be achieved if crystals of the latter form as conglomerates [that is, the (+)-enantiomer preferentially associates with other (+)-isomers while the (−)-enantiomer also preferentially associates with its own kind]. Racemic sodium ammonium tartrate is a conglomerate and this allowed Pasteur to pick out crystals of the (+)- and (−)-enantiomers because the crystals also have an enantiomeric relationship. However, it is not possible to use such a painstaking method for the isolation of large quantities of material. Fortunately other methodology is available. For example, α-methyl-3′,4′-dihydroxy-L-phenylalanine (α-methyl-L-dopa) (Fig. 14.11) is obtained in optically active form by circulating a supersaturated solution of the racemate through two crystallization chambers that contain seed crystals of the respective enantiomers (Fig. 14.12).

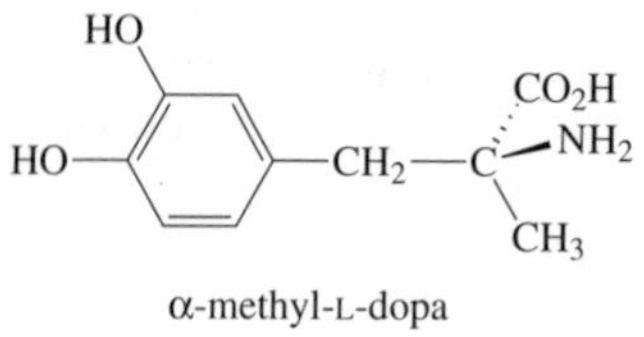

α-methyl-L-dopa

Fig. 14.11

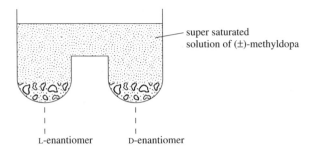

Fig. 14.12 Diagrammatic representation of the resolution of racemic methyldopa.

A related method involves crystallizing of one of the enantiomers of a racemate, while keeping the other in a supersaturated state. This process has been used for the preparation of large quantities of the antifungal agent chloramphenicol (Fig. 14.13).

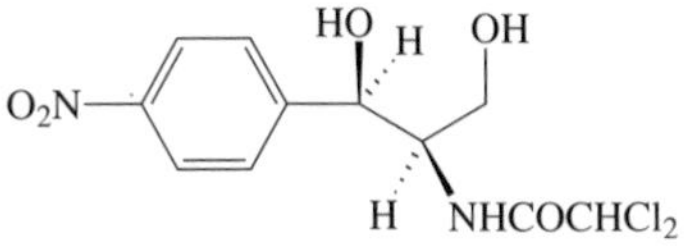

chloramphenicol

Fig. 14.13

In the technique known as 'crystallization induced asymmetric transform-ation' the target molecule should have a racemizable chiral centre. The Merck Company (New Jersey, USA) applied this methodology to synthesize an optically active 3-aminobenzodiazepinone (Fig. 14.14). (Benzodiazepines are famous for their sleep-inducing and anxiety-relieving properties.) Thus, the amine **8** was reacted with (+)-camphorsulfonic acid [(+)-CSA] in the presence of 3 mole per cent of an aldehyde such as benzaldehyde. The *S*-enantiomer of the 3-aminobenzodiazepinone forms a less soluble salt with (+)-CSA than does the *R*-enantiomer, and this salt then crystallizes out of solution. Equilibration of the *S*- and *R*-enantiomers of the benzodiazepinone is aided by the reversible formation of a Schiff's base with the aldehyde which makes the proton next to the carbonyl group of the amide moiety more acidic. Racemization of the Schiff's base **9** is effected by a small amount of the free amine. As a result of this process the desired product was obtained essentially optically pure (>99% e.e.) in 92% yield.

To resolve a racemate through diastereoisomer formation two principle conditions should be fulfilled:

(a) The resolving agent should be inexpensive, easily recovered without loss of chiral integrity, preferably available as both enantiomers and should react readily and completely with the racemate.

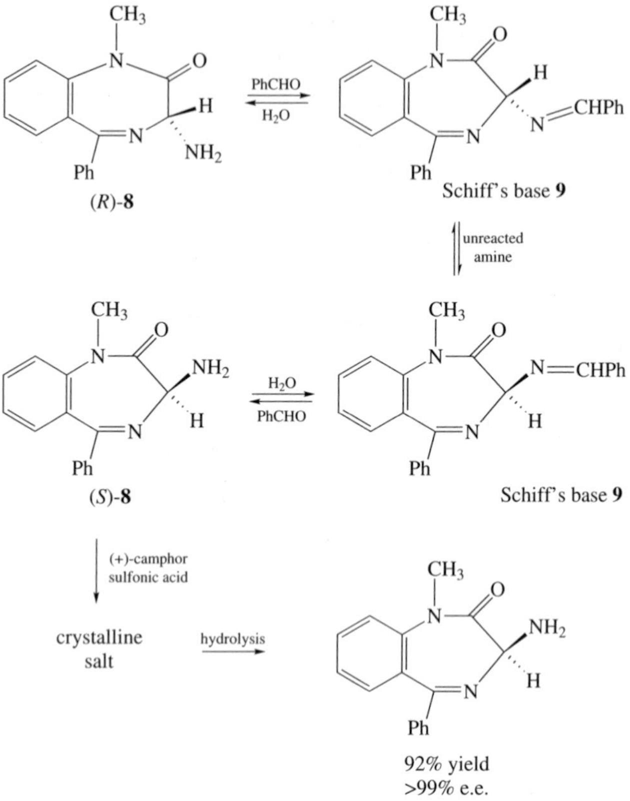

Fig. 14.14

(b) The diastereoisomeric products should be easily separated by crystallization, distillation or chromatography.

The most common method of resolving a racemate [for example, a racemic acid (±)-**A**] requires combination with a chiral reagent [such as an optically active base (+)-**B**]. Diastereoisomers are formed and then separated. The separated diastereoisomers (in this case the salts) are then decomposed and separation of the enantiomers has been effected. This method was pioneered by Pasteur in 1853, who found that (±)-tartaric acid could be resolved using optically active, naturally occurring bases (alkaloids) (Fig. 14.15).

Similarly, (±)-*N*-benzyloxycarbonylalanine (Z-Ala) can be reacted with (1*R*,2*S*)-(−)-ephedrine (Fig. 14.16) to give (−)-Z-Ala·(−)-ephedrine as the less soluble salt. Employment of (+)-ephedrine as the resolving agent gives (of course! see Chapter 7, question 7.3) (+)-Z-Ala·(+)-ephedrine as the less soluble salt. Since, from a practical point of view, it is less easy to obtain the more soluble salt in a diastereoisomerically pure state, the availability of both enantiomers of the resolving agent is advantageous.

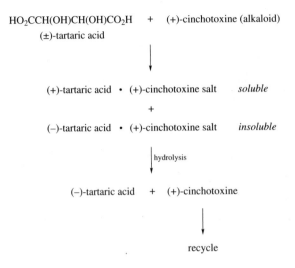

HO$_2$CCH(OH)CH(OH)CO$_2$H   +   (+)-cinchotoxine (alkaloid)
(±)-tartaric acid

(+)-tartaric acid · (+)-cinchotoxine salt   *soluble*

+

(−)-tartaric acid · (+)-cinchotoxine salt   *insoluble*

hydrolysis

(−)-tartaric acid   +   (+)-cinchotoxine

recycle

Fig. 14.15

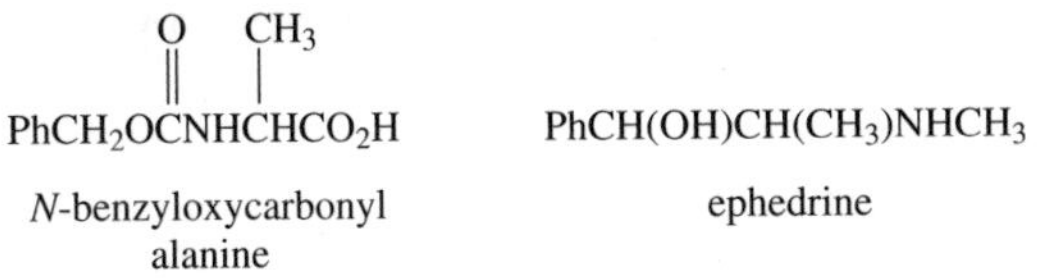

O   CH$_3$
‖ |
PhCH$_2$OCNHCHCO$_2$H   PhCH(OH)CH(CH$_3$)NHCH$_3$

N-benzyloxycarbonyl   ephedrine
alanine

Fig. 14.16

Racemic bases may be resolved in a similar way using acidic resolving agents. For example, the Andeno Chemical Company have developed such a process for the production of the unnatural amino acid D-(+)-phenylglycine (PG) (Fig. 14.17), using (+)-camphor-10-sulfonic acid, on a scale of 1000 tons per annum. Easy racemization of the unwanted enantiomer is often a requirement for the economic viability of such a process. The large quantity of D-phenylglycine is needed for coupling to 6-aminopenicillanic acid *en route* to ampicillin (Fig. 14.5).

The formation of diastereoisomers by covalent bond formation was discussed earlier: the resolution of racemic butan-2-ol by forming a diastereoisomeric mixture of esters using 2-chloropropanoic acid followed by fractional distillation or chromatography, then hydrolysis, was described in Chapter 2. However, you should be aware that fractional crystallization of diastereoisomeric salts is easier in practice than fractional distillation or chromatography, particularly on a large scale.

If the compound to be resolved does not have an acidic or basic functional group as a 'handle' then it is often helpful to introduce one. For example, oct-1-yn-3-ol is resolved by forming a hemiphthalate ester and using optically active phenethylamine (Fig. 14.18). Using a similar strategy, the ketone

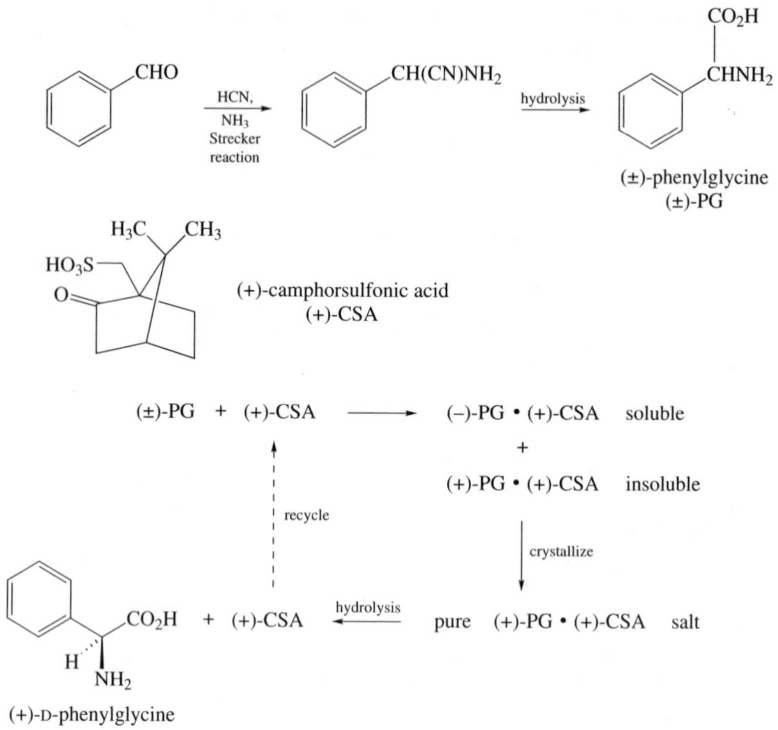

Fig. 14.17

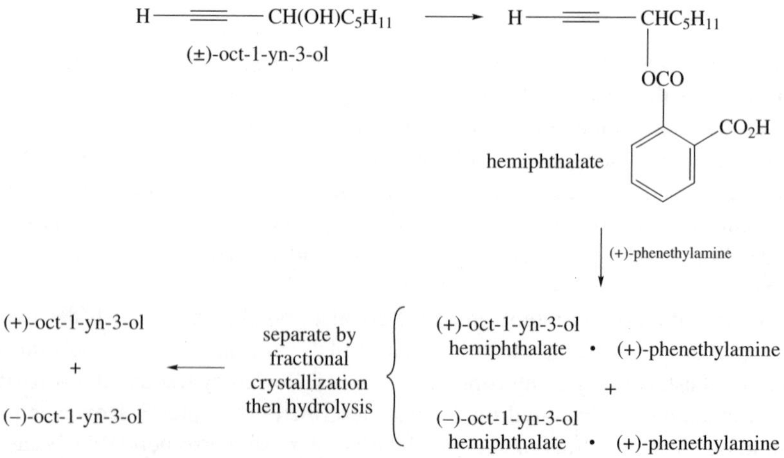

Fig. 14.18 Resolution of oct-1-yn-3-ol.

bicyclo[3.2.0]hept-2-en-6-one (Fig. 14.19) has been resolved by the bisulfite addition compound and using optically active phenethylamine as the resolving agent.

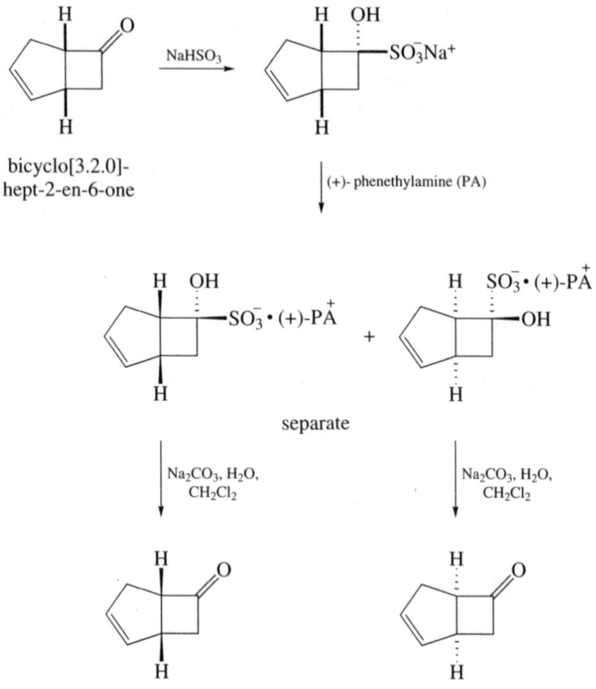

Fig. 14.19 Resolution of bicyclo[3.2.0]hept-2-en-6-one.

Q. 14.3 Shown below are a series of common resolving agents **A–E**. Indicate which one(s) is (are) suitable for the resolution of racemic carboxylic acids, amines, ketones, dienes.

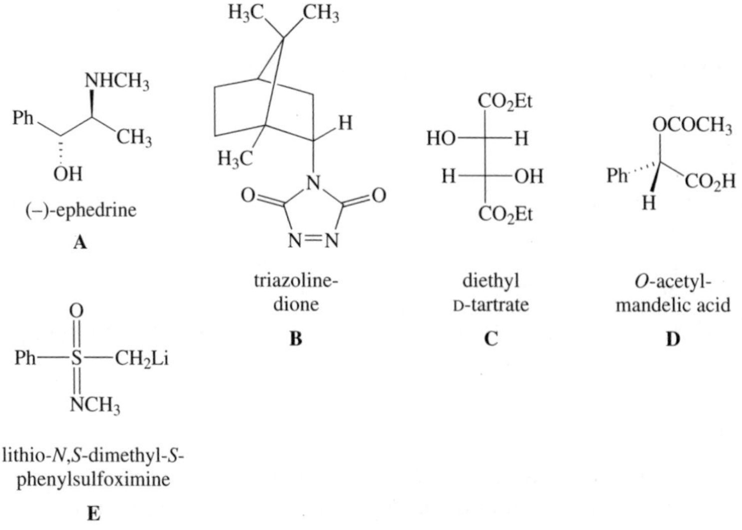

You should remember that enantiomers can be separated by chromatography (Chapter 6). As different types of chromatography columns with chiral stationary phases become cheaper and more readily available, their use on a preparative scale has become more commonplace. High pressure liquid chromatography (HPLC) over columns containing cyclodextrins (see Chapter 13) or amino acids such as phenylglycine attached to the solid support have become popular.

Answers to questions

Q. 14.1　The acetate acts to stabilize the carbocation at the adjacent position; attack by water then leads to the observed disposition of substituents.

NGP = neighbouring
　　　group participation

Q. 14.2　In glucose the configurations are as follows:
　C3 – S
　C4 – R
　C5 – R
In deoxymannojirimycin:
　C3 – R
　C4 – R
　C5 – R

The change in stereodescriptor is due to the replacement of oxygen by nitrogen at C2.

Q. 14.3

A is used to resolve racemic acids such as *N*-acetylamino acids.
B is used to resolve chiral dienes by a Diels–Alder reaction using the triazolinedione as the dienophile.

C is used to resolve ketones by intermediate formation of diastereoisomeric acetals.

D is used to resolve amines through formation of a diastereoisomeric salt.

E is the least obvious reagent. It has been used to resolve ketones, for example 2-substituted cyclohexanones:

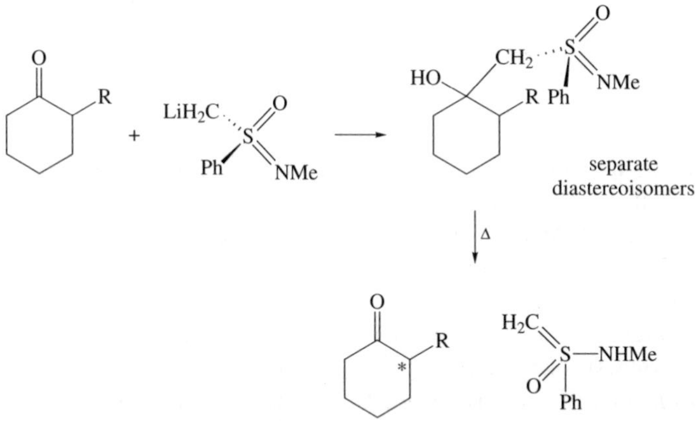

The structure of the ketone must be such that the addition of the organometallic reagent should occur from only one face, to avoid unwanted mixtures on creation of the third stereogenic centre. 2-*tert*-Butylcyclohexanone has been resolved in this way.

15 Asymmetric synthesis

The preparation of chiral compounds has become one of the most important and intellectually challenging areas of organic synthesis, not least in the pharmaceutical industry. As touched upon earlier (Chapter 14), biological systems are very sensitive to the chirality of substances with which they interact and so practical methods for the production of single stereoisomers are of great value in terms of higher yields, fewer waste products and ease of purification. For example, the synthesis of a chiral drug in its racemic form is not efficient if only one stereoisomeric form is active, since 50% of the final product will be useless and may require further treatment in the form of separation, possible recycling or safe disposal.

Asymmetric synthesis is the name given to that area of organic chemistry in which the preparation of specific stereoisomers is targeted and which doesn't rely on the elimination of one of a pair of stereoisomers or on the isolation of one stereoisomer from a mixture of several. There are many research groups active in this area and the wealth of asymmetric synthesis methods described in the chemical literature is beyond the scope of this book. However, in this chapter we outline some of the more popular strategies for the stereocontrolled preparation of some chiral compounds, and discuss the events that occur to make this stereocontrol possible.

The three aspects of asymmetric synthesis considered here are the use of chiral auxiliaries, use of chiral reagents and use of chiral ligands and catalysts.

15.1 Chiral auxiliaries

A chiral auxiliary is an agent that facilitates the transformation of an achiral starting material into a chiral product. During this transformation the auxiliary becomes chemically bound to the achiral substrate, forming a chiral intermediate, but does not appear in the final product. However, during the time that the auxiliary is attached it directs the course of further reactions by making available one preferred pathway for attack by the added reagents.

The general sequence of events is given in Fig. 15.1 where the achiral starting material is a carboxylic acid, the chiral auxiliary is represented by A*, E is the reagent and C* is the new stereogenic centre. The two features that characterize an ideal chiral auxiliary are (i) a functional group that permits the ready attachment and removal of the auxiliary before and after the stereocontrolled reaction, respectively, and (ii) a sterically demanding structure. In practice, it is the latter condition that provides the element of stereocontrol and which is best illustrated by considering specific examples. Two such examples are presented here: C2-functionalization of propanoic acid and a Diels–Alder reaction.

$$RCH_2CO_2H \xrightarrow[\text{auxiliary}]{\text{append}} RCH_2COA^* \xrightarrow[\text{(ii) E}^+]{\text{(i) base}} R\underset{\underset{E}{|}}{C}{}^*HCOA^* \xrightarrow[\text{auxiliary}]{\text{remove}} R\underset{\underset{E}{|}}{C}{}^*HCO_2H \;+\; A^*$$

Fig. 15.1

The C2 functionalization of propanoic acid is a seemingly simple reaction. However, it can be used to illustrate the subtleties involved in the use of different chiral auxiliaries. The order of events is as follows: either the chiral auxiliary is attached to the propanoic acid by reacting with the carboxy group or the propanoic acid unit is constructed on the auxiliary. The resulting acyl derivative, on treatment with a base, undergoes enolization and the enolate then reacts in a stereocontrolled manner with an electrophile to give a new stereogenic centre at C2.

15.1.1 Using an iron carbonyl as a chiral auxiliary: synthesis of Captopril

Many sufferers of hypertension are likely to be familiar with the name Captopril. It is a widely prescribed drug which effectively relieves the symptoms of high blood pressure. Its mode of action is to prevent the formation of angiotensin II, a protein that occurs in mammalian cells and which is a potent vasoconstrictor.

The structure of Captopril consists of an *N*-acylated pyrrolidinecarboxylic ester (Fig. 15.2) and contains two stereogenic centres both of which are *S*. One synthesis of Captopril involves the use of a cyclopentadienyl(triphenylphosphine)iron complex on which is attached the propanoic acid unit. Butyl lithium is then added and transforms the acyl carbonyl group into the corresponding lithium enolate. The geometry about the double bond is *E*, that is to say the methyl group is on the same side as the enolate oxygen. From Fig. 15.2 you should be able to see that the iron complex also fulfils the second criterion of being sterically demanding: the ligands around the iron are in fixed positions and particularly important is one of the phenyl groups of the PPh$_3$ substituent. This group effectively blocks the approach of the electrophile (in this case *tert*-butylsulfanylmethyl bromide) from the *Re* face of C2*. Therefore, the reaction has to take place at the *Si* face, during which the bromine atom is displaced from the electrophile and a new *R* stereogenic centre[†] is created.

---

*For a recap of the usage of *Re* and *Si* for trigonal centres, see Chapter 8.
[†]At this point the stereogenic centre is *R* but it becomes *S* when the chiral auxiliary is removed. This is because the former contains iron (of high atomic number) which is later removed.

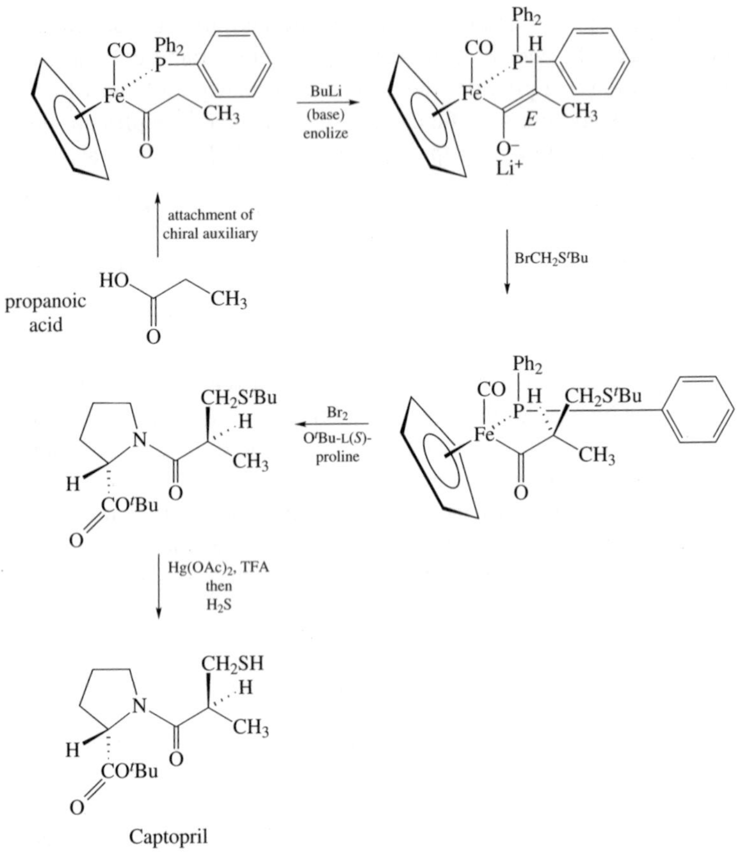

propanoic acid

Captopril

Fig. 15.2

The second stereogenic centre is introduced by displacing the chiral auxiliary with *O-tert*-butylproline. The latter reagent is an easily obtained, naturally occurring α-amino acid derivative, whose single stereogenic centre occurs exclusively in the *S* form. The cyclic amide thus formed now has the two requisite *S* stereogenic centres and requires only the removal of the *tert*-butyl group attached to the sulfur, using standard methodology, to become Captopril.

15.1.2 Using heterocycles as chiral auxiliaries: alkylation and hydroxyalkylation of propanoic acid

(*S*)-Prolinol is a useful chiral auxiliary. It reacts with activated propanoic acid to form the amide shown in Fig. 15.3. Treatment of this amide with the base lithium diisopropylamide (LDA) causes the carbonyl group to form the (*Z*)-enolate at the same time as the hydroxylic hydrogen is replaced by a lithium atom. The result is a complex in which the hydroxymethyl group of the chiral auxiliary acts as an anchor maintaining the enolate structure by interacting, or chelating, in its lithiated form with the enolate OLi group.

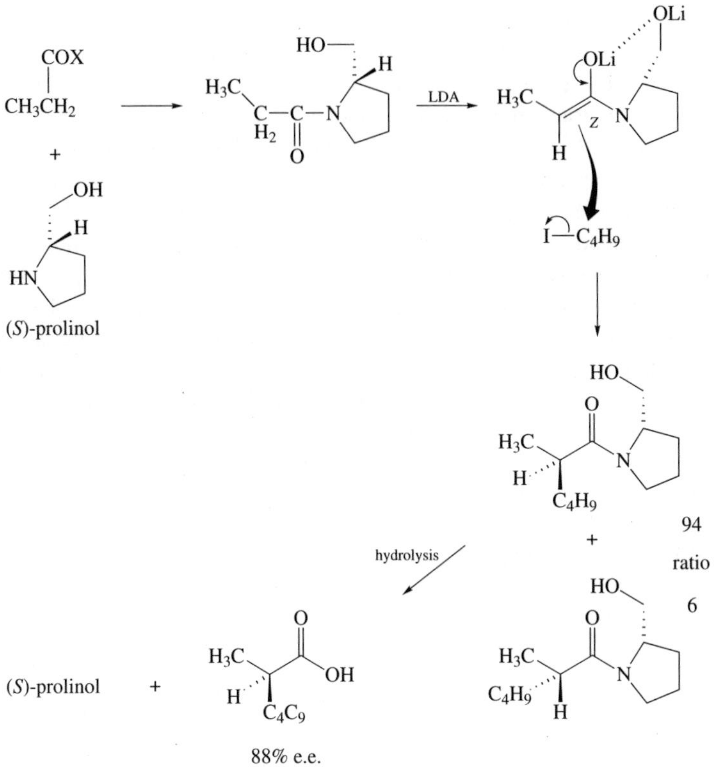

Fig. 15.3

The structure is now set up to undergo electrophilic addition, in this example with butyl iodide. The stereochemical outcome of the entire transformation hinges on which side of the double bond the electrophilic attack takes place. Examination of both faces of C2, the site of the electrophilic attack, will show that the approach to the *Re* face is hindered by the presence of the chiral auxiliary. On the other hand, there is nothing to prevent the electrophile approaching the *Si* face which is what actually happens. C2 is then butylated and a new stereogenic centre (*R*-configuration) is created. It should be emphasized, however, that this reaction is not 100% stereoselective; although approach of the electrophile to the *Re* face is hindered, it is not entirely prevented and there is a small degree of reaction by this route leading to the *S*-product. In practice, the ratio 2*R*:2*S* is 94:6.

The high degree of stereocontrol of this reaction, therefore, emanates from the chirality of the auxiliary. If (*R*)-prolinol were used as the chiral auxiliary the hydroxymethyl group would point in the opposite direction, that is, it would hinder the approach of the electrophile to the *Si* face and force it to approach the *Re* face thus leading to the 2*S*-product as the major diastereoisomer.

The final step is hydrolysis, which regenerates the chiral auxiliary (which can be recycled) and gives the final product, (*R*)-2-butylpropanoic acid [(*R*)-2-

methylhexanoic acid is the conventional name]. The overall reaction gives an enantiomeric excess (Chapter 6) of 88%.

4-Isopropyloxazolidin-2-one is also a useful chiral auxiliary. In alkylation reactions it functions in much the same way as prolinol. Taking (*S*)-4-iso-propyloxazolidin-2-one as an example, we see that it contains appropriate functionality to become attached to propanoic acid, *via* the nitrogen (Fig. 15.4), and, as with the previous example, treatment with a lithium base (LDA or BuLi) leads to the formation of a lithium enolate. The geometry about the double bond is again *Z* (the methyl group on the same side as the enolate oxygen) and the chelation is very similar, being this time between the enolate oxygen, the lithium and the ring carbonyl oxygen. The isopropyl group at C4 blocks the approach of the alkylating agent (ethyl iodide) to the *Re* face of C2, leading to almost exclusive reaction at the *Si* face and concomitant creation of an *R* stereogenic centre (92% e.e.).

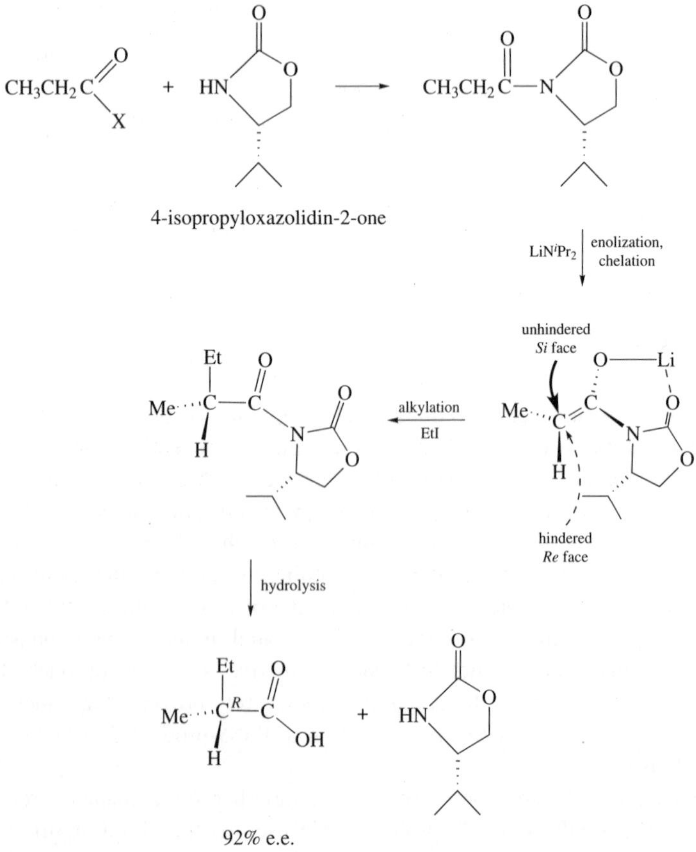

4-isopropyloxazolidin-2-one

Fig. 15.4

Interestingly, the same chiral auxiliary on the same substrate causes a reversal of the above product stereochemistry when the electrophile is changed to an aldehyde instead of an alkyl halide. Figure 15.5 illustrates the C2-hydroxy-methylation of propanoic acid and the chelation agent necessary for this is a dialkylborane moiety (BBu_2). The alkylborane serves not only as a chelating reagent for the enolate, but also as an activating agent for the aldehyde.

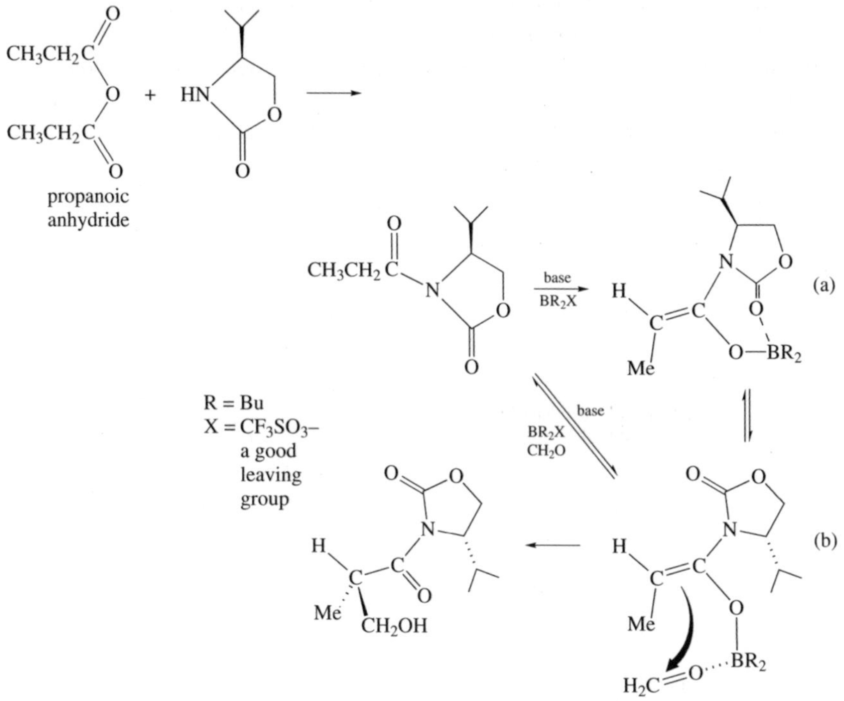

Fig. 15.5

The enolized structure can adopt one of two possible conformations. One, denoted (a) in Fig. 15.5, is similar to that described in the preceding example, in that the boron becomes covalently bound to the enolate oxygen and chelates with the ring carbonyl oxygen. This is a poor arrangement in terms of further reaction, however, as aldehydes are insufficiently activated to bring about electrophilic attack and the boron is unable to assist; its activating capability is taken up by the chiral auxiliary. An equilibrium exists, however, between this and the arrangement denoted (b) in which the boron is released from chelation with the auxiliary and becomes free to activate the aldehyde sufficiently to bring about reaction.

The stereochemical significance of this transformation is profound: once the chelation between the boron and ring carbonyl is broken, unfavourable dipole–dipole interactions cause the oxazolidinone to rotate about the C—N bond so as to afford maximum separation between the O-BR$_2$ and the ring amide carbonyl group. The result of this is that, unlike the alkylation example (which requires no additional activation) it is the *Si* face of C2 that is protected from electrophilic attack by the isopropyl group and reaction therefore takes place at the *Re* face and the new stereogenic centre is *S*.

Further complications arise when the aldehyde is one other than formaldehyde. When, for example, the reaction is carried out with benzaldehyde, the final product is 2-hydroxy-2-phenylpropanoic acid (Fig. 15.6). The transformation from starting material to final product has therefore created two new stereogenic centres. Of the four possible stereoisomers (*RR, RS, SR, SS*) this elegant reaction system gives predominantly only one and so a little further exploration of the reaction mechanism is worthwhile here.

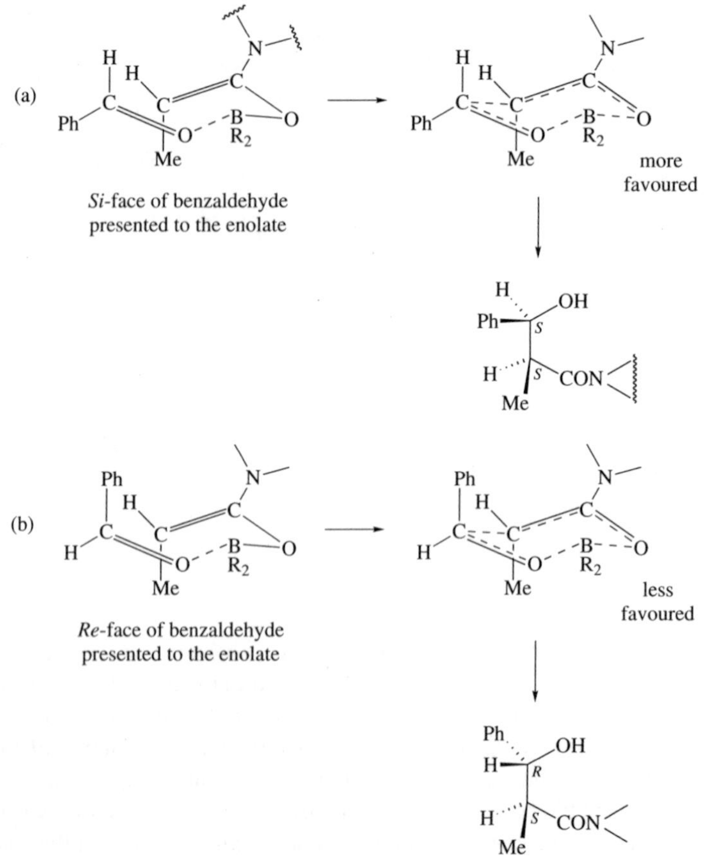

Fig. 15.6

Returning to our reaction and Fig. 15.6, we know that the C2 *Si* face is effectively blocked by the isopropyl group of the chiral auxiliary and so the benzaldehyde, coordinated to the boron, is directed towards the C2 *Re* face. However, the carbonyl group of benzaldehyde is also stereoheterotopic: it has an

Re and an *Si* face. There are therefore two possible approaches of the benzalde-hyde to the enolate: one in which the *Re* face is presented for reaction and the other in which the *Si* face is presented. The deciding factor is what happens to the phenyl group in the transition state.

The proposed transition state for this reaction is a loose 'chair' with the six members comprising the boron, C1, C2 and the oxygen of the enolate, and the carbon and oxygen of the aldehyde, as shown in Fig. 15.6. The dotted lines out-line the chair shape and represent interactions that are sufficiently strong to maintain the transition state (Zimmerman–Traxler transition state) long enough to bring about the electrophilic reaction.

You will recall from Chapter 1 that the chair conformation is an energetically favourable arrangement of atoms for cyclohexane systems and this also holds true for transition states. Remember also that large bulky groups exert a prefer-ence for less crowded equatorial occupancy, and the reaction proceeds so as to place the phenyl group of benzaldehyde in an equatorial position. Inspection of Fig. 15.6 will show that this is achieved when the *Re* face of C2 reacts with the *Si* face of the benzaldehyde. Otherwise the corresponding *Re–Re* interaction leads to less preferred axial positioning of the phenyl group. The methyl group on C2 has to be axial in the transition state (also shown in Fig. 15.6) because of the *Z*-geometry of the double bond obtained during the prior enolization step. The two new stereogenic centres at C2 and C3 are *S* and *S*, respectively, and the final product, after hydrolysis, is (2*S*,3*S*)-3-hydroxy-3-phenylpropanoic acid. The chiral auxiliary can then be removed by hydrolysis to allow the isolation of the corresponding hydroxy acid.

Q. 15.1 In general, the reaction of carbonyl compounds $R^1COCH_2R^2$ with non-nucleophilic bases $LiNR^3_2$ gives a mixture of *Z*- and *E*-enolates, under kinetic control. *Z*-Enolates are favoured if R^1 is bulky (e.g. Ph). *E*-Enolates are favoured if R^3 is bulky (*sec*- or *tert*-alkyl). Given that the lithium cation transfers from nitrogen to oxygen and a hydrogen atom transfers from the methylene group to nitrogen, draw six-membered chair-like transition states to explain the observed results.

15.1.3 Camphor-derived chiral auxiliaries: Diels–Alder reaction

Camphor-based chiral auxiliaries are widely used in Diels–Alder chemistry, one reason for their usefulness being the fact that both geometrically possible stereoisomeric forms of camphor (Fig. 15.7) are readily available. The follow-ing example is the preparation of (*R*)-cyclohex-3-enecarboxylic acid in which the chiral auxiliary is a sultam derived from (1*S*,4*S*)-camphor.

Before we come to the part played by the chiral auxiliary, we should just remind ourselves of the basic reactions (Fig. 15.8). The dienophile acrylic acid undergoes cycloaddition to buta-1,3-diene to give a cyclohexene derivative

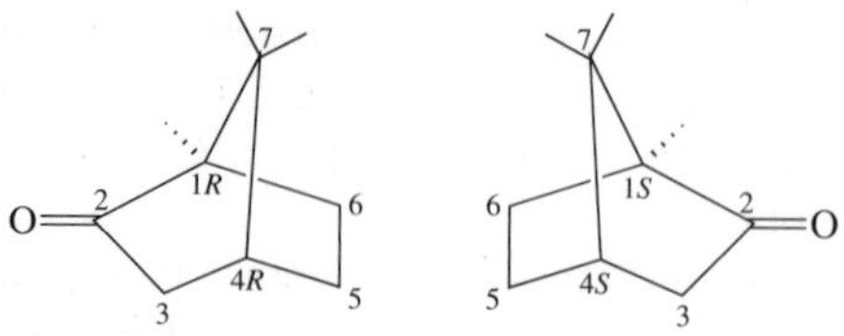

(1R,4R)-camphor   (1S,4S)-camphor

Fig. 15.7

containing one new stereogenic centre. Whether this stereogenic centre is *R* or *S* depends on which side of the dienophile the diene approaches: if it approaches the side C2*Re*, the new stereogenic centre will be *S*, whereas approach to the side C2*Si* will give an *R* stereogenic centre.

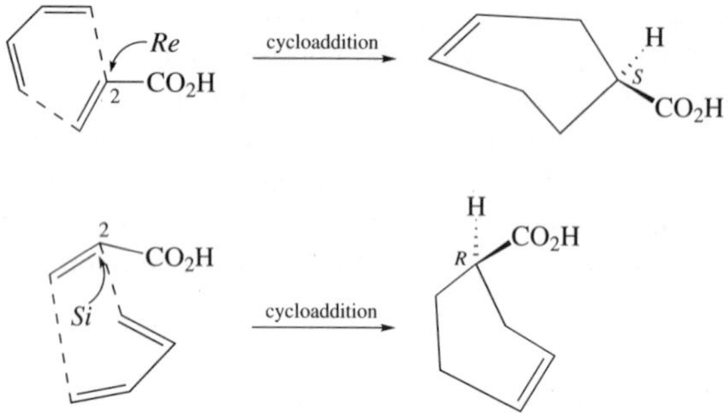

Fig. 15.8

What the chiral auxiliary does is to render one of these approaches to the diene virtually impossible and force the reaction to occur almost entirely from one face only. It does this by being attached, *via* a sultam nitrogen (Fig. 15.9), to the carboxy group of the dienophile. The conformation of the dienophile is then locked by the addition of ethylaluminium chloride to the reaction mixture. This aluminium reagent is a Lewis acid, which coordinates to the carbonyl oxygen of the dienophile and one of the sultam oxygens, thereby preventing rotation about the CO—N bond and at the same time activating the dienophile towards addition.

As you will see from Fig. 15.9, the sheer size and bulk of the camphor derivative causes virtually complete obstruction of the *Re* face of C2 and so the diene must instead approach the *Si* face. The ensuing cycloaddition gives a cyclohexene in which the stereogenic centre is *R* as predicted. Lithium hydroxide promoted hydrolysis of this initial product releases the chiral auxiliary, which can be used again and the final product is (*R*)-cyclohex-3-enecarboxylic acid. In fact, so effective is this chiral auxiliary, the diastereomeric excess of the initially formed adduct is 97%.

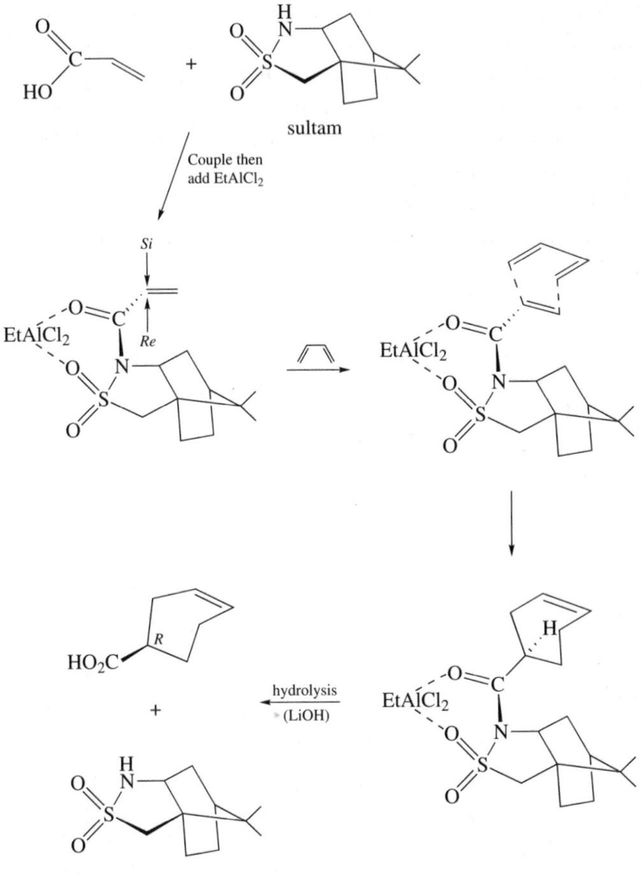

Fig. 15.9

15.2   Chiral reagents

An asymmetric reagent will often react with an achiral substrate to give a pre-
ponderance of one chiral product. Classic examples include the hydration of a
carbon–carbon double bond through asymmetric hydroboration and conversion
of a ketone into a chiral secondary alcohol (asymmetric reduction). We shall
focus on each of these reactions in turn.

15.2.1 *Asymmetric hydroboration reactions*

Before concentrating on asymmetric hydroboration we need to review, briefly,
the chemistry of boron hydrides. Alkenes react with borane (BH_3) and simple
alkylboranes (such as R_2BH; R = simple alkyl) through *cis*-addition of the boron

unit and the hydrogen atom across the carbon–carbon double bond (Fig. 15.10). As indicated in this figure, although the C—B bond and the new C—H bond are made simultaneously, formation of the C—B bond is slightly more advanced than the formation of the C—H bond, which accounts for the build up of partial positive charge at one of the carbon centres and overall anti-Markownikoff addition. Oxidation of the organoborane gives the corresponding alcohol (overall addition of H_2O). Borane can react with two molecules of alkene (Fig. 15.11); steric hindrance often makes the third hydroboration a relatively slow process.

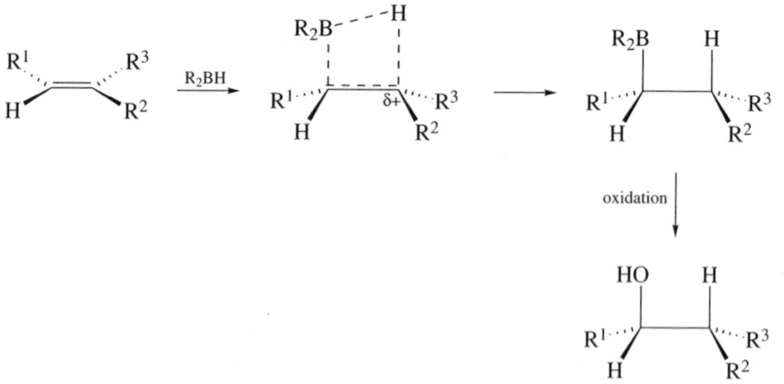

Fig. 15.10

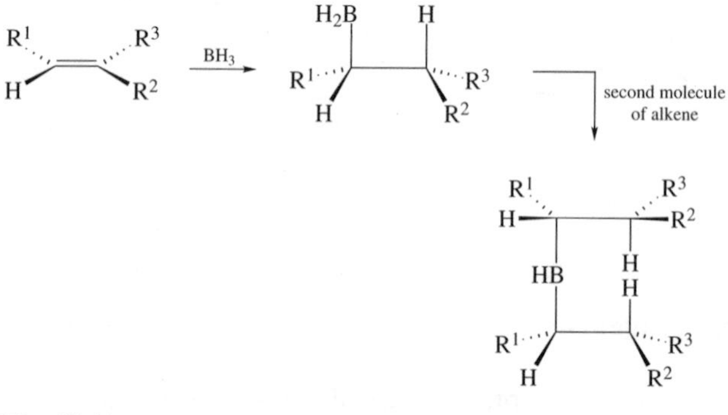

Fig. 15.11

Finally in this review section it should be noted that, for cyclic alkenes, face-selective hydration takes place, the reagent approaching the alkene towards the more exposed face (Fig. 15.12).

Armed with the above information it should be easy to understand the reaction of borane with the chiral, naturally occurring compound called pinene. Dipinanylborane (Fig. 15.13) is available in both enantiomeric forms and is a useful chiral hydroboration agent. For example (+)-dipinanylborane reacts with a prochiral diene such as a 5-substituted cyclopentadiene at the less hindered

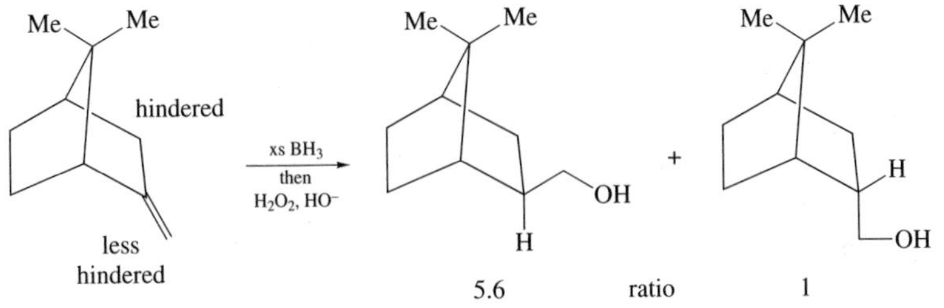

Fig. 15.12

face (Fig. 15.14) to give a mixture of two diastereoisomeric boranes in the ratio 98:2; treatment of the boranes with alkaline peroxide then gives the corresponding alcohols. The observed enantiomeric excess (96%) of the (*R*)-cyclopenten-3-ol arises from the difference in the energies of the diastereoisomeric transition states for the alkene addition step, the transition state leading to the *R*-alcohol being much lower in energy.

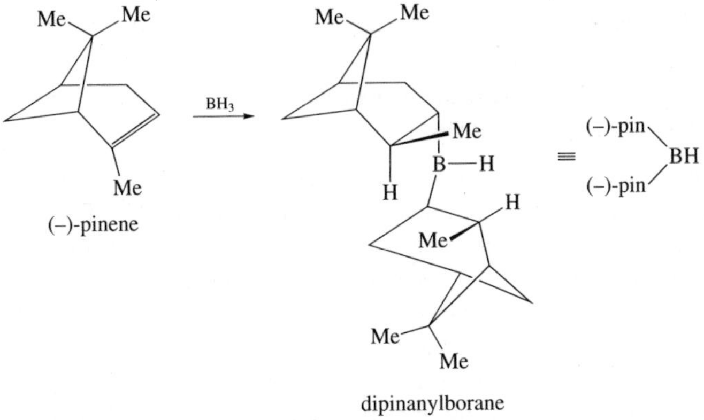

Fig. 15.13

Q. 15.2 The alkene **A** is hydroborated through the conformation shown below which minimizes unfavourable interactions between the alkene methyl group and the substituents at the allylic position. Predict the configuration of the product after hydroboration using BH$_3$ and oxidation.

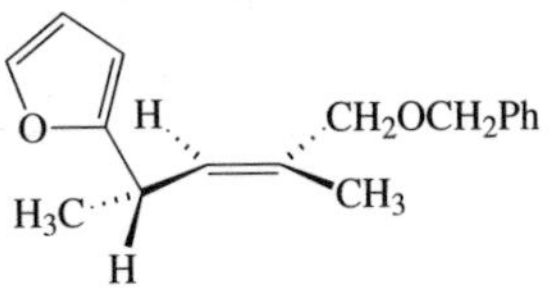

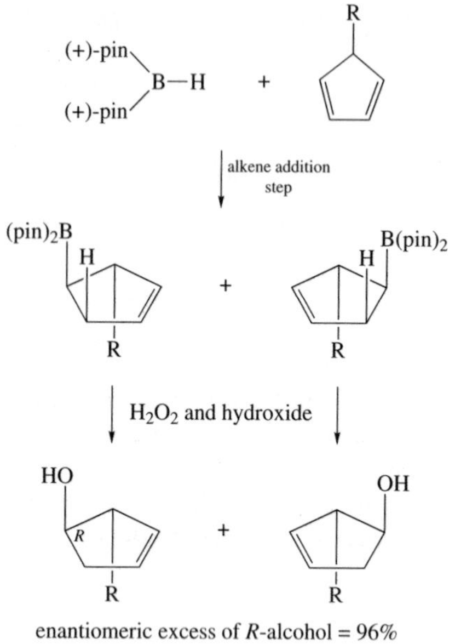

enantiomeric excess of R-alcohol = 96%

Fig. 15.14

15.2.2 *Asymmetric reducing reagents related to lithium aluminium hydride and sodium borohydride*

Reduction of an unsymmetrical ketone R^1COR^2 with sodium borohydride yields a racemic mixture of the secondary alcohol $R^1CH(OH)R^2$. This is because sodium borohydride is achiral and approach of the reductant to the carbonyl carbon atom from the *Re* or *Si* face is equally facile (Chapter 2). On the other hand, a chiral hydride donor reacts with a ketone R^1COR^2 through two transition states that have a diastereoisomeric relationship and hence different energies (Fig. 15.15). Since the donation of hydride ion is essentially irreversible this energy difference for the transition states is reflected in the different amounts of R- and S-alcohols formed in the process.

One of the most commonly used asymmetric reducing agents is 'BINAL-H' (Fig. 15.16), formally derived from equimolar quantities of optically pure binaphthol, ethanol and lithium aluminium hydride. The reagent is particularly well suited for the reduction of unhindered aralkyl ketones and acyclic conjugated enones. One example is shown in Fig. 15.16.

On the other hand, asymmetric reduction of simple ketones such as butanone and 4,4-dimethylpentan-2-one (DIMPO) is best achieved using (R,R)- or (S,S)-dimethylborolane (Fig. 15.17). In general the RR-reducing agent gives the R-alcohol and the SS-reagent gives the S-product.

15.3 Chiral catalysts

In Fig. 15.17 a stoichiometric amount of a borolane was used to effect the asymmetric reduction of a ketone. Further studies provided a catalytic version of this

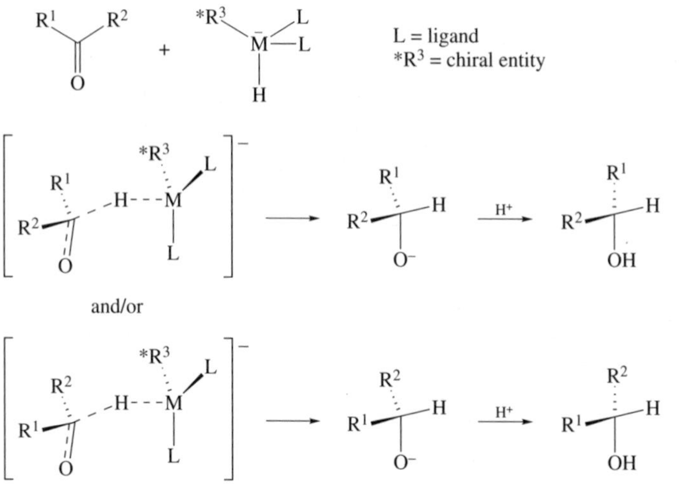
L = ligand
*R3 = chiral entity
and/or

Fig. 15.15

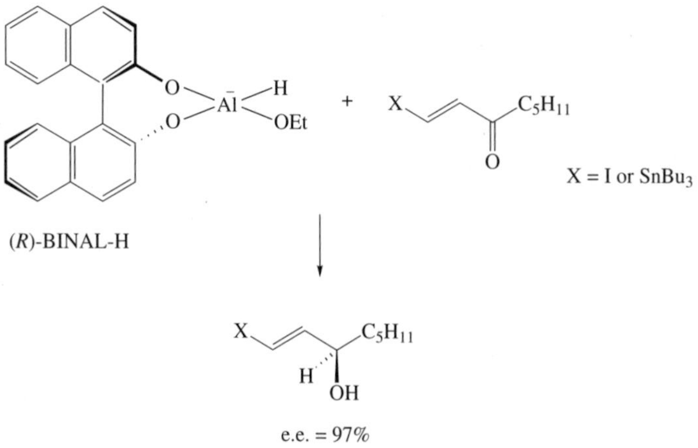
X = I or SnBu3
(R)-BINAL-H
e.e. = 97%

Fig. 15.16

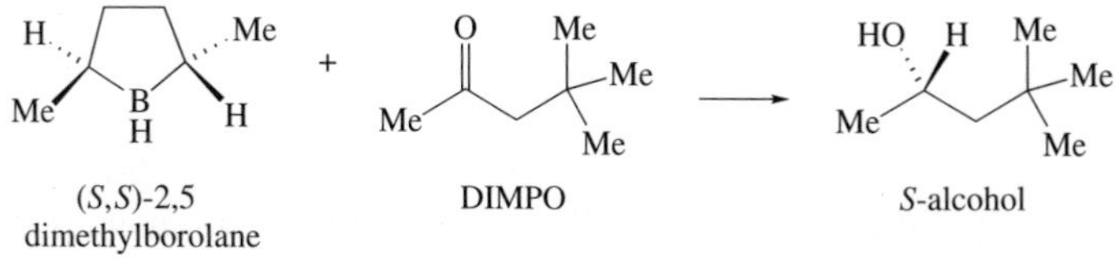
(S,S)-2,5
dimethylborolane
DIMPO
S-alcohol

Fig. 15.17

transformation. Thus a small amount of the oxazaborolidine **1** in conjunction with a stoichiometric amount of borane gives a method for the enantioselective reduction of a wide range of ketones, including alkyl aryl ketones and dialkyl

> **Q. 15.3** (*S*)-BINAL-H reduces acetophenone to give (*S*)-1-phenylethanol (95% e.e.). Given that (i) hydrogen atom transfer takes place from the aluminium centre to the carbonyl carbon atom and (ii) the lithium ion coordinates with the carbonyl oxygen atom and the oxygen of the ethoxy unit, draw a six-membered chair-like transition state which accounts for the observed stereochemistry of the major product.

ketones (Fig. 15.18). For example, acetophenone (PhCOCH$_3$) is reduced by 1/BH$_3$ to (*R*)-1-phenylethanol. The mechanism for this reaction is as follows: borane coordinates to the nitrogen atom in the catalyst (Fig. 15.18). The carbonyl oxygen atom of acetophenone then coordinates to the electrophilic boron and hydrogen atom transfer takes place through a six-membered transition state such that the larger substituent (Ph) attached to the carbonyl carbon atom is remote from the heterocyclic complex.

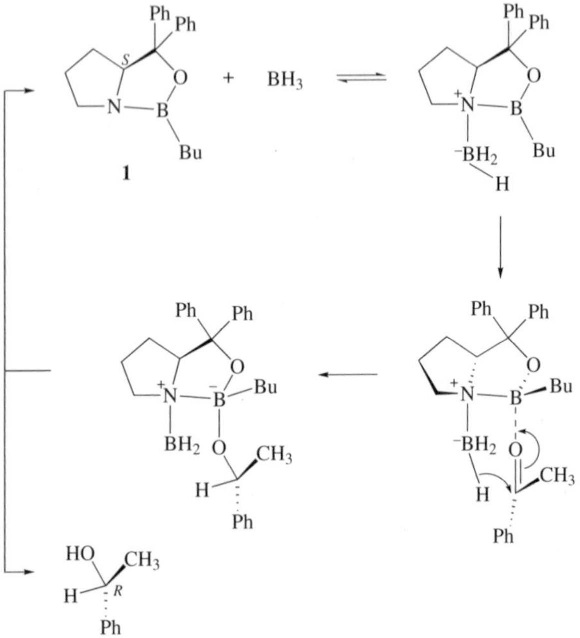

Fig. 15.18

The excellent catalytic properties of the oxazaborolidine additive lies in the fact that it brings the two reactants (BH$_3$ and the carbonyl compound) into close proximity in a stereocontrolled manner.

Two alkene oxidation processes, epoxidation and dihydroxylation (Fig. 15.19), provide examples of the use of other extremely important asymmetric catalysts.

Before launching into a discussion of the asymmetric epoxidation reaction some important background information is necessary. This oxidation reaction

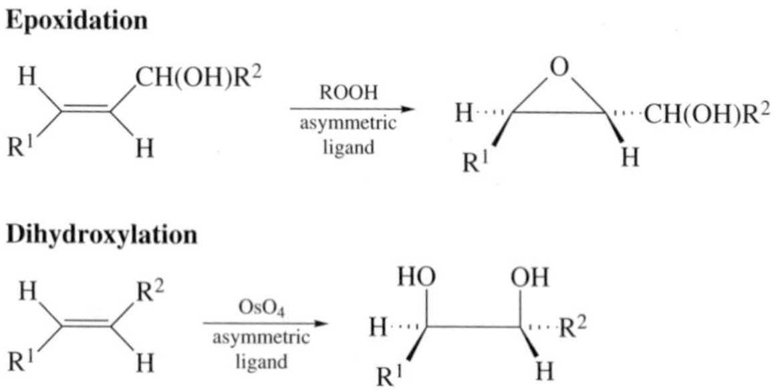

Epoxidation

Dihydroxylation

Fig. 15.19

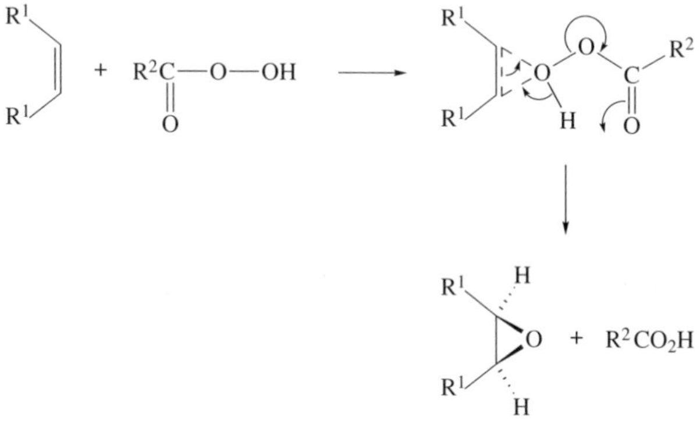

Fig. 15.20

extends the well known derivatization of an alkene to afford an epoxide, a transformation which can be accomplished using a peracid (Fig. 15.20; see also Chapter 10). In the case of chiral cyclic alkenes attack by the reagent will take place from the less-hindered face, *except* when a hydroxy group is directly adjacent to the olefin. Here, the OH acts as a neighbouring group which directs the reagent to the nearby face of the alkene through intermolecular hydrogen bonding (the Henbest effect). Figure 15.21 shows the effect of peracid on two 3-substituted cyclohex-1-enes. Epoxidation occurs on the opposite face to that occupied by the acetate group (for steric reasons) whereas the allylic hydroxy group guides the incoming oxidant to the nearby face of the alkene unit.

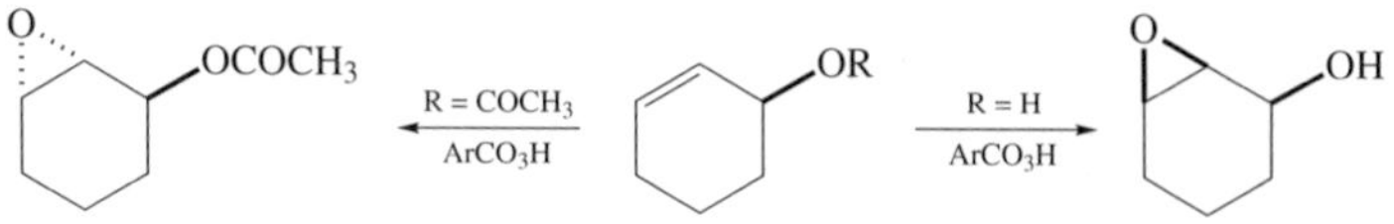

Fig. 15.21

Alkyl hydroperoxides (ROOH) also oxidize alkenes under catalysis by a transition metal. Being relatively stable, *tert*-butylhydroperoxide is often used as the preferred oxidant (Fig. 15.22). The catalysts are commonly based on

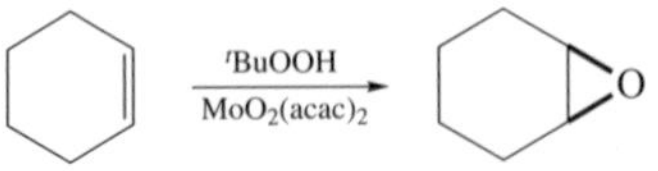

Fig. 15.22

molybdenum, vanadium or titanium as the central metal. In general, more highly substituted alkenes react faster than less substituted alkenes.

The oxidation of allylic alcohols using *tert*-butyl hydroperoxide and a transition metal is highly selective, leading to the epoxy alcohol with the hydroxy group on the same side as the epoxy moiety (Fig. 15.23). Reaction involves

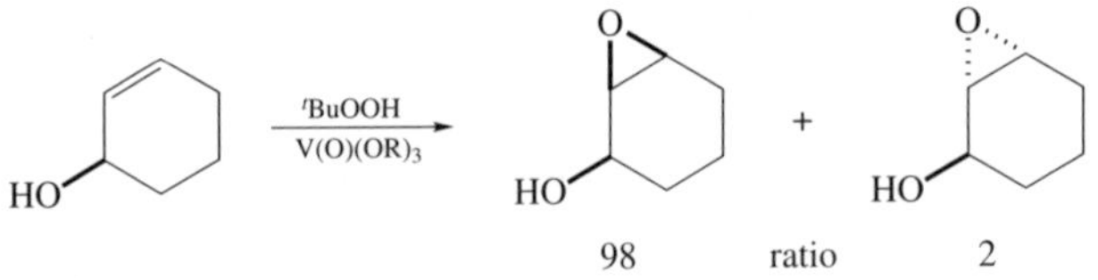

Fig. 15.23

initial coordination of the metal to both the allylic alcohol and the hydroperoxide, followed by displacement at the peroxy group by the alkene unit. Thus in the above case the vanadium(V) species gives a reactive complex *via* displacement of two alkoxy ligands as described in Fig. 15.24.

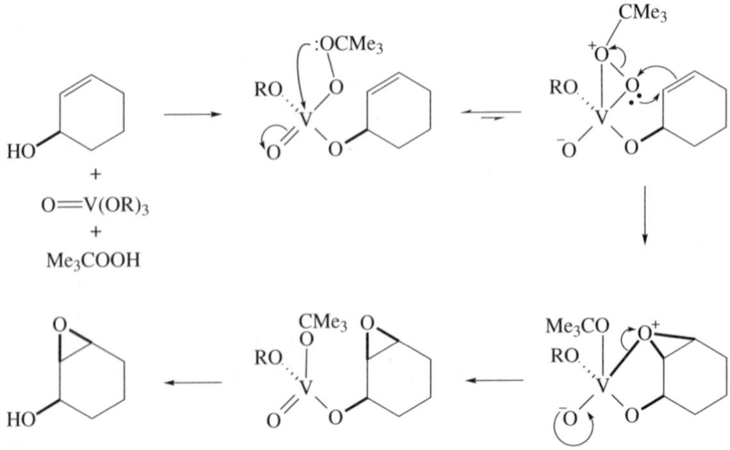

Fig. 15.24

This *diastereoselective* epoxidation reaction was made *enantioselective* by providing the transition metal with a chiral ligand. This field has been developed most elegantly and extensively by Sharpless in the USA. As an example, reaction of geraniol with *tert*-butyl hydroperoxide, catalysed by titanium(IV)

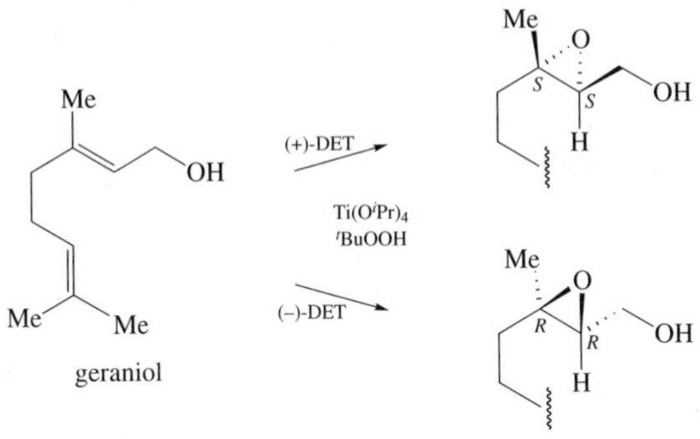

Fig. 15.25

tetraisopropoxide $Ti(O^iPr)_4$ in the presence of L-(+)- or D-(–)-diethyl tartrate (DET) gave the (2S,3S)- or the (2R,3R)-enantiomer of the corresponding epoxide respectively (Fig. 15.25).

The tartrate is believed to displace two isopropoxy groups from the tetra-isopropoxide [Fig. 15.26; L-(+)-tartrate is shown]. Displacement of two more isopropoxy groups by the allylic alcohol (in this case $R^1CH=CHCH_2OH$) and the peroxide sets up the preferred disposition of the alkene and the oxidant.

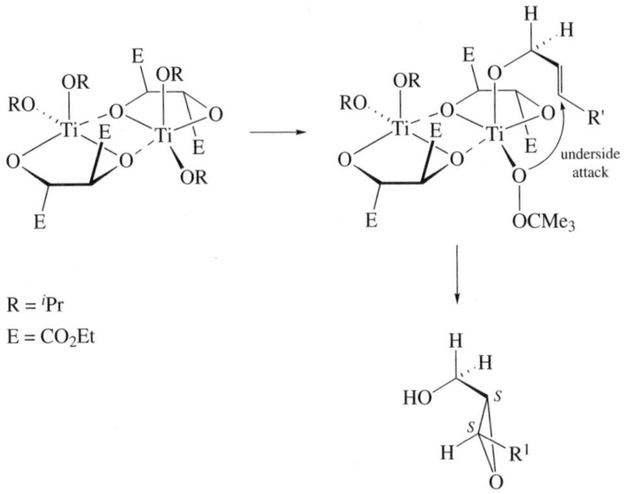

$R = {}^iPr$
$E = CO_2Et$

Fig. 15.26

Similar arguments explain why the two enantiomers of a chiral allylic alcohol are oxidized at different rates (Fig. 15.27): kinetic resolution will take place such that if the reaction is taken to the halfway point, one enantiomer of the allylic alcohol will be present in excess. The rationale for the greater rate of reaction of one enantiomer can be appreciated by inspection of the two intermediates featured in Fig. 15.27. In one the group R^2 at the stereogenic centre in the alkene points towards the tartrate residue, giving rise to unfavourable interactions. The preferred intermediate has the group R^2 in an unhindered situation. Obviously, if the kinetic resolution is perfect only one enantiomer of the allylic alcohol will remain at the halfway stage. In order to reverse the sense of the kinetic resolution, unnatural (−)-tartrate should be employed as the ligand. The inclusion of molecular sieves in the reaction-pot greatly improves the epoxidation process by removing minute quantities of water which would otherwise deactivate the system. This adaptation allows the Sharpless epoxidation to be truly catalytic in terms of the involvement of the titanium species.

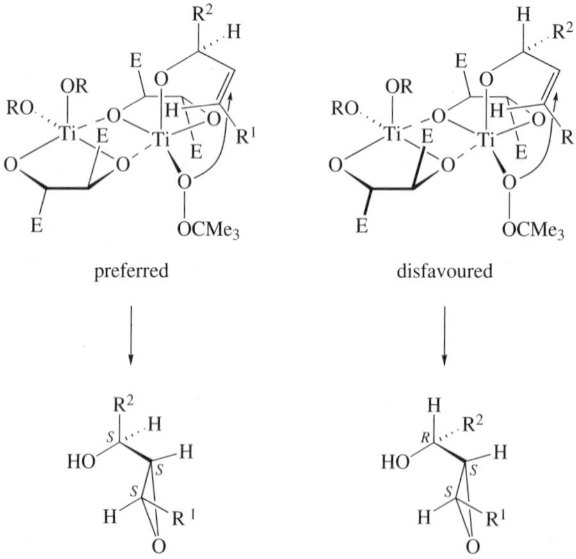

preferred   disfavoured

R^2 = alkyl
R = ^{i}Pr
E = CO_2Et

Fig. 15.27

Overall, the allylic alcohol (chiral or achiral) is oxidized by a peroxide in the presence of a transition metal and a chiral ligand to give an oxirane (epoxide) with two new stereogenic centres. The epoxide moiety is a reactive unit and suffers nucleophilic attack (usually S_N2 with fracture of one of the C—O bonds) making these three-membered ring compounds exceedingly important in organic synthesis.

Osmium tetraoxide oxidizes alkenes to *cis*-diols via the intermediacy of a cyclic osmate ester (Chapter 10). The reaction can be catalysed by amines, and derivatives of naturally occurring cinchona alkaloids have been shown to be particularly effective ligands in the promotion of *asymmetric* bis-hydroxylation reactions. Note that it was only after a great deal of painstaking research that suitable effective ligands were found; it was not clear, initially, that certain alkaloids would be the sought-after ligands.

Incorporation of these selected alkaloid derivatives into the reaction means that, for a substrate such as (*E*)-stilbene, one diastereoisomer of the osmate ester is preferred to the other, leading to the preferential formation of one enantiomer of the diol (Fig. 15.28). In the favoured diastereoisomer the bulky phenyl groups of the substrate are oriented away from the aryl groups (Ar) on the azabicyclo[2.2.2]octane moiety of the alkaloid. Note that, once again, an immense amount of work had to be undertaken to optimize the ligand by variation of the aromatic groups that are attached to the central core structure.

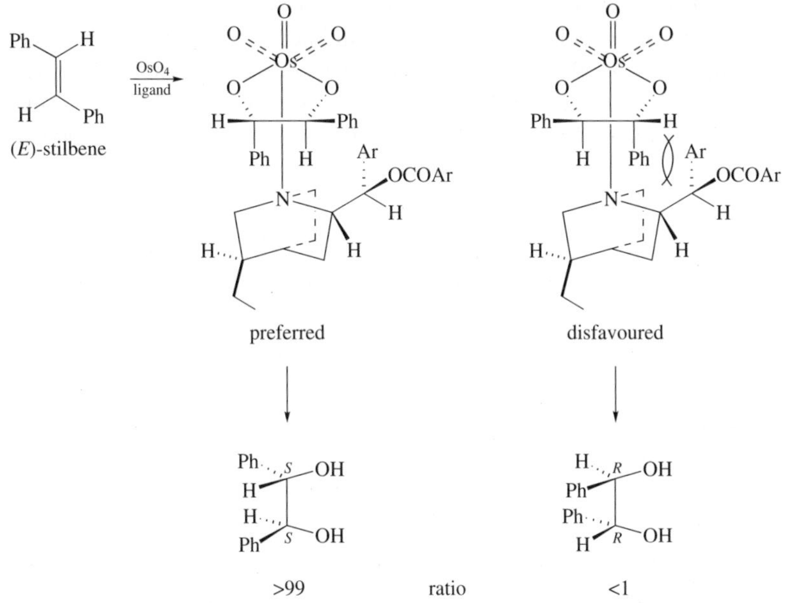

Fig. 15.28

It is noteworthy that only a small amount of osmium tetraoxide and ligand are needed if the metal oxide is recycled by using *N*-methylmorpholine *N*-oxide (NMO) as the terminal oxidant (Fig. 15.29).

In addition to the many man-made catalysts that are available, including those of the above types, nature provides the chemist with another, often complementary, range of catalysts called enzymes. Enzymes are proteins (molecular weight ≥ 30 000 daltons) composed of chains of L-amino acids. Folding of these chains provides a three-dimensional catalytic system (Chapter 12). Reactions can be promoted on natural and unnatural substrates.

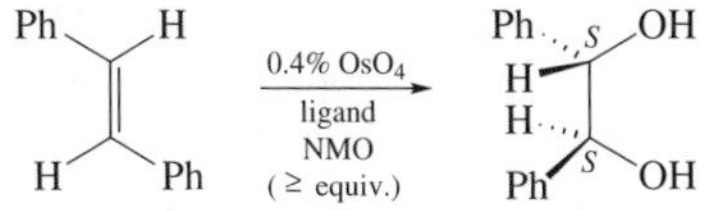

Fig. 15.29

Some of the simplest enzymes to use in the laboratory are hydrolase enzymes, a set of proteins that catalyse the hydrolysis of esters, amides, etc. (Fig. 15.30). Since enzymes are chiral and the 'active site' of the enzyme is often at the bottom of a cavity or cleft within this chiral environment, you should not be surprised to learn that they catalyse a given reaction at different rates for the two enantiomers of a particular substrate. Take the first example shown in Fig. 15.30, the hydrolysis of the two enantiomers of butyl 2-bromopropanoate. These two enantiomers are transformed at different rates using the hydrolase enzyme (the lipase from the fungus *Candida cylindracea*) because the acyl–enzyme complexes (and, importantly, the tetrahedral intermediates leading to these complexes) derived from the two enantiomers have a diastereoisomeric relationship, one to the other (Fig. 15.31). The fact that one ester is processed faster than the other leads to a kinetic resolution: (*R*)-bromopropanoic acid is formed preferentially, leaving an enantiomeric excess of the *S*-ester, provided, of course, that the hydrolysis is not taken to completion.

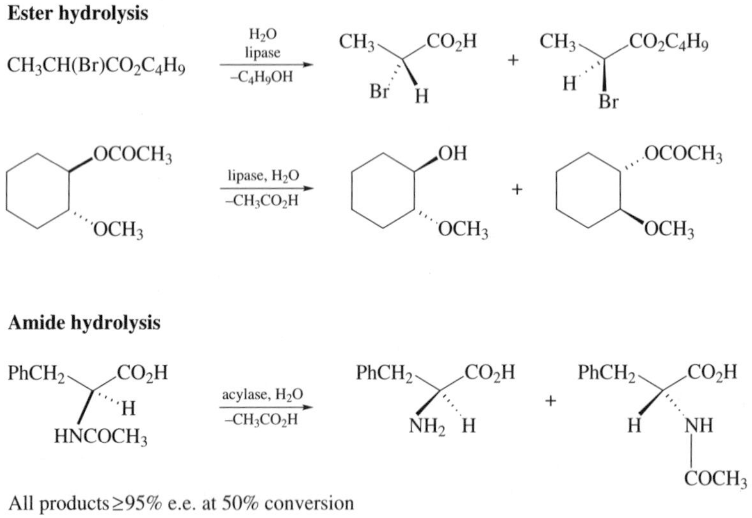

Ester hydrolysis

Amide hydrolysis

All products ≥95% e.e. at 50% conversion

Fig. 15.30

Reduction reactions can also be catalysed by enzymes. For example, a prochiral ketone can be reduced to an optically active secondary alcohol (Fig. 15.32). Note that the enzyme does not provide the hydrogen atoms for the reduction process (it is only the catalyst). Instead the hydrogen atoms are provided by the relevant cofactor which is usually nicotinamide adenine dinucleotide (NADH) or nicotinamide adenine dinucleotide phosphate (NADPH). These are fairly complicated natural macromolecules but in both cases a nicotinamide moiety is the working part of the cofactor (Fig. 15.33); the 1,4-dihydropyridine system is oxidized to the corresponding pyridinium species as the substrate is reduced. The enzymes catalysing these transformations are called dehydrogenase enzymes. This name serves to emphasize that these proteins are catalysts

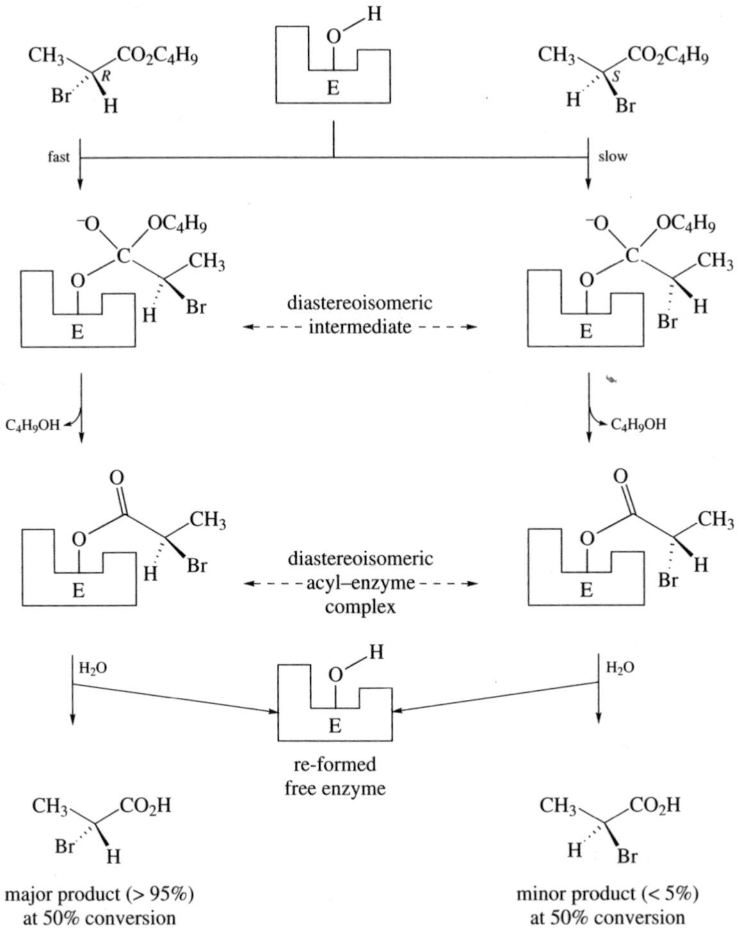

Fig. 15.31

that will increase the rate of both the reduction *and* the oxidation (dehydrogenation) processes. The direction taken by the reaction will depend on factors such as the pH of the solution. (One important function of mammalian dehydrogenase enzymes is detoxification of the ethanol in the system after ingestion of alcoholic beverages!) Reduction reactions can be accomplished using whole cells such as common bakers' yeast (Fig. 15.34). In this case both the enzyme and the cofactor are provided by the organism. The oxidized cofactor is reduced back to NADH as part of the metabolism of the whole cell. Other reduction reactions involving diketones and carbon–carbon double bonds can be accomplished using whole cells and/or enzymes (Fig. 15.35).

To emphasize the complementarity of natural and man-made catalysts it should be pointed out that the asymmetric hydrogenation of other carbon–carbon double bond systems can often be accomplished effectively using chiral

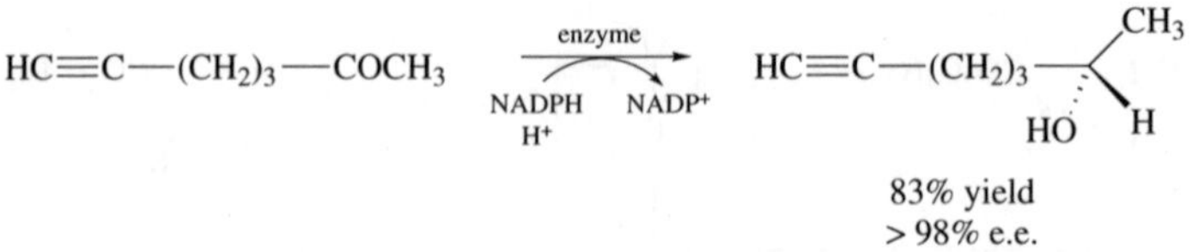

Fig. 15.32

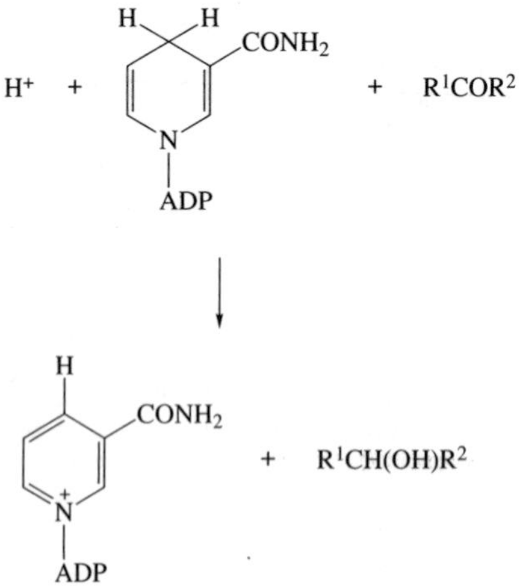

Fig. 15.33

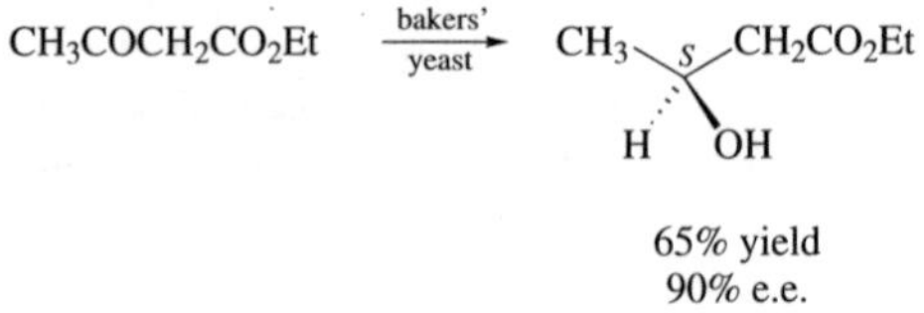

Fig. 15.34

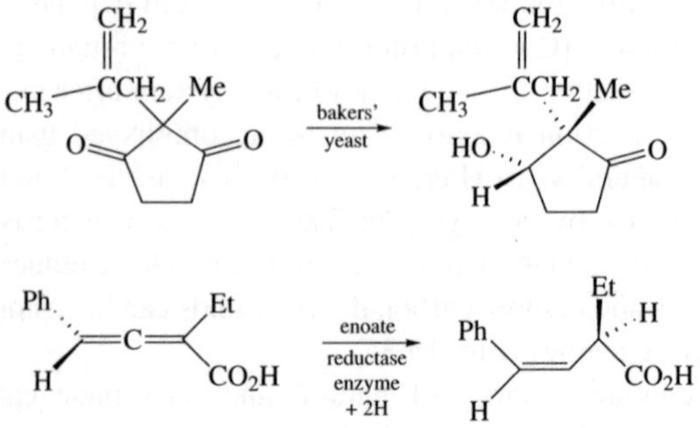

Fig. 15.35

bisphosphine–rhodium complexes [based on the chiral bisphosphine ligand DIPAMP (Chapter 4)] rather than enzymes. This rhodium-based catalyst can form two diastereoisomeric complexes **2** and **3** (Fig. 15.36) with a substrate such as an achiral enamide. The latter complex is hydrogenated more rapidly to give an alkylrhodium hydride which decomposes with release of the saturated product and regeneration of the catalyst. This strategy in asymmetric synthesis has been important in the synthesis of some amino acid derivatives. Figure 15.36 illustrates the preparation of a derivative of (*S*)-alanine.

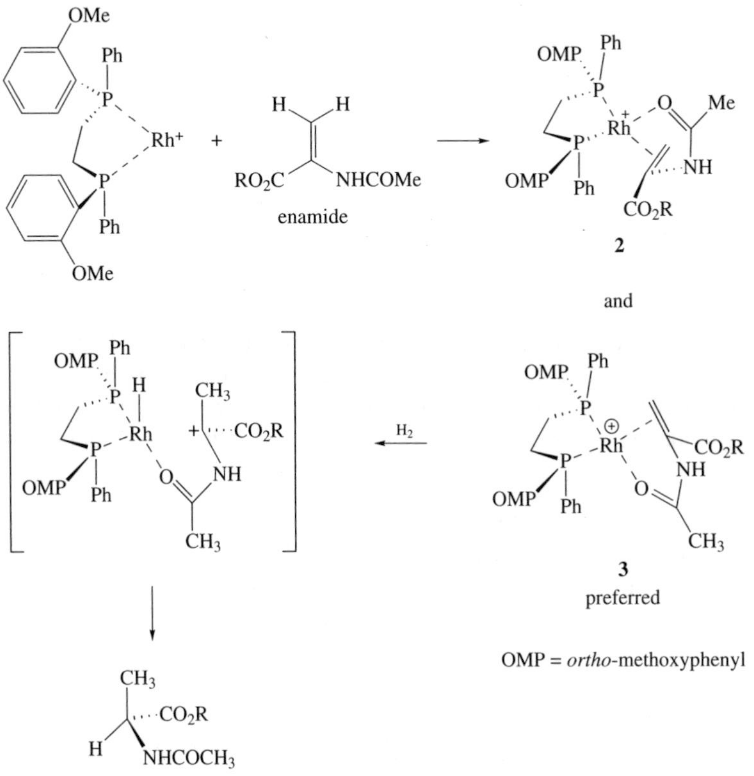

Fig. 15.36

Answers to questions

Q. 15.1 Proton abstraction goes through two possible transition states. The transition state leading to the (*E*)-enolate is generally disfavoured when R^1 is bulky due to the close approach of groups R^1 and R^2. The transition state leading to the (*Z*)-enolate is disfavoured when R^3 is very bulky because the 1,3-transannular interactions between R^2 and R^3 become dominant.

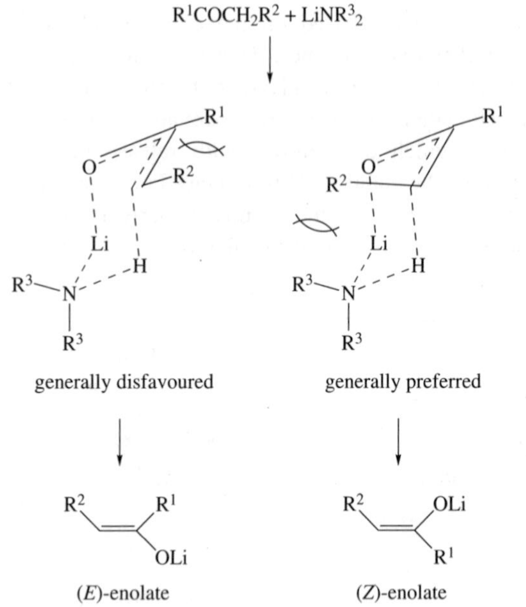
R¹COCH₂R² + LiNR³₂
generally disfavoured
generally preferred
(E)-enolate
(Z)-enolate

Q. 15.2 The attack on the alkene by BH_3 takes place from the 'bottom' face of the molecule to avoid unfavourable interactions with the furan ring. The incoming hydrogen atom is attached to the more substituted end of the double bond. After oxidation the predominant product has the (1*S*,2*R*,3*R*)-configuration.

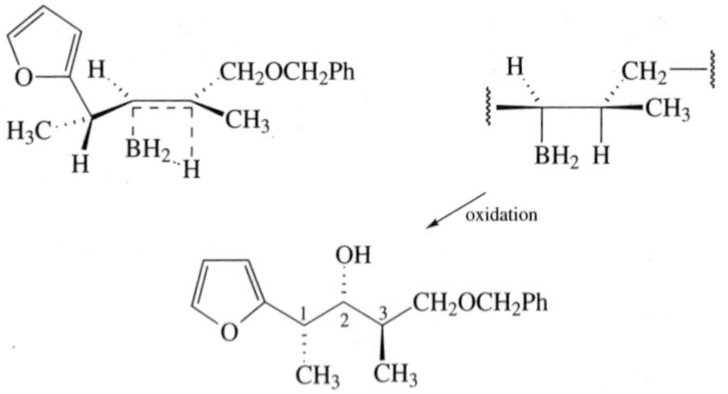
oxidation

Q. 15.3 The preferred transition state is shown below.

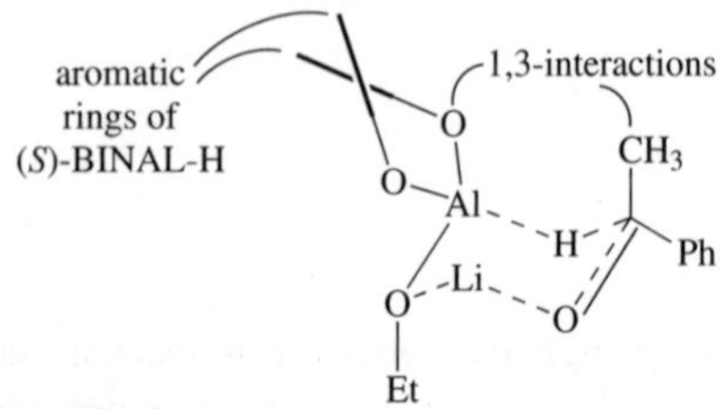
aromatic rings of (S)-BINAL-H
1,3-interactions

The π-electron-rich phenyl group occupies a *pseudo*-equatorial position to avoid unfavourable transannular 1,3-interactions with the non-bonding electrons surrounding the oxygen atom of the reagent.

16 Total asymmetric syntheses of prostaglandin F$_{2\alpha}$ and compactin

For reasons alluded to in the previous chapters, the preparation of optically pure compounds is a great challenge to a synthetic organic chemist and there are a number of different strategies that can be employed. Whatever tactic is chosen, the preparation of an unnatural compound in optically active form is often quite a lot more expensive than the synthesis of the corresponding (±)-racemate since the route may have to employ an expensive optically active starting material or there might be extra steps needed to resolve a particular synthetic building block (synthon) into an optically pure form. We have seen that the approaches to optically active synthons can be categorized as those requiring: classical resolution processes; use of stoichiometric amounts of optically active (preferably reusable) agents and reagents; use of optically active natural products from the chiral pool; and the use of chiral catalysts, including enzymes. The different strategies listed above can be exemplified using two target molecules, namely prostaglandin F$_{2\alpha}$ (PGF$_{2\alpha}$) and compactin.

16.1 Syntheses of prostaglandin F$_{2\alpha}$ (PGF$_{2\alpha}$)

PGF$_{2\alpha}$ is a member of a family of naturally occurring compounds that have attracted a huge amount of interest because of their diverse range of biological activities. A small number of prostaglandin derivatives (prostanoids) have been marketed as useful drugs (cytoprotective agents) and as veterinary aids (controlling the oestrus cycles of horses and cattle).

PGF$_{2\alpha}$ represents a useful model to describe the various approaches to chiral synthesis, since the molecule is relatively simple but does contain five chiral

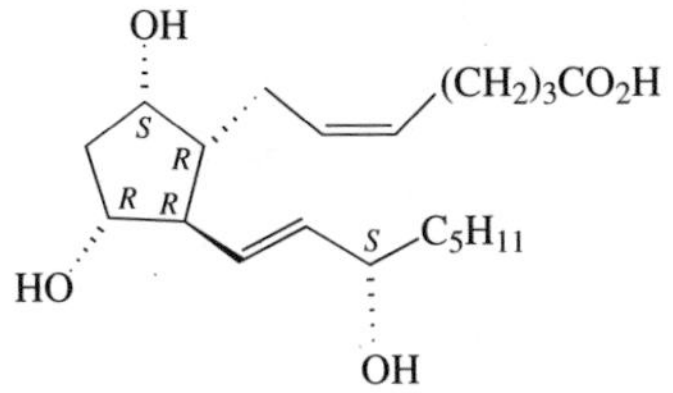

1

Fig. 16.1 PGF$_{2\alpha}$

centres (designated *R* or *S* in formula **1**, Fig. 16.1), four around the five-membered ring and one in the lower side-chain. The different approaches are discussed in turn.

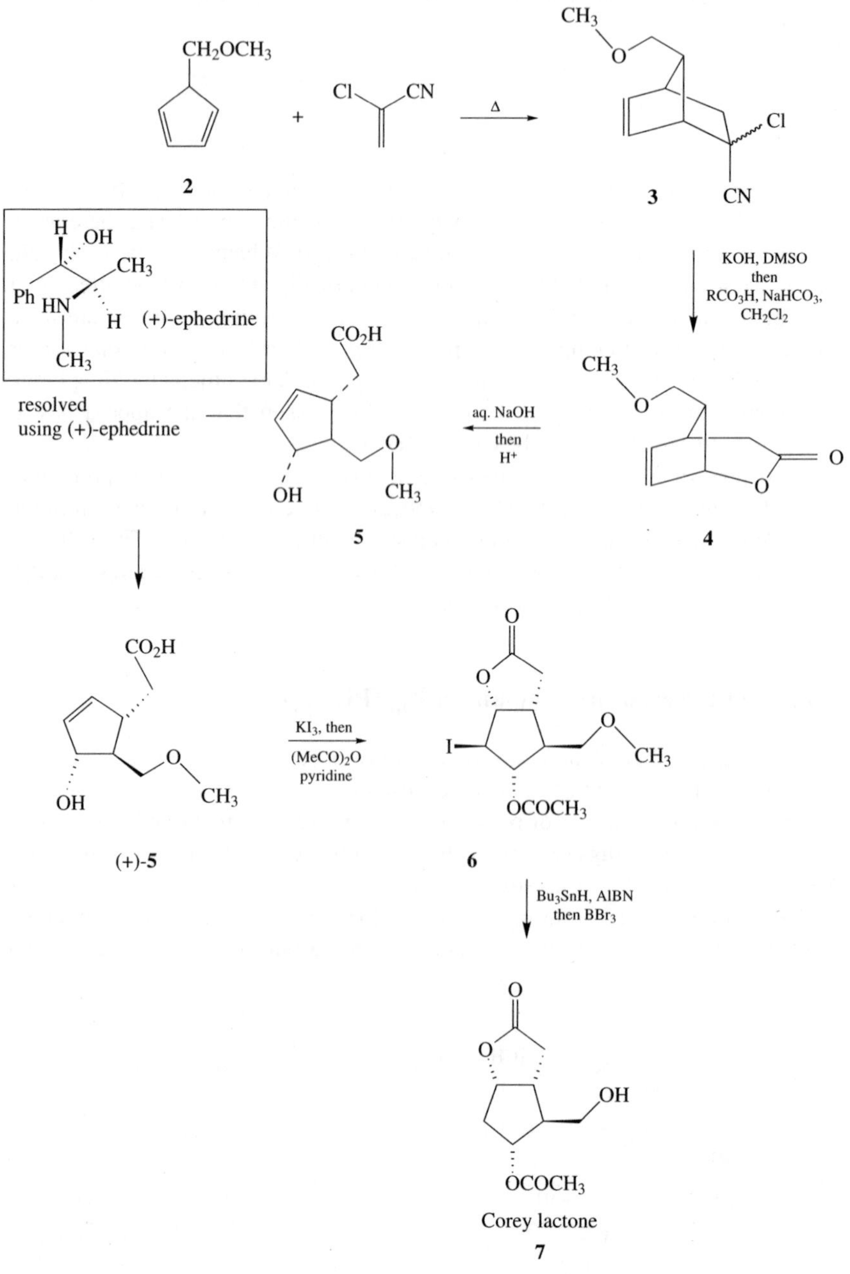

Fig. 16.2

16.1.1 Classical resolution of a PGF$_{2\alpha}$ intermediate

Optically pure PGF$_{2\alpha}$ was synthesized first by E.J. Corey *et al.* by classical resolution of a key intermediate (Fig. 16.2). Thus the alkylated cyclopentadiene **2** underwent a Diels–Alder reaction with 2-chloroacrylonitrile to afford the adduct **3** from which the lactone **4** was obtained after hydrolysis to the ketone (*via* the cyanohydrin) and Baeyer–Villiger oxidation. Hydrolysis of **4** gave the hydroxy-acid **5**. The racemic acid **5** was resolved with the optically active base, (+)-ephedrine. Treatment of (+)-**5** with potassium triiodide followed by protection of the hydroxy group gave the lactone **6**. The iodine atom was removed with tri-*n*-butyltin hydride under radical-initiating conditions: ether cleavage then gave the (+)-lactone **7** in good overall yield.

Q. 16.1 The Baeyer–Villiger oxidation involves the conversion of a ketone (cyclic ketone) into an ester (lactone) using peracid:

$$R^1COR^2 + R^3CO_3H \rightarrow R^1CO_2R^2 + R^3CO_2H$$

Suggest a mechanism for this reaction, given that the initial attack is by a peracid oxygen on the carbonyl carbon of the ketone.

The (+)-lactone was converted into PGF$_{2\alpha}$ using standard methodology (Fig. 16.3). Thus, the (+)-lactone **7** was oxidized with Collins' reagent to give the aldehyde **8**, which on treatment with the appropriate phosphonate under Wadsworth–Emmons–Wittig conditions gave the *E*-enone **9**. In Corey's original route, reduction of the keto function of **9** with zinc borohydride gave a 1:1 diastereoisomeric mixture of alcohols from which the *S*-isomer **11** was separated by chromatography. Subsequent improvements involving the *p*-biphenylyl carbamoyl protecting group (**10** → **12**) and use of hindered borohydrides (for example, lithium tri-*s*-butylborohydride) gave a more stereoselective reduction, resulting in an improved ratio (89:11) of 15*S*:15*R* isomers.

Lactone **11** or **12** was readily converted into PGF$_{2\alpha}$ *via* partial reduction using diisobutylaluminium hydride (DIBAL) to give a lactol and Wittig reaction using a non-stabilized ylide to incorporate the *Z*-alkene moiety in the upper, α-oriented side-chain.

16.1.2 Using chiral auxiliaries and optically active reagents

In an attractive approach where optical activity is incorporated at an early stage, Corey developed the use of the chiral acrylic ester **13** in a stereo-controlled Diels–Alder reaction. Cycloaddition of **13** with 5-benzyl-oxymethylcyclopentadiene **14** catalysed by aluminium(III) chloride gave the optically active *endo-anti*-norbornene **15** (Fig. 16.4). Asymmetric induction by the (*S*)-(–)-pulegone-derived acrylate **13** exceeded that shown by other optically active acrylates (cf. Chapter 15). The norbornene **15** was converted into an enolate with lithium diisopropylamide (LDA) and oxidized with

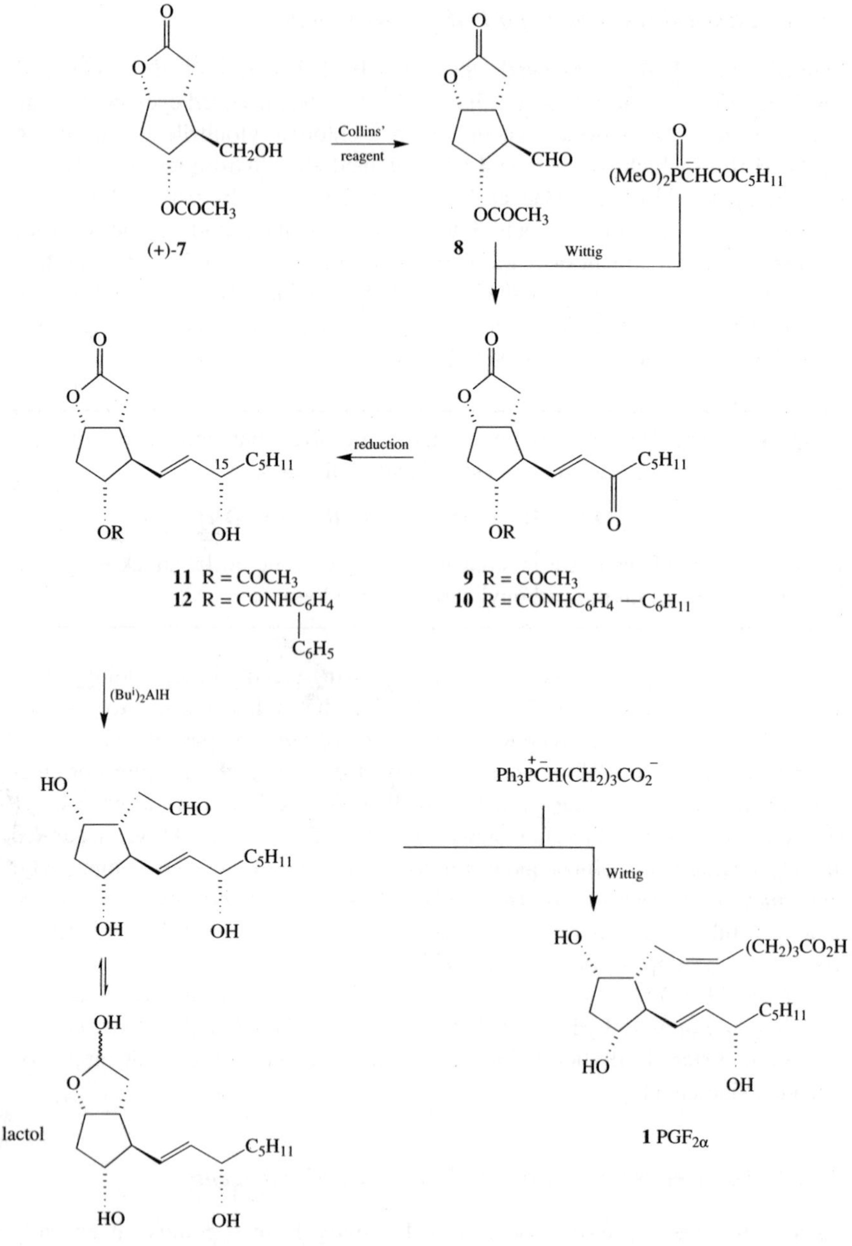

Fig. 16.3

molecular oxygen in the presence of triethyl phosphite to afford an isomeric mixture of the hydroxy esters **16**. Reduction with lithium aluminium hydride gave a diol which on oxidation with sodium metaperiodate gave the ketone **17**.

The ketone **17** was oxidized with alkaline hydrogen peroxide to give the corresponding unsaturated δ-lactone which was converted into the Corey lactone derivative using methodology outlined previously (see **4 → 7** in Fig. 16.2).

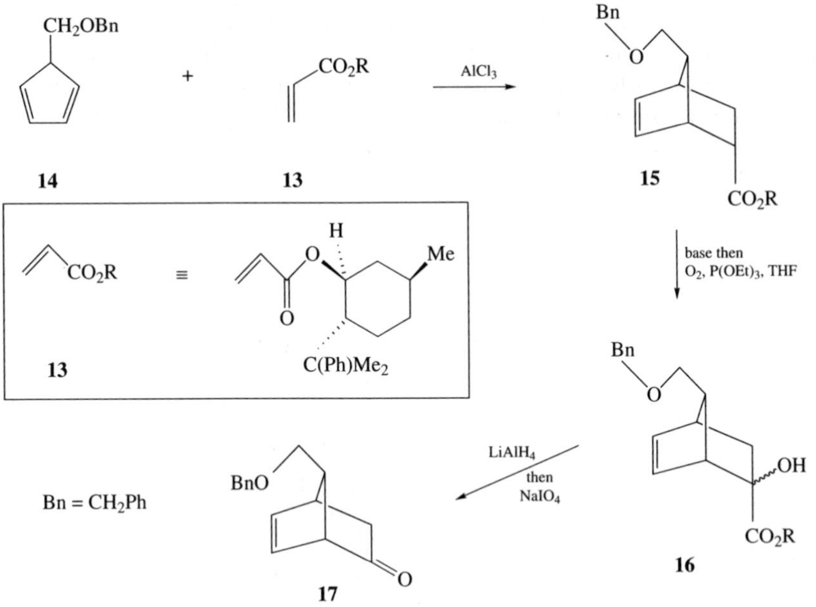

Fig. 16.4

2-Oxabicyclo[3.3.0]oct-6-en-3-one **18** (Fig. 16.5) is another versatile intermediate in the preparation of prostaglandins and syntheses of **18** have been reported which employ optically active reagents. Thus, cyclopentadienylsodium

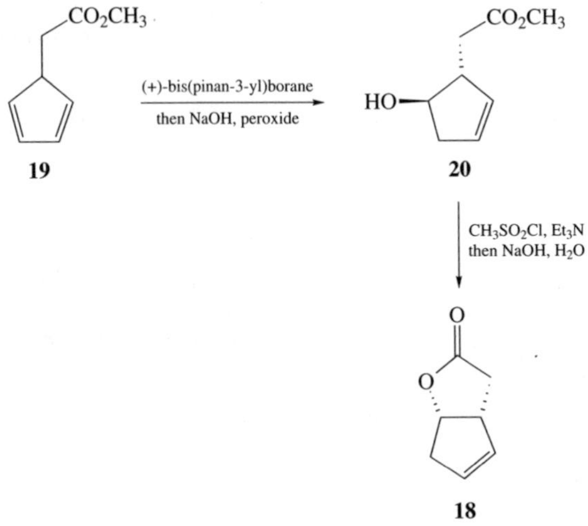

Fig. 16.5

was treated at −78°C with methyl bromoacetate to afford the ester **19**. Hydroboration *in situ* with (+)-di(pinan-3-yl)borane followed by oxidation with hydrogen peroxide afforded the hydroxy ester **20** in 45% yield. The optical purity was assessed to be about 95% e.e. Obviously one of the diastereo-isomerically related transition states in the hydroboration reaction is much preferred to the other (Chapter 15). The hydroxy ester **20** was converted into optically pure lactone **18** *via* the corresponding mesylate (see Fig. 16.7 for further derivatization of the lactone **18**).

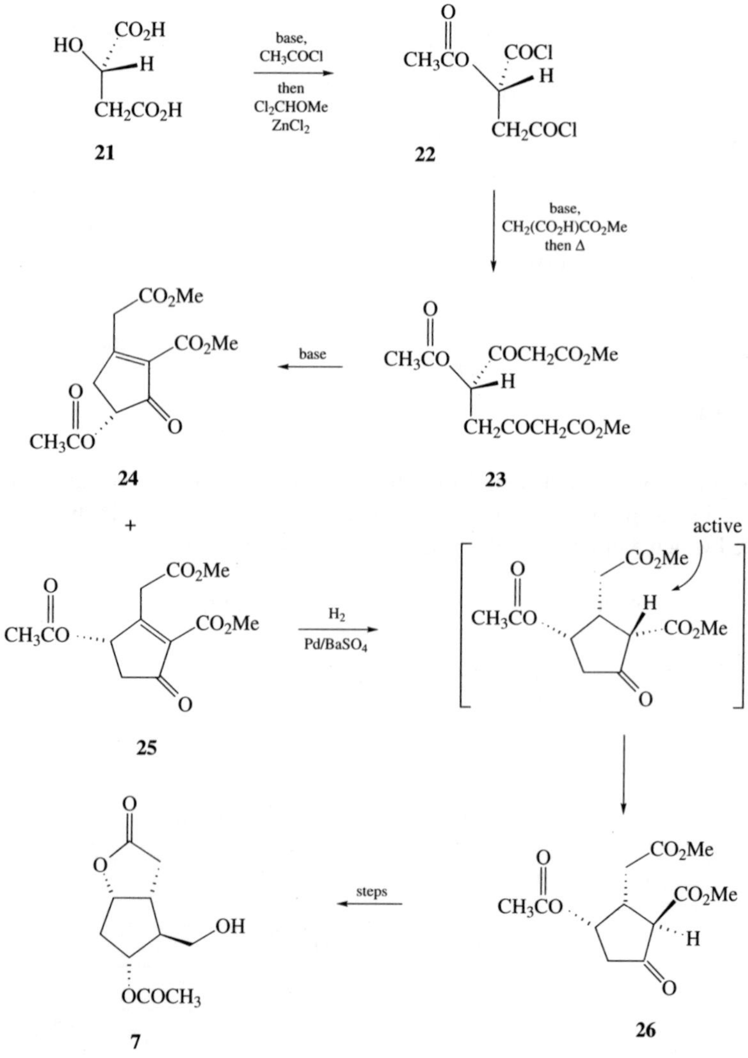

Fig. 16.6

16.1.3   Synthesis of PGF$_{2\alpha}$ starting with a natural product from the 'chiral pool'

The synthesis of prostaglandins from natural products is illustrated by the synthesis of the Corey lactone from (S)-malic acid **21** (Fig. 16.6). (S)-Malic acid **21** was treated with acetyl chloride to afford (S)-(–)-2-acetoxysuccinic anhydride which, in turn, gave the bis-acid chloride **22** on reaction with dichloromethyl methyl ether in the presence of zinc chloride. Reaction with two equivalents of methyl hydrogen malonate anion, followed by decarboxylation, gave the triester **23**.

Cyclization using triethanolamine as base gave two cyclopentenones **24** and **25**, which were separated. The required product **25** (85% yield) was reduced to furnish, after equilibration, the thermodynamically more stable trisubstituted cyclopentanone **26**. The cyclopentanone **26** was converted into the Corey lactone **7** using standard techniques.

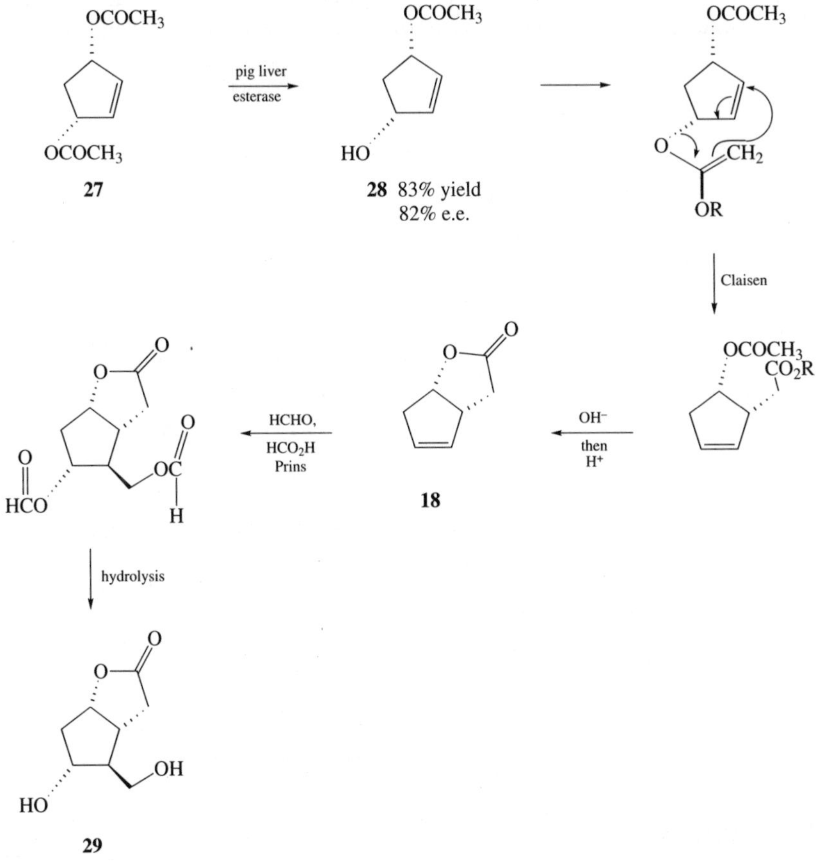

Fig. 16.7

16.1.4 Synthesis of primary prostaglandins using enzymes as chiral catalysts

Lipase or esterase catalysed hydrolysis of the prochiral diacetate **27** gave the hydroxy ester **28**. This hydroxy ester was converted into the optically active lactone **18** by way of a Claisen rearrangement (Fig. 16.7). The lactone **18** can be converted into the Corey lactone intermediate **29** *via* a Prins reaction.

Q. 16.2   Discuss the stereochemistry of the product resulting from the Prins reaction on lactone **18** (Fig. 16.7).

Alternatively, the alcohol **28** can be converted into the enone **30**. Conjugate addition of an organometallic reagent **31** is stereocontrolled by the protected hydroxy group. The product from the conjugate addition is held as the metal enolate **32** and an electrophile (propargylic iodide, **33**) is added. Attack by the electrophile on the enolate takes place from the side opposite to the eight carbon

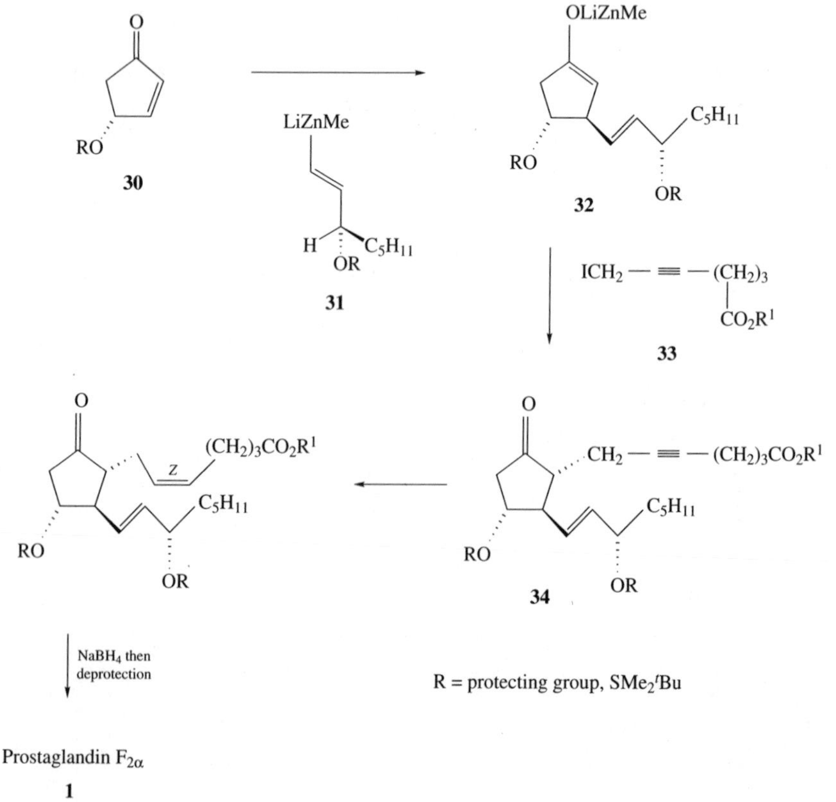

R = protecting group, $SMe_2{}^tBu$

Prostaglandin $F_{2\alpha}$

1

Fig. 16.8

side chain to give the ketone **34**. Controlled hydrogenation of the triple bond using a poisoned catalyst gave the Z-alkene. Reduction of the carbonyl group using a hydride donor takes place at the more exposed *Re* face to give the *S*-alcohol, which upon deprotection gives prostaglandin F$_{2\alpha}$ (Fig. 16.8).

Comparison of the routes to prostaglandin F$_{2\alpha}$ shown in Figs. 16.2/3 and Fig. 16.8 highlights a difference in the method of obtaining the required stereo-chemical configuration of the secondary alcohol group at C15. In the former routes the correct configuration is induced by modifying the C15 ketone using a bulky reducing agent in the presence of an influential unit attached to the pre-existing chiral centre at C11. In the latter route the chiral centre in the lower side chain is produced in the correct form in the eight carbon unit; this unit is sub-sequently incorporated into the embryonic prostanoid. The relevant eight carbon unit, a cuprate reagent, is prepared from the iodoalkene **35** which is, in turn, derived from the alkynol **36**.

The methods available for producing the *S*-alkynol **36** represent a microcosm of the present technology for the preparation of optically active compounds. In the first method the racemic alkynol is converted into the hemiphthalate and resolved using optically pure phenethylamine by formation of diastereoisomeric salts and fractional crystallization of the less soluble component (Fig. 16.9 and Chapter 14). Alternatively, the readily available ketone **37** can be reduced using (*S*)-BINAL-H to give the required *S*-alcohol (Chapter 15). The third strategy involves conversion of the racemic heptyn-3-ol into the corresponding acetate **38** and kinetic resolution using a hydrolase enzyme. Using readily available lipases the *S*-enantiomer is hydrolysed preferentially to give (*S*)-**36** and recovered (*R*)-acetate, (*R*)-**38**.

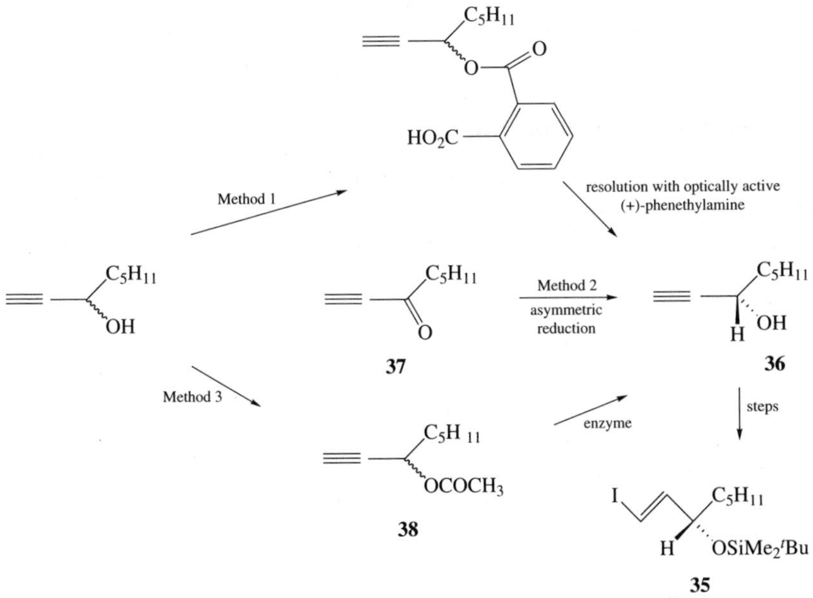

Fig. 16.9

Q. 16.3 Prostacyclin (prostaglandin-I_2, PGI_2) is a naturally occurring compound that has the important biological activity of inhibiting the aggregation of blood platelets (such aggregated clumps can subsequently

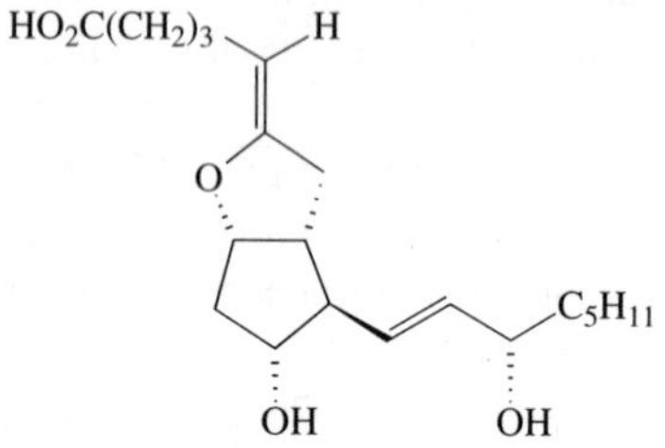

HO2C(CH2)3
H
O
C5H11
OH
OH

PGI_2

disengage from artery walls to cause thrombosis and stroke). In the laboratory prostacyclin can be made from prostaglandin-$F_{2\alpha}$. The key steps involve iodoetherification and dehydroiodination as shown below. Show how the ring-closure and E2 elimination steps give the desired *Z*-geometry about the enol ether.

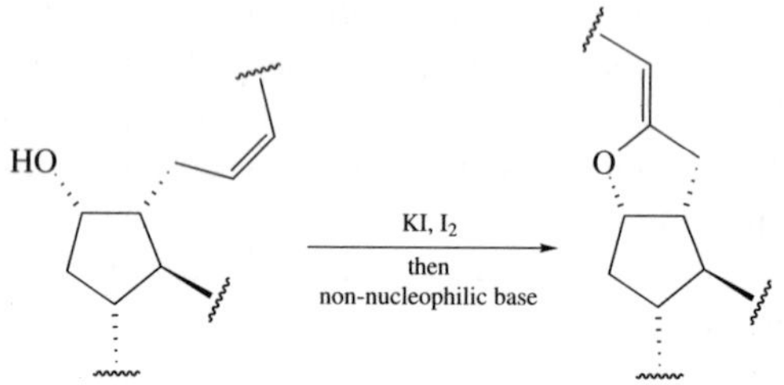

HO
KI, I2
then
non-nucleophilic base
O

16.2 Synthesis of compactin

Compactin **39** and mevinolin **40** are naturally occurring compounds that act as potent cholesterol-lowering agents in mammals. The molecular architecture of the latter compound is incorporated in a top-selling drug (marketed under the

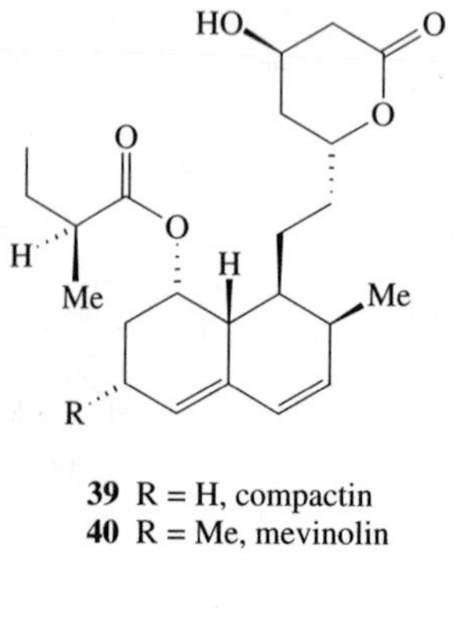

HO
O
O
O
O
H
H
Me
Me
R

39 R = H, compactin
40 R = Me, mevinolin

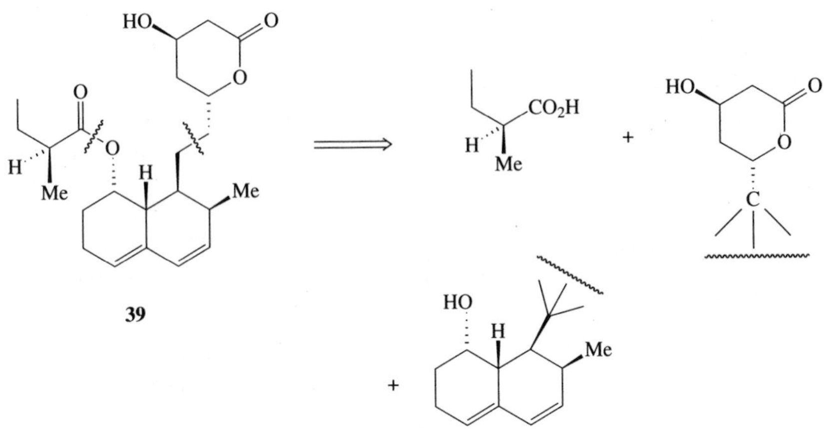

Fig. 16.10

trade name Lovastatin®) that effectively helps patients with high-cholesterol levels by inhibiting one of the key enzymes in the biosynthetic pathway to cholesterol in humans [namely, hydroxymethylglutaryl-Coenzyme A reductase (HMG-CoA reductase)].

The synthesis of compactin described by Burke and Heathcock constructs the three segments of the natural compound (Fig. 16.10) using classical resolution, an optically active reagent and a chiral auxiliary obtained by modification of

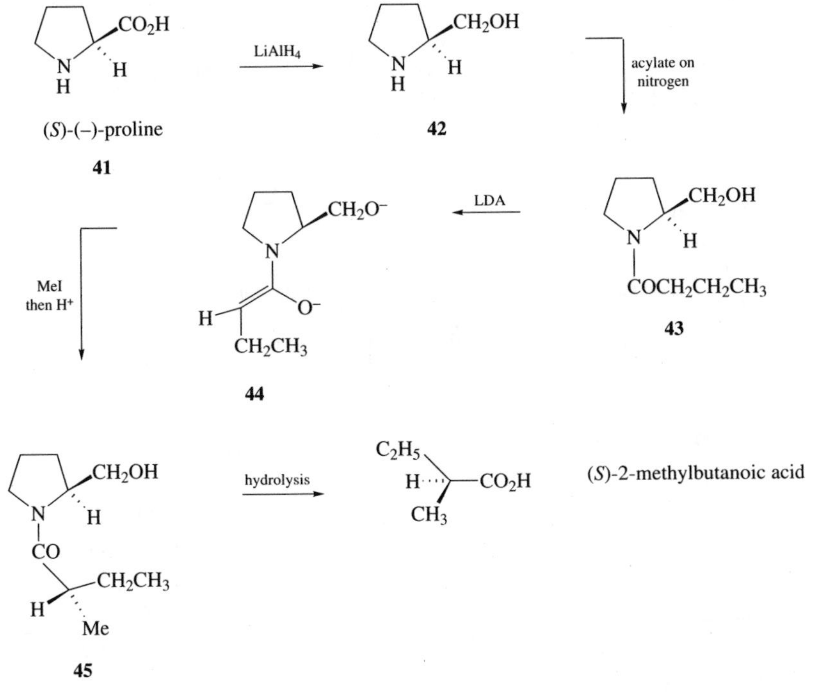

Fig. 16.11

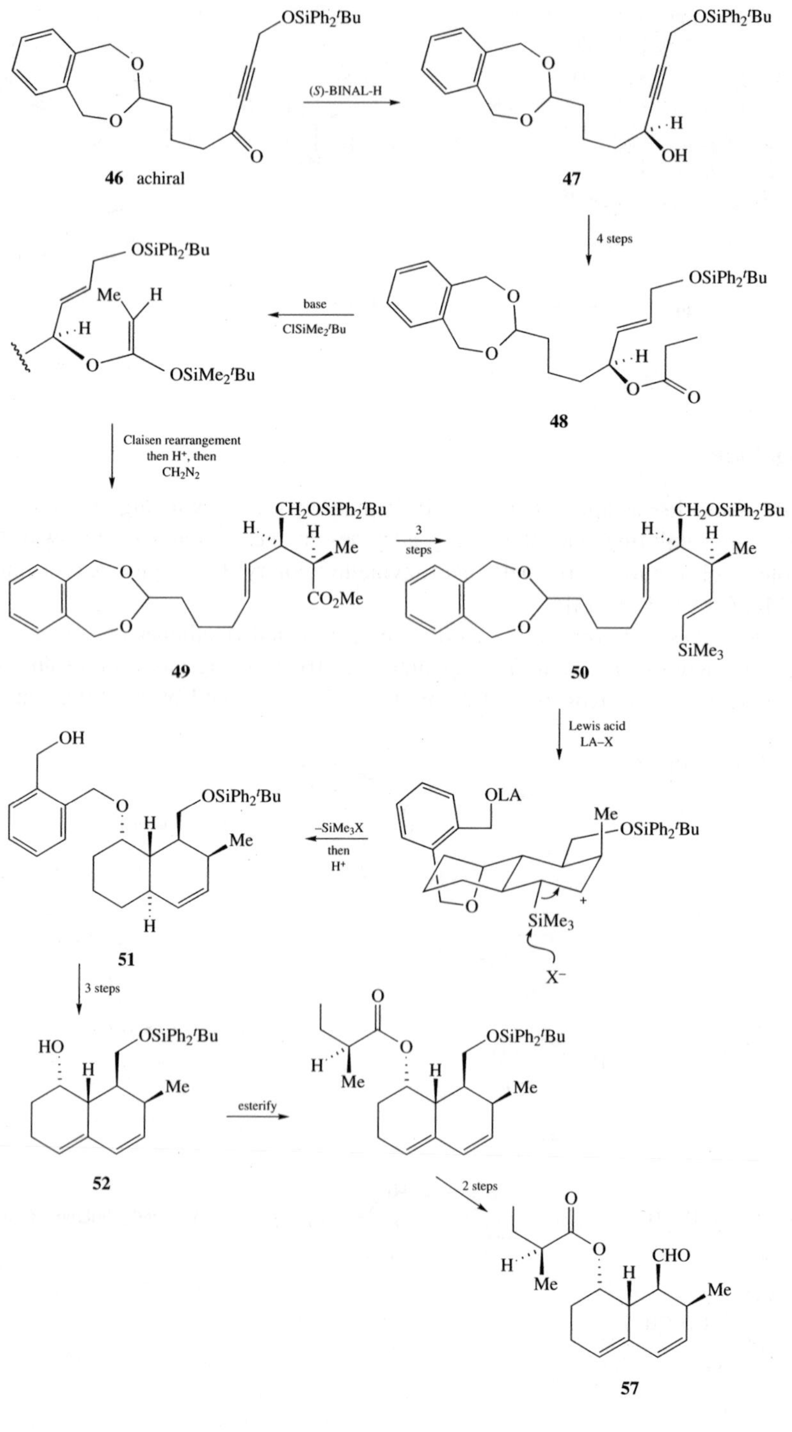
OSiPh2tBu
(S)-BINAL-H
46 achiral
47
4 steps
base
ClSiMe2tBu
48
Claisen rearrangement then H+, then CH2N2
CH2OSiPh2tBu
CO2Me
49
3 steps
SiMe3
50
Lewis acid LA–X
OLA
–SiMe3X then H+
51
3 steps
52
esterify
2 steps
CHO
57

Fig. 16.12

material from the chiral pool. The synthesis also has a number of elegant features that demonstrate the modern invention and use of stereocontrolled reactions in organic synthesis and it seems eminently fitting to end the book on this note.

The smallest fragment used to make compactin is prepared using (*S*)-(–)-proline **41** to provide the chiral auxiliary (Fig. 16.11). This natural amino acid is reduced to prolinol **42** and the amino group is acylated to give compound **43**. Deprotonation of the active methylene group *and* the hydroxy unit gives the dilithio species **44** with the enolate in the Z-configuration (see Chapter 15). Alkylation takes place from the less hindered lower face of the molecule (the *Si* face of the reacting carbon centre) to give the diastereoisomer **45** as the major product (about 12:1). Hydrolysis produces (*S*)-2-methylbutanoic acid (84% e.e.).

The hexahydronaphthalene portion of the natural compound has four stereogenic centres and needs to be prepared in single-enantiomer form. The route to the required bicyclic unit involves three steps that are of paramount importance in terms of stereochemical control (Fig. 16.12).

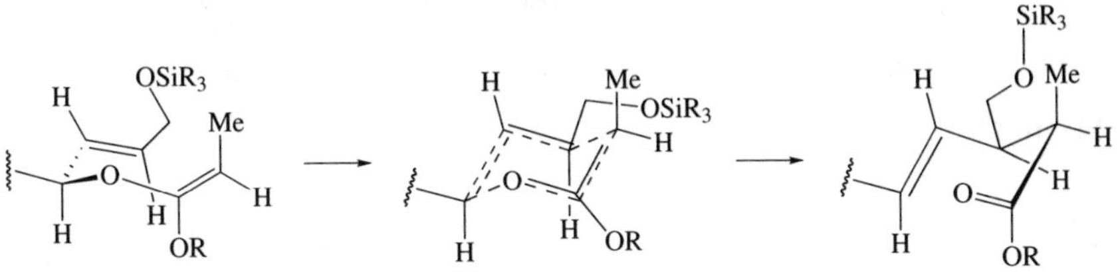

Fig. 16.13

First the ketone **46** was reduced using (*S*)-BINAL-H (Chapter 15) to introduce a chiral centre into the molecule. The product obtained was the optically active secondary alcohol **47** which was subsequently transformed to give the ester **48**.

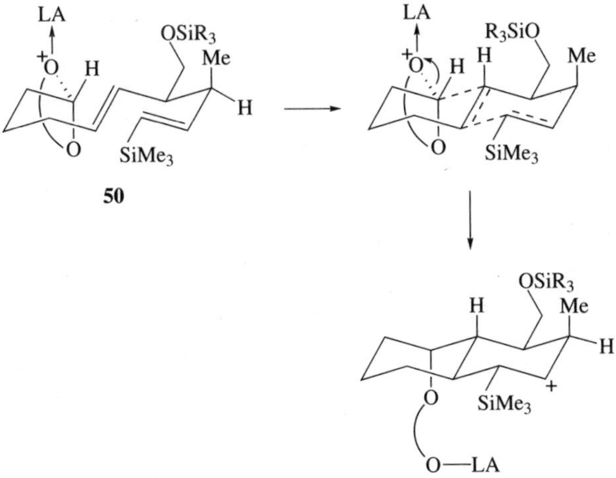

Fig. 16.14

Formation of the silyl enol ether followed by Claisen rearrangement (Fig. 16.13) furnished the compound **49** (after methylation); two stereogenic centres of the target compound are set up in the process. Further chemical steps led to the diene

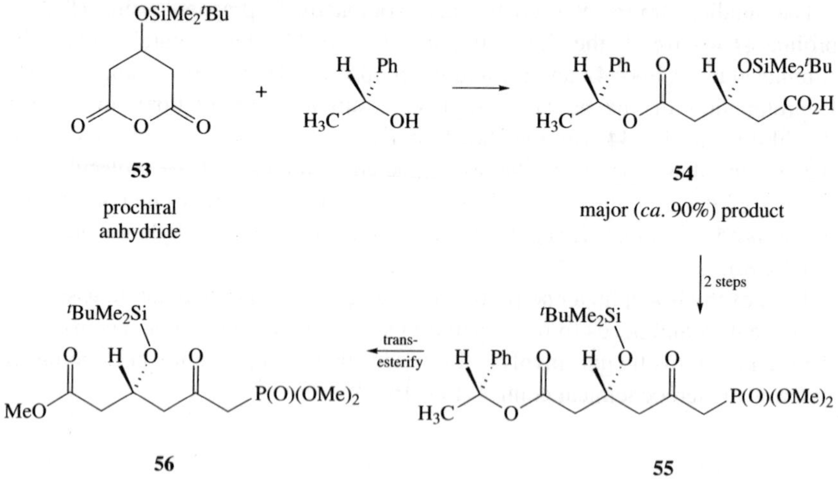

OSiMe$_2{}^t$Bu
+
H
Ph
H$_3$C
OH
53
prochiral
anhydride
54
major (*ca.* 90%) product
2 steps
^{t}BuMe$_2$Si
trans-
esterify
MeO
P(O)(OMe)$_2$
56
55

Fig. 16.15

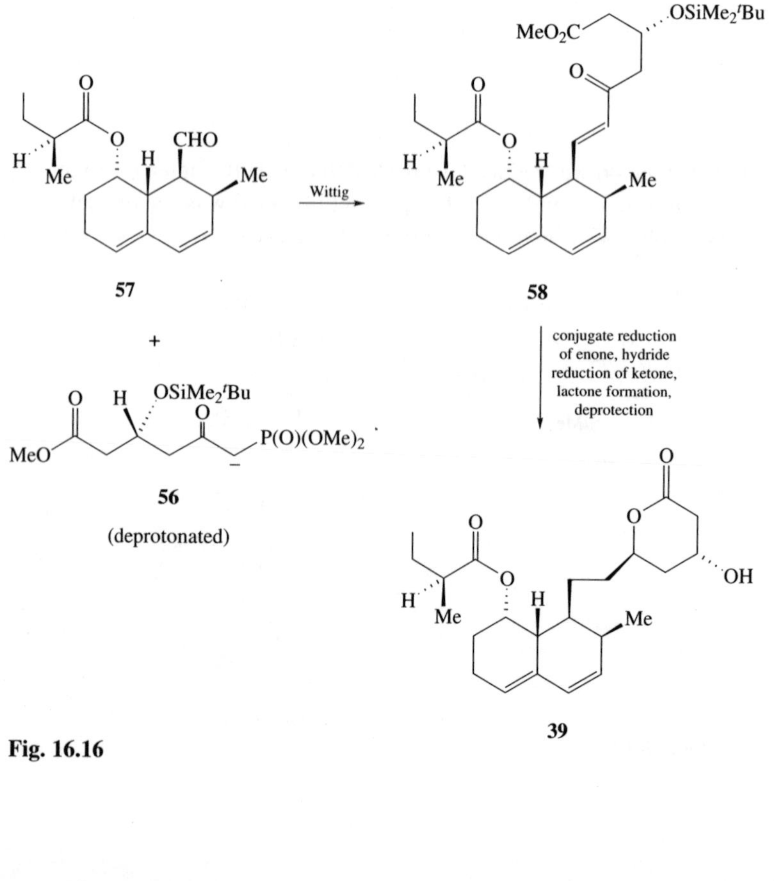

CHO
Me
H
Wittig
57
58
MeO$_2$C
OSiMe$_2{}^t$Bu
+
56
(deprotonated)
conjugate reduction
of enone, hydride
reduction of ketone,
lactone formation,
deprotection
OH
39

Fig. 16.16

50 and treatment of this acetal with Lewis acid (LA) (Fig. 16.14) gave a bicyclic compound **51** (in a reaction reminiscent of the Johnson cyclization that transformed a polyene into a steroidal compound, Chapter 12) with exquisite stereocontrol. Removal of the residual benzyl linkage and bromination of the alkene unit present in compound **51** followed by elimination of two molecules of HBr gave the bicyclo[4.4.0]decadiene **52**.

The third and final portion of compactin was prepared by opening the anhydride **53** with (*R*)-1-phenylethanol, the latter acting as a chiral auxiliary (Fig. 16.15). The alcohol attacks one of the carbonyl groups selectively (ratio 8:1) to give mainly the carboxylic acid **54**. This acid was transformed into the Wittig reagent **55**. At this stage the chiral auxiliary is removed and replaced by a methyl group to give the ester **56**.

The stage was then set to couple the three fragments of compactin. First the alcohol **52** was coupled with (*S*)-2-methylbutanoic acid, before the C1 side chain was modified to provide the aldehyde group in compound **57** (Fig. 16.12). Reaction between the stabilized ylide derived from **56** and the aldehyde **57** gave an *E*-alkene **58** which, after conjugate reduction, ketone group reduction and deprotection, gave compactin **39** (Fig. 16.16).

This synthesis provides one of the most beautiful examples of the use of chiral agents and reagents, as well as the strategic employment of stereocontrolled transformations, in the preparation of an important natural product.

Answers to questions

Q. 16.1   The Baeyer–Villiger oxidation takes place by way of the Craigee tetrahedral intermediate **A**. Migration of a chiral group R^2 takes place with retention of configuration.

$$R^1CR^2 + R^3CO_3H \xrightarrow{\text{(i)}}$$

A

$$\downarrow \text{(ii)}$$

$$R^1CO_2R^2 + R^3CO_2H$$

(i) nucleophilic attack by R^3CO_3H then H^+ shift
(ii) rearrangement and then H^+ shift

Q. 16.2   The protonation of formaldehyde under acidic conditions gives the cation $(H_2C{=}OH)^+$. Electrophilic attack on the alkene unit of lactone **18** takes place from the less-hindered face.

Attack by the attendant nucleophile HCO_2H takes place at the less-hindered position. The primary alcohol is esterified under the acidic conditions.

Q. 16.3   The iodo-etherification step proceeds through the initial formation of an iodonium ion: the hydroxy group acts as the intramolecular nucleophile giving overall 5-*exo-tet*, *trans*-addition across the alkene. The base-catalysed elimination reaction proceeds by way of an E2 mechanism with an anti-periplanar arrangement of the departing atoms. The correct stereochemistry of PGI_2 is obtained as a result.

Appendix 1

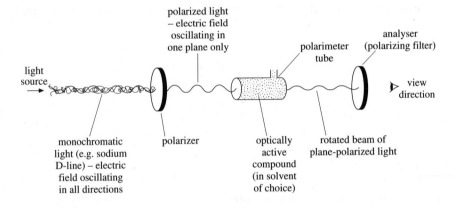

The degree of rotation of the plane-polarized light is measured by adjusting the analyser. The analyser is set so that no light passes through when the polarimeter tube is filled with pure solvent. The angle through which the analyser has to be rotated to block the passage of light after the sample of optically active compound in the solvent has been placed in the polarimeter tube gives the value of α.

Appendix 2

Transformations of D-glyceraldehyde that do not affect the stereogenic centre (*).

$$
\begin{array}{c}
\text{CHO} \\
\text{H}\!\!-\!\!\overset{}{\underset{*}{|}}\!\!-\!\!\text{OH} \\
\text{CH}_2\text{OH}
\end{array}
\xrightarrow{\text{HgO}}
\begin{array}{c}
\text{CO}_2\text{H} \\
\text{H}\!\!-\!\!\overset{}{\underset{*}{|}}\!\!-\!\!\text{OH} \\
\text{CH}_2\text{OH}
\end{array}
$$

D-glyceraldehyde   D-glyceric acid

↓ HCN

$$
\begin{array}{c}
\text{CN} \\
\text{H}\!\!-\!\!|\!\!-\!\!\text{OH} \\
\text{H}\!\!-\!\!\underset{*}{|}\!\!-\!\!\text{OH} \\
\text{CH}_2\text{OH}
\end{array}
\quad + \quad
\begin{array}{c}
\text{CN} \\
\text{HO}\!\!-\!\!|\!\!-\!\!\text{H} \\
\text{H}\!\!-\!\!\underset{*}{|}\!\!-\!\!\text{OH} \\
\text{CH}_2\text{OH}
\end{array}
$$

(i) hydrolysis (CN→CO₂H)
(ii) oxidation (CH₂OH→CO₂H)

$$
\begin{array}{c}
\text{CO}_2\text{H} \\
\text{H}\!\!-\!\!|\!\!-\!\!\text{OH} \\
\text{H}\!\!-\!\!\underset{*}{|}\!\!-\!\!\text{OH} \\
\text{CO}_2\text{H}
\end{array}
\qquad
\begin{array}{c}
\text{CO}_2\text{H} \\
\text{HO}\!\!-\!\!|\!\!-\!\!\text{H} \\
\text{H}\!\!-\!\!\underset{*}{|}\!\!-\!\!\text{OH} \\
\text{CO}_2\text{H}
\end{array}
$$

meso-tartaric   D-tartaric acid
acid

While the usage of D-tartaric acid and L-tartaric acid is commonplace, it is not to be recommended. The conversion of D-glyceraldehyde to D-tartaric acid is described above; however other chemical conversions link L-tartaric acid and D-glyceraldehyde (see below).

$$
\begin{array}{c}
\text{CO}_2\text{H} \\
\text{H}\!\!-\!\!|\!\!-\!\!\text{OH} \\
\text{HO}\!\!-\!\!|\!\!-\!\!\text{H} \\
\text{CO}_2\text{H}
\end{array}
\xrightarrow[\text{(ii) Ac}_2\text{O, base}]{\text{(i) esterify}}
\begin{array}{c}
\text{CO}_2\text{Me} \\
\text{H}\!\!-\!\!|\!\!-\!\!\text{OAc} \\
\text{HO}\!\!-\!\!|\!\!-\!\!\text{H} \\
\text{CO}_2\text{Me}
\end{array}
\xrightarrow[\substack{\text{(ii) Zn/HCl} \\ \text{(iii) hydrolysis}}]{\text{(i) SOCl}_2}
\begin{array}{c}
\text{CO}_2\text{H} \\
\text{H}\!\!-\!\!|\!\!-\!\!\text{OH} \\
\text{H}\!\!-\!\!|\!\!-\!\!\text{H} \\
\text{CO}_2\text{H}
\end{array}
\xrightarrow[\text{half-amide}]{\text{prepare}}
$$

L-tartaric
acid

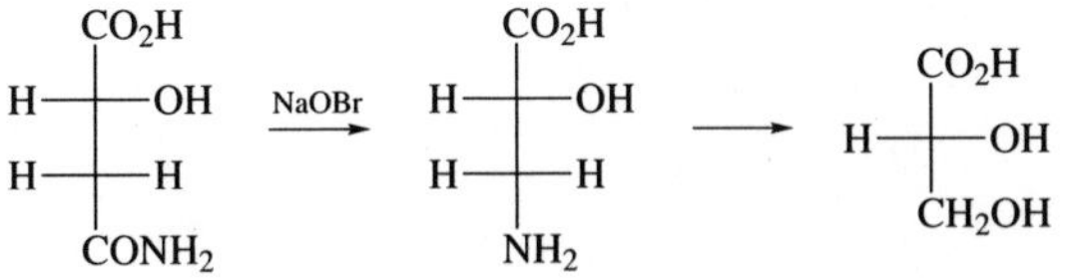

$$\begin{array}{ccc}
\begin{array}{c}
CO_2H \\
H\!\!-\!\!\!\!\overline{}\!\!\!\!-\!\!OH \\
H\!\!-\!\!\!\!\overline{}\!\!\!\!-\!\!H \\
CONH_2
\end{array}
&
\xrightarrow{\text{NaOBr}}
&
\begin{array}{c}
CO_2H \\
H\!\!-\!\!\!\!\overline{}\!\!\!\!-\!\!OH \\
H\!\!-\!\!\!\!\overline{}\!\!\!\!-\!\!H \\
NH_2
\end{array}
\end{array}$$

$$\longrightarrow
\begin{array}{c}
CO_2H \\
H\!\!-\!\!\!\!\overline{}\!\!\!\!-\!\!OH \\
CH_2OH
\end{array}$$

D-glyceric
acid
(see above)

To avoid any possibility of confusion, it is much better to refer to naturally occurring (+)-tartaric acid as (R,R)-tartaric acid and to limit the D,L nomenclature strictly to amino acids and carbohydrates $OHC\!-\!(CHOH)_n\!-\!CH_2OH$.

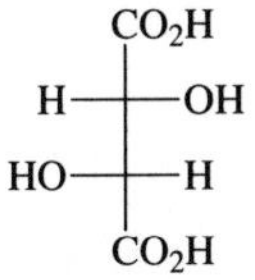

$$\begin{array}{c}
CO_2H \\
H\!\!-\!\!\!\!\overline{}\!\!\!\!-\!\!OH \\
HO\!\!-\!\!\!\!\overline{}\!\!\!\!-\!\!H \\
CO_2H
\end{array}
\qquad\qquad
\begin{array}{c}
CO_2H \\
HO\!\!-\!\!\!\!\overline{}\!\!\!\!-\!\!H \\
H\!\!-\!\!\!\!\overline{}\!\!\!\!-\!\!OH \\
CO_2H
\end{array}$$

(R,R)-tartaric acid (S,S)-tartaric acid
(+)-tartaric acid   (−)-tartaric acid

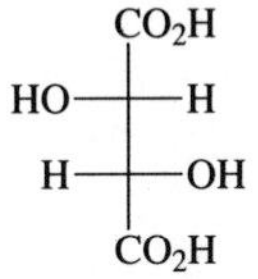

Index